Direct Thermochemical Liquefaction for Energy Applications

Woodhead publishing series in energy

Direct Thermochemical Liquefaction for Energy Applications

Edited by

Lasse Rosendahl

ELSEVIER

WP
WOODHEAD
PUBLISHING
An imprint of Elsevier

Woodhead Publishing is an imprint of Elsevier
The Officers' Mess Business Centre, Royston Road, Duxford, CB22 4QH, United Kingdom
50 Hampshire Street, 5th Floor, Cambridge, MA 02139, United States
The Boulevard, Langford Lane, Kidlington, OX5 1GB, United Kingdom

Notices
Knowledge and best practice in this field are constantly changing. As new research and experience broaden our understanding, changes in research methods, professional practices, or medical treatment may become necessary.

Practitioners and researchers must always rely on their own experience and knowledge in evaluating and using any information, methods, compounds, or experiments described herein. In using such information or methods they should be mindful of their own safety and the safety of others, including parties for whom they have a professional responsibility.

To the fullest extent of the law, neither the Publisher nor the authors, contributors, or editors, assume any liability for any injury and/or damage to persons or property as a matter of products liability, negligence or otherwise, or from any use or operation of any methods, products, instructions, or ideas contained in the material herein.

Library of Congress Cataloging-in-Publication Data
A catalog record for this book is available from the Library of Congress

British Library Cataloguing-in-Publication Data
A catalogue record for this book is available from the British Library

ISBN: 978-0-08-101029-7 (print)
ISBN: 978-0-08-101025-9 (online)

For information on all Woodhead publications
visit our website at https://www.elsevier.com/books-and-journals

Publisher: Joe Hayton
Acquisition Editor: Raquel Zanol
Editorial Project Manager: Mariana Kuhl
Production Project Manager: Omer Mukthar
Cover Designer: Victoria Pearson

Typeset by SPi Global, India

Contents

List of contributors

H. Bergem SINTEF Materials and Chemistry, Trondheim, Norway

P. Biller Aarhus University, Aarhus, Denmark

D. Castello Aalborg University, Aalborg, Denmark

W.-T. Chen University of Illinois at Urbana-Champaign, Urbana, IL, United States

J.K.R. Guerrero Steeper Energy Canada Ltd, S.W. Calgary, AB, Canada

H.J. Heeres University of Groningen, Groningen, The Netherlands

D. Ho Western University, London, ON, Canada
Trojan Technologies, London, ON, Canada

S.B. Iversen Steeper Energy ApS, Hørsholm, Denmark

C.U. Jensen Steeper Energy ApS, Hørsholm, Denmark

S. Karatzos Steeper Energy Canada Ltd, S.W. Calgary, AB, Canada

R. Lødeng SINTEF Materials and Chemistry, Trondheim, Norway

L. Nazari Western University, London, ON, Canada

G. Olofsson Steeper Energy ApS, Hørsholm, Denmark

A. Pattiya Mahasarakham University, Kamriang, Maha Sarakham, Thailand

M.B. Ray Western University, London, ON, Canada

L. Rosendahl Aalborg University, Aalborg, Denmark

D. Santoro Western University, London, ON, Canada
Trojan Technologies, London, ON, Canada

S. Sarathy Western University, London, ON, Canada
Trojan Technologies, London, ON, Canada

I. Sintamarean Aalborg University, Aalborg, Denmark

S.S. Toor Aalborg University, Aalborg, Denmark

R.H. Venderbosch Biomass Technology Group BV, Enschede, The Netherlands

C. (Charles) Xu Western University, London, ON, Canada

W. Yin University of Groningen, Groningen, The Netherlands

Y. Zhang University of Illinois at Urbana-Champaign, Urbana, IL, United States

Preface

In the quest for sustainable alternatives to fossil hydrocarbon fuels, biomass direct thermochemical liquefaction (DTL) technologies are emerging as promising and viable pathways, potentially capable of producing necessary volumes of relevant fuel types and doing so with both environmental and financial sustainability.

Several recent studies, for example, [1,2], show that DTL (hydrothermal liquefaction—HTL—and pyrolysis) are indeed very competitive, capable of delivering at production costs only some 25%–50% higher than average costs of fossil equivalents. At the same time, DTL technologies show very high GHG capping potential [3], due to efficient processing and the ability to utilise waste and residual streams and aquatic biomass. Without presuming knowledge of future carbon or other tax schemes or even future pricing of fossil fuels, this indicates a favourable business proposition for this type of fuels in a carbon-constrained, environmentally conscious future—a future that is not necessarily decades away. Currently, significant effort is put into liquefaction research and development at a global scale, manifesting itself in increased publication rates and dedicated funding going into R&D efforts at national and international levels, such as through the Horizon 2020 programme of the EU. Technologies are now being taken beyond 'simple' boiler fuel production by fundamental research into fuel design and applications, and plans to supplement existing piloting/demonstration activities such as the Empyro plant in Hengelo, NL, [4] are announced on different continents: Queensland, Australia, by Southern Oil [5] and in Europe in Norway by Statkraft [6]. ASTM approval for liquefaction pathway fuels is ongoing, with hydrotreated depolymerised cellulosic jet (HDCJ) being certified by UOP and others, and it could be argued that catalytic hydrothermolysis (CH, by ARA, Chevron, and others) is essentially what happens when fatty acids (present in a range of biomass and organic waste streams) are coprocessed in a hydrothermal liquefaction process. All these efforts are crucial in setting up the mechanisms taking these fuels to the market, thus defining business propositions and market opportunities for investors and technology providers.

A number of key features set DTL technologies positively aside from currently available bioderived fuels and other technology pathways under development:

– *Nonfood/feed input streams*: DTL technologies are inherently feedstock flexible and do not rely on isolation of specific macrocomponents (typically carbohydrates or starch) before processing. Rather than being a problem, lignin is a valuable source of desired product compounds. Wet and dry feedstock streams can be efficiently handled by DTL technologies, and input streams can be combined from a variety of sources according to availability, economics, or other considerations. Residual and waste streams from society or from other bioprocessing can be efficiently integrated.

- *Multiple component product stream*: Unlike bioethanol/biomethanol/biobutanol, bio-diesel (FAME/FAEE), sugar-derived jet fuels and Fischer-Tropsch fuels, DTL products are multicomponent, not single-molecule fuels. This enables DTL products to address the entire range of fuels available today, defined by boiling point ranges and hydrocarbon characteristics (i.e. paraffins and aromatics). For jet fuels, currently the most challenging in terms of approval, this means that full replacement is an option, rather than a blend limit less than 50%, which is the case for single-molecule bioderived jetfuels. Furthermore, it provides an extensive source for platform chemicals and high-value carbon-containing products, not just single-platform molecules.
- *Drop-in-type fuels*: DTL-derived fuels are potentially drop-in fuels; in the definition of IEA Task 39, 'drop-in' biofuels are defined as "liquid bio-hydrocarbons that are functionally equivalent to petroleum fuels and are fully compatible with existing petroleum infrastructure" [7], targeting full replacement of fossil equivalents. The main product from a DTL platform is a liquid biofuel intermediate, a so-called biocrude or biooil, which can be individually processed by almost standard refinery operations into fully compatible, functionally equivalent fossil replacements. Thus, the drop-in point can be chosen according to a number of metrics—scale, process pathway, and, not least, value proposition. This opens for a significant opportunity of repurposing existing hydrocarbon infrastructure, potentially saving both implementation cost and time (Fig. 1).
- *Smart energy system integration potential*: Most DTL technologies have a natural link to other energy infrastructures through an inherent need for larger or smaller external hydrogen sources in order to fully meet drop-in criteria of zero oxygen fuels. Today, most hydrogen is produced at refineries making an integration relatively simple, but in the future, shaving fluctuating peaks in the electrical grid by large-scale electrolysis of seawater may become a sustainable path to hydrogen. DTL plants could then serve as hydrogen sinks to these units, effectively storing electrical energy by chemical bonding in liquid fuels.

The purpose of this book is to give an insight into relevant, up to date, state-of-the-art DTL technologies, applications, and perspectives through contributions from both academia, research institutions, and commercial technology developers. It covers both major DTL technology groups, pyrolysis and hydrothermal liquefaction, going

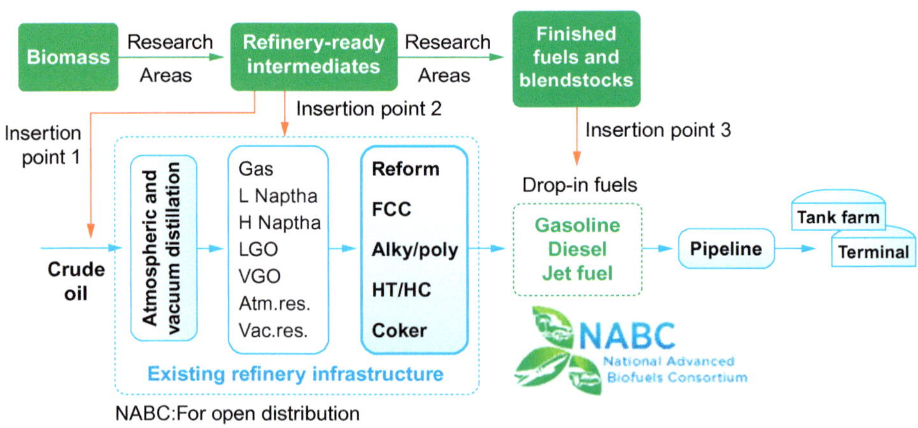

Fig. 1 Drop-in points as defined by the National Advanced Biofuels Consortium (www.nabcprojects.org).

into different biomass feedstock scenarios, process layouts, and product upgrading schemes and identifies bottlenecks and challenges, as well as potentially high-reward focus areas. It is intended for an audience with technical insight but can be read by students and professionals alike and serve as a source of technical data, for example, for the LCA and legislative communities.

L. Rosendahl
Aalborg University, Aalborg, Denmark

References

[1] De Jong S, Hoefnagels R, Faaij A, Slade R, Mawhood R, Junginger M. The feasibility of short-term production strategies for renewable jet fuels—a comprehensive techno-economic comparison. Biofuels Bioprod Biorefin 2015;9:778–800. https://doi.org/10.1002/bbb.1613.
[2] Biomass Direct Liquefaction Options. TechnoEconomic and life cycle assessment. Technical report PNNL-23579, for US DOE contract DE-AC05-76RL01830, Richland, Wa: Pacific Northwest National Laboratory; 2014.
[3] Michel J. WP6: synthesis of results and recommendations. Final report. CORE-Jetfuel, grant no FP7-605716, www.core-jetfuel.eu; 2016. [Accessed 6 July 2017].
[4] btg-btl: Empyro project website. www.empyro.nl. [Accessed 6 July 2017].
[5] Queensland Government. New Gladstone plant a boost for biofuel industry in Queensland. Media statement, The Queensland Cabinet and Ministerial Directory; 2017. http://statements.qld.gov.au/Statement/2017/6/1/new-gladstone-plant-a-boost-for-biofuel-industry-in-queensland. [Accessed 6 July 2017].
[6] Reuters. Norway's Statkraft to make biofuel from wood chips and other waste, https://mobile-reuters-com.cdn.ampproject.org/c/mobile.reuters.com/article/amp/idUSKBN19B2C7; 2017. [Accessed 6 July 2017].
[7] Karatzos S, McMillan JD, Saddler JN. The potential and challenges of drop-in biofuels. 2014. A report by IEA bioenergy task 39 report T39-T1, July 2014.

Part One

Pyrolysis

Fast pyrolysis

1

A. Pattiya
Mahasarakham University, Kamriang, Maha Sarakham, Thailand

1.1 Introduction

1.1.1 *Background*

It is widely accepted that the problems related to environmental pollutions, global warming, domestic energy shortage, economic crisis, and dependency of crude oil import are rooted from the unbalanced utilisation of fossil fuels for electricity generation, transportation, industries, and agriculture. The use of fossil fuels leads to the increase of pollutants and greenhouse gases such as sulphur oxides (SO_X), nitrogen oxides (NO_X), and carbon oxides (CO_X) in the atmosphere. A potential solution to alleviate these problems is to employ renewable resources such as biomass in place of fossil fuels. There are several types of biomass available. The appropriate ones should be nonfood. Otherwise, other problems related to the shortage of food supply would occur. Nonfood biomass such as agricultural and industrial residues is considered as a potential raw material for alternative energy production.

Biomass can be directly used or converted to other forms of fuels. It can be converted by a technology called 'fast pyrolysis' for liquid biofuel production. The liquid product is known as 'bio-oil'. Bio-oil can also be called pyrolysis liquid, pyrolysis oil, pyrolytic oil, pyrolysis tar, bio-crude, bio-crude oil, biofuel oil, wood liquid, liquid wood, wood oil, liquid smoke, wood distillates, pyroligneous tar, and pyroligneous acid [1–4].

Bio-oil can be directly used as a burner fuel or further upgraded to transportation fuels. Fast pyrolysis technology has been investigated and developed for >40 years. Currently, many countries around the world are involved in the development of industrial-scale fast pyrolysis plants. These include Canada, the Netherlands, China, Germany, Finland, the United Kingdom, the United States, and Spain. Many other countries such as Brazil, Ukraine, France, Australia, Belgium, Denmark, Greece, Italy, Malaysia, India, Republic of Korea, Vietnam, Japan, and Thailand have also investigated fast pyrolysis at research and development level. This chapter aims at providing an overview of biomass fast pyrolysis and upgrading routes for the production of liquid biofuel.

1.1.2 *Process*

Fast pyrolysis is a moderate temperature (400–600°C) thermochemical conversion process in which biomass is rapidly heated with a heating rate higher than 1000°C/s in the absence of any oxidising agent. Under these conditions, biomass decomposes producing pyrolysis vapour and solid residue, which is called char

Direct Thermochemical Liquefaction for Energy Applications. https://doi.org/10.1016/B978-0-08-101029-7.00001-1

or biochar. Part of the pyrolysis vapour is condensed to a dark brown liquid that has a heating value of about half of conventional fuel oil [5], whereas the rest is noncondensable gaseous product. Depending on biomass composition and process conditions, pyrolysis reactions can be generally classified as a simultaneous mix of dehydration, depolymerization, fragmentation, rearrangement, repolymerization, and condensation [6].

The essential features of a fast pyrolysis process in order to achieve maximum liquid yields are (i) very high heating and heat transfer rates that require a finely ground biomass feed to reduce the mass and heat transfer limitations, (ii) carefully controlled temperature of around 500°C and vapour phase temperature of 400–450°C, (iii) short vapour residence time of typically <2 s, and (iv) rapid cooling of the pyrolysis vapours to give the bio-oil product [7].

Fast pyrolysis can be carried out in a wide range of reactor types such as fluid bed, ablative, circulating fluid bed, entrained flow, rotating cone, transported bed, and vacuum moving bed. It is not the intention to review the reactor configurations here, since an extensive and excellent review already exists [5]. Nevertheless, one of the most common reactors, a fluidised-bed type reactor is briefly described here as this type of reactor is known to give a very high liquid yields of ~80% from very finely divided (~0.1 mm) clean aspen poplar wood in a bubbling fluidised-bed bench-scale reactor with an extremely short vapour residence time of <1 s [8].

Fig. 1.1 shows a conceptual fluidised-bed fast pyrolysis system. The system starts with a dryer to obtain biomass with moisture content below 10 wt% in order to control the water content of bio-oil product because water in the biomass feedstock would inevitably end up in the bio-oil. Drying is not always necessary, especially if the biomass is already naturally dried. Since drying also consumes energy or heat, the heating source can be obtained by combustion of char and/or gaseous products. After drying, the biomass feedstock is ground to particle size of <5 mm in order to achieve

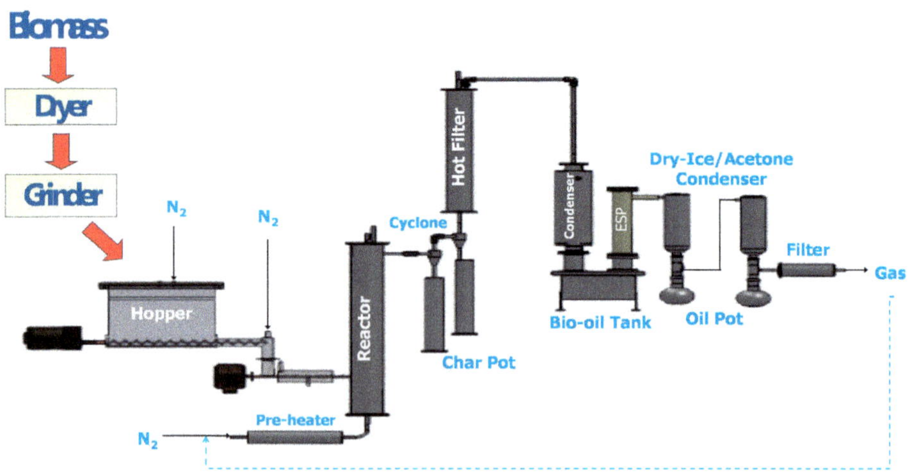

Fig. 1.1 Fast pyrolysis process.

high heating and heat transfer rate. In the fluidised-bed pyrolysis reactor, sand, a typical fluidising medium, is preheated to a temperature of around 450–550°C. It is important that the initial bed temperature prior to biomass feeding should be slightly higher than expected reaction temperature since it is known that the heat of reaction for the fast pyrolysis process is marginally endothermic [9]. The heat for pyrolysis reactor can be derived from combustion of char and/or noncondensable gases. In fast pyrolysis, both the sensible heat required for biomass to raise its temperature to the pyrolysis temperature and the heat of the pyrolysis reactions are estimated to be ~1–2 MJ/kg of biomass containing ~10% moisture [10]. Nitrogen is generally used as a carrier and fluidising gas to start up the process. After that, the recycle gas stream is used. The dried and ground biomass particles are conveyed to the reactor by a screw or pneumatic feeder.

Once a particle is heated, primary pyrolysis vapour is evolved. This vapour includes water, aerosols, condensable vapour, and noncondensable gases, all of which are derived from the degradation by fragmentation and depolymerisation of the original macrobiopolymers including cellulose, hemicellulose, and lignin as well as the decomposition of organic extractives such as tannins, fatty acids, and resins. It is important to note that certain degradation products are of very high molecular weight and they exist in gas phase as liquid aerosols rather than vapour. The pyrolysis vapour together with aerosols and solid residues is quickly entrained out of the reactor passing a cyclone where most of the solid char is separated and the vapour is cooled and collected by a bio-oil recovery unit. This unit can be composed of a quench column and an electrostatic precipitator (ESP). The use of ESP is possible due to the polar character of the pyrolysis vapour [11]. In addition, low-temperature cooling may be useful for collection of low-boiling-point chemicals such as acetic acid and formaldehyde. This may be done by the use of dry-ice condenser for lab-scale unit. For large scale, economic consideration is needed to trade off the chemical recovery and the cost for low-temperature cooling. If only ambient temperature is used for liquid product recovery, it is likely that some light volatiles would be in the gas phase. This will increase the overall heating value of the gaseous product. Combustion of the gas for process heat is then recommended.

Apart from the fluidised-bed reactor, fast pyrolysis can be carried out in a wide range of other reactor configurations such as circulating fluid bed (Fig. 1.2A, auger; Fig. 1.2B, rotating cone; Fig. 1.2C, free-fall or entrained flow; and Fig. 1.2D, ablative and vacuum moving bed).

The advantages of fast pyrolysis for bio-oil production are simplicity of the process, atmospheric pressure operation, main product being in the form of liquid that can be readily stored and transported, estimated low production cost, high thermal efficiency, and low fossil fuel inputs leading to CO_2 neutral fuel [16]. The overall energy balance of biomass fast pyrolysis can give 60%–70% efficiency to liquids with low environmental emissions due to low sulphur and nitrogen contents of the original biomass [17]. Bio-oil fuels generate >50% lower NO*X* emissions than diesel oil in a gas turbine, and no SO*X* emissions are generated [9]. Although bio-oil is regarded as the main product from fast pyrolysis, solid char and noncondensable gases can also be considered as valuable by-products.

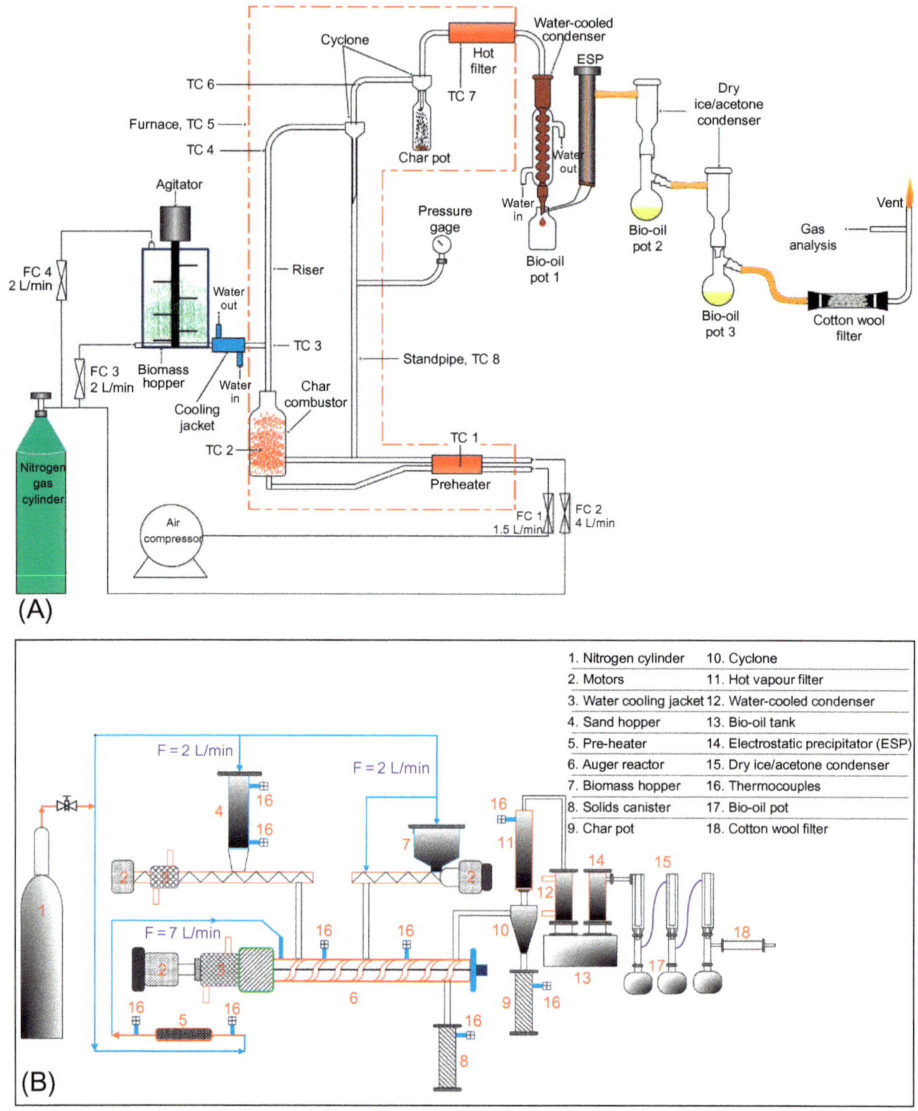

Fig. 1.2 See the legend in next page

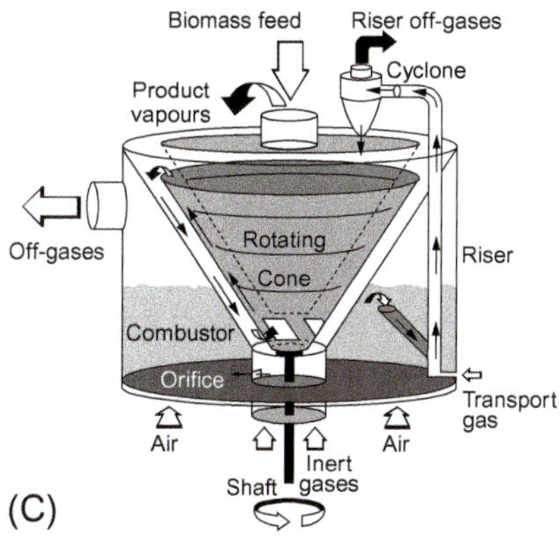

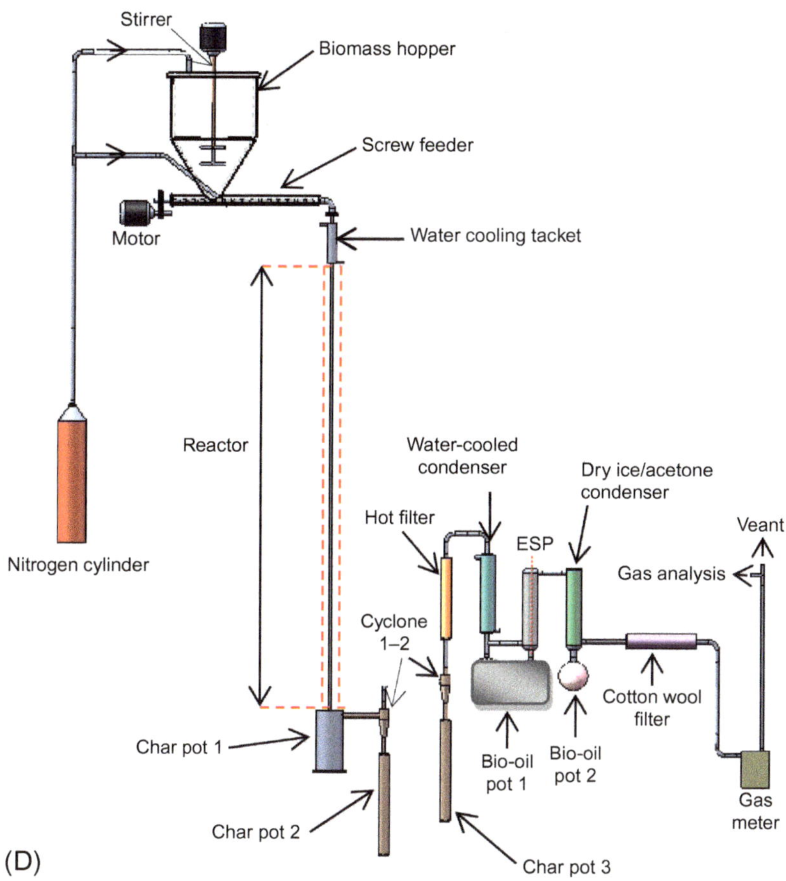

Fig. 1.2 Examples of reactor configurations: (A) circulating fluidised-bed reactor [12], (B) twin-screw (auger) reactor [13], (C) rotating cone reactor [14], and (D) free-fall reactor [15].

1.2 Fast pyrolysis products

1.2.1 Bio-oil

Bio-oil is a dynamic mixture of water, char fines, and compounds (monomers, oligomers, polymers, or fragments) derived from cellulose, hemicellulose, and lignin macropolymers of biomass, which include sugars, acids, alcohols, aldehydes, ketones, furans, esters, phenols, guaiacols, syringols, and multifunctional compounds, such as hydroxyacetic acids, hydroxyaldehydes, and hydroxyketones. These organic materials present in bio-oil are constantly reacting to move towards chemical equilibrium [18]. The actual composition of a bio-oil is a complex function of feedstock, pyrolysis technique, char removal system, condensation system, and storage conditions [19]. Bio-oil has nearly the same elemental composition as the biomass feedstock [20], thus having a high oxygen content, which makes it different from petroleum oils. Since bio-oil contains large proportion of organic acids such as acetic and formic acids, it has low pH value (typically 2–4).

Bio-oil can be separated into water-soluble and water-insoluble fractions. The former contains light organic compounds, while the latter contains mainly large molecules derived from lignin, which are known as pyrolytic lignin. The content of water-insoluble fraction of bio-oil varies from about 20 to 40 wt% [21]. Pyrolytic lignin is composed mainly of tri- and tetramers of lignin subunits (hydroxyphenyl, guaiacyl, and syringyl units), has an average molecular weight between 650 and 1300 g/mol [22], and represents almost 80% of the original content of biomass lignin [11]. According to the thermal ejection theory postulated by Piskorz et al. [23], lignin oligomers found in pyrolytic lignin are formed by direct mechanical expulsion from biomass particles as a result of a partial cracking of lignin molecules. This theory is supported by a study of bio-oil nature by using small-angle neutron scattering (SANS) technique [24] by which bio-oils were found to be nanostructured fluids constituted by a complex continuous phase and nanoparticles mainly formed by the association of units of pyrolytic lignins. The aggregation or polymerisation of these units during storage produces branched structures, which are responsible for bio-oil ageing. The polymerisation continues until the heaviest lignin-rich fraction separates out of the matrix as a viscous sludge [24]. The consequence of polymerisation reactions can be noticed from the increase in the average molecular weight and the viscosity of bio-oil over time as it is known that the bio-oil viscosity is directly correlated with the average molecular weight (Mw) and the water-insoluble fraction of bio-oil [18,25]. The increase of bio-oil Mw during storage can also occur by condensation or dehydration reactions of carbohydrate-derived constituents such as aldehydes and ketones [9], which produce water as a by-product. Therefore, an increase in water content of bio-oil during storage is apparent. This could ultimately lead to the phase separation of the bio-oil with heavy sludge at the bottom and light aqueous liquid at the top.

Due to significant amounts of the oxygenated compounds including the lignin derivatives, bio-oils have relatively low heating values, are relatively unstable, have high viscosity, have low volatility, and are corrosive and incompatible with the less polar petroleum oils [26]. Therefore, bio-oil should be upgraded in order to improve its properties prior to utilisation. Possible bio-oil upgrading options are discussed in Section 1.4.

1.2.1.1 Bio-oil properties

Since bio-oils have different properties from petroleum crude oil or fuel oil due to their oxygenated origin, different standard requirements are applied for bio-oil. Table 1.1 summarises the standard requirements for pyrolysis liquid biofuel covering two grades—G and D. Grade G is intended for use in industrial burners equipped to handle the pyrolysis liquid biofuels meeting the requirements listed for Grade G in the table and is not intended for use in residential heaters, small commercial boilers, engines, or marine applications. Grade D is intended for use in commercial/industrial burners requiring lower solids and ash content that are equipped to handle the pyrolysis liquid biofuels meeting the requirements listed for Grade D in the table and is not intended for use in residential heaters, engines, or marine applications not modified to handle these types of fuels.

According to Table 1.1, the gross heat of combustion or higher heating value (HHV) of bio-oil is required to be >15 MJ/kg. It is measured by ASTM D240 using a bomb calorimeter. The HHV of typical bio-oil is 16–19 MJ/kg [3,6,27], which is less than half of those of petroleum oil. The main reason for the low heating value of bio-oil is due to its high oxygen and water contents. In order to increase the heating value of bio-oil, the water content can be reduced by, for example, controlling the condensation temperature in the process. By increasing the cooling temperature, water and some light components such as acids and aldehydes can be removed or fractionated into different fraction. However, one should bear in mind that if significant amount of water is removed from bio-oil, its viscosity could be increased accordingly. The ASTM D7544-12 limits the kinematic viscosity of bio-oil to 125 mm^2/s. Therefore, removal of water from bio-oil in order to increase its heating value should consider the

Table 1.1 Standard specification for pyrolysis liquid biofuel (ASTM D 7544-12)

Property	Test method	Grade G	Grade D
Gross heat of combustion, MJ/kg, min	D240	15	15
Water content, % mass, max	E203	30	30
Pyrolysis solid content, % mass, max	D7579	2.5	0.25
Kinematic viscosity at 40°C, mm^2/s, max	D445[a]	125	125
Density at 20°C, kg/dm^3	D4052	1.1–1.3	1.1–1.3
Sulphur content, % mass, max	D4294	0.05	0.05
Ash content, % mass, max	D482	0.25	0.15
pH	E70	Report	Report
Flash point, °C, min	D93, procedure B	45	45
Pour point, °C, max	D97	−9	−9

[a] Without filtering.

impact on the viscosity changes. However, if the viscosity of bio-oil is too high for application such as pumping or atomising, a simple technique of solvent addition should be applied to reduce its viscosity. This technique is discussed as an option for bio-oil upgrading in Section 1.4.2. Another technique that can be applied to increase the bio-oil heating value is to remove the oxygen from bio-oil, which can be performed by the help of catalyst either during the pyrolysis processing or after bio-oil is produced. The former technique is called catalytic pyrolysis, and the latter is catalytic cracking or hydrotreating technique.

As mentioned earlier, water in bio-oil affects the heating value and viscosity. Therefore, the ASTM requires that bio-oil should have water content below 30 wt% measured according to ASTM E203 using volumetric Karl Fischer titration technique. Typical bio-oil contains around 15–30 wt% water, whereas petroleum fuels hardly contain water due to their nonpolar nature.

Solid content of bio-oil is also important and affects its applications since solids are detrimental to vessel or equipment parts. Solids in bio-oil are derived mainly from char fines, which are the residues of biomass decomposition. In the case of high ash content feedstock, ash will end up in char after the reactions. Therefore, ash content of bio-oil should also be measured together with the solids. The solids and ash contents are the only two properties of bio-oils that differentiate the Grades G and D. Grade G is defined as the liquid biofuel intended for use in industrial burners equipped to handle the pyrolysis liquid biofuels and is not intended for use in residential heaters, small commercial boilers, engines, or marine applications. Grade D is defined as the liquid biofuel intended for use in commercial/industrial burners requiring lower solids and ash content and which are equipped to handle the pyrolysis liquid biofuels and is not intended for use in residential heaters, engines, or marine applications not modified to handle these types of fuels. Grade D requires a maximum solid content of 0.25 wt% and ash content of 0.15 wt%, whereas Grade G requires maximum 2.5 wt% solid content and 0.25 wt% ash content. The solid and ash contents of bio-oil can be reduced in the process by hot gas filtration, which is discussed in more details in Section 1.4.1 as a technique for bio-oil upgrading.

1.2.1.2 Application of bio-oil

The bio-oils can be used either as whole bio-oil, fractionated bio-oil, or extracted specific chemicals. The potential applications of the bio-oils include fuels in boilers, engines, and turbines for heat and power generation. Bio-oils were previously tested for electricity and heat production in boilers, furnaces, and combustors [28], diesel engines [29,30], and gas turbines [31,32].Bio-oil can be converted to transportation biofuel by upgrading processes. Since bio-oil contains various chemical compounds, some of them are valuable. It is therefore possible to produced specialty chemicals by chemical extraction or reactions of bio-oil such as preservatives, liquid smoke, resin precursors, additives in fertilising and pharmaceutical industries, flavouring agents (such as glycolaldehyde) in food industries, acetic acid, hydroxyacetaldehyde, levoglucosan, levoglucosenone, and maltol [33–36] [3]. Fig. 1.3 summarises the overview of bio-oil potential applications.

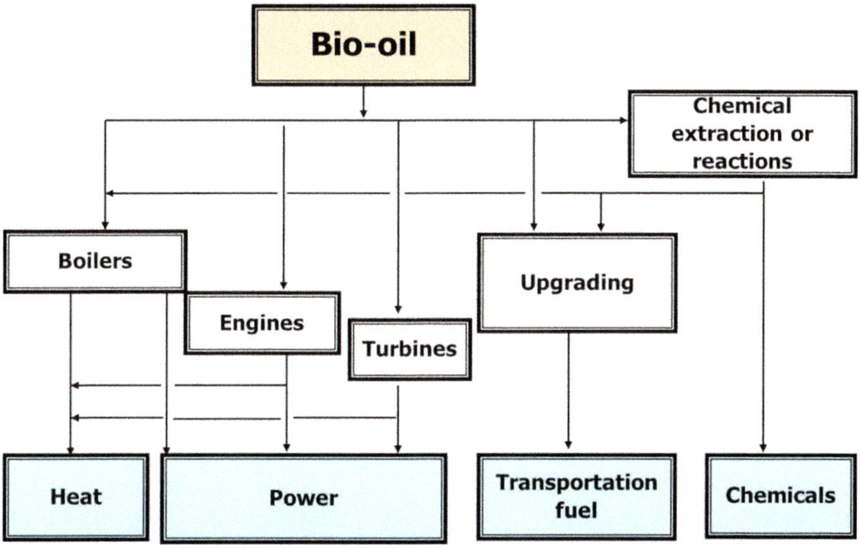

Fig. 1.3 Potential applications of bio-oil [37].

1.2.2 Char

Char is the solid residue of pyrolysed biomass particles. Char is normally removed from the vapour stream by cyclone separation. It contains high proportion of carbon with small quantities of oxygen and hydrogen. It also contains most of the inorganics present in the original biomass. Therefore, it may be used for soil amendment, thus capturing the carbon back into the earth. Since the char product has heating value (HHV) of around 23 MJ/kg, it can be used as an energy source for process heat. The char may also be used for production of hydrogen or syngas by steam reforming or thermal cracking processes [33,38]. The char could be used as a solid fuel in boilers or could be cofired with conventional fossil fuels. Consequently, char can be regarded as a valuable by-product. The char may also be physically or chemically activated to improve its microscopic surface structure so as to be used as filtration and adsorption of organic and inorganic pollutants [39,40].

1.2.3 Gas

Noncondensable gases from biomass fast pyrolysis are composed mainly of carbon dioxide (CO_2), carbon monoxide (CO), hydrogen (H_2), methane (CH_4), ethene or ethylene (C_2H_4), ethane (C_2H_6), propene or propylene (C_3H_6), and propane (C_3H_8). Additionally, if the liquid collection unit is not 100% effective, some light volatile could be in the gaseous stream such as pentane, benzene, toluene, xylenes, and acetaldehyde. Since pyrolysis gas contains significant amount of carbon dioxide along with methane and some other combustible gases, it might be used as a fuel for industrial combustion purposes [33]. It is suggested by Bridgwater [41] that the most effective use for the pyrolysis gas is as a fluidising medium or carrier gas for a fluid bed reactor and it can also be used in-plant for providing process heat.

1.3 Parameters affecting pyrolysis products

There are many parameters that affect fast pyrolysis product yields and properties. It is not the intention of this section to review all of them. Nevertheless, certain key parameters are discussed in this section including feedstock composition, pyrolysis temperature, vapour residence time, carrier gas, and char separation system.

1.3.1 Feedstock composition

Biomass is composed of water, cellulose, hemicellulose, lignin, organic extractives, and inorganics. The types and proportion of all of these components influence fast pyrolysis processing and product yields and properties.

1.3.1.1 Water

Water or moisture content of biomass affects the fast pyrolysis processing as additional heat is required to vapourise the water and additional cooling load is required to condense the water back into liquid product. Initial water in biomass will end up in bio-oil. If biomass with high moisture content is pyrolysed, bio-oil would have high water content, which affects the heating value, viscosity, pH, and density of the bio-oil. Although the viscosity is lowered and the pH is increased due to high water content, the heating value and the density (especially the energy density) are decreased. Therefore, typical criteria for biomass fast pyrolysis require that the biomass moisture content should be below 10 wt%.

1.3.1.2 Cellulose, hemicellulose and lignin

Among the macropolymer components of biomass, hemicellulose is the easiest to decompose followed by cellulose. Hemicellulose decomposed at 220–350°C, while cellulose decomposed mainly at 315–400°C [42–45]. However, lignin is the most difficult to decompose as it is the most stable component and the weight loss occurred in a wide range of temperature from 160°C to 900°C with high solid residue [44,45] with the main decomposition temperature range of 300–550°C [46]. Therefore, cellulose and hemicelluloses mainly contribute to the bio-oil product, whereas lignin yields larger proportion of solid char [47]. Fast pyrolysis of hemicellulose gives water, methanol, formic, acetic, propionic acids, hydroxyl-1-propanone, hydroxyl-1-butanone, 2-methylfuran, 2-furfuraldehyde, dianhydroxylopyranose, and anhydroxylopyranose as main products [48]. Cellulose gives levoglucosan as the main pyrolysis product [42,43]. Although most difficult to decompose, fast pyrolysis of lignin can give bio-oil yield of 45 with 27.6 wt%daf organic liquid yield [49].

Compared with cellulose and hemicellulose, lignin seems to significantly affect bio-oil properties. A clear, positive, dependence between lignin content in the feedstock and water-insoluble fraction (pyrolytic lignin) of the bio-oil has been reported [50]. Pyrolytic lignin is speculated to be responsible for the high viscosity, high average molecular weight, and low stability of bio-oil. Biomass with higher proportion of lignin may increase the average molecular weight and viscosity of bio-oil [51].

Ghetti et al. [52] studied the effect of biomass lignin content on burning characteristics of bio-oil products using thermogravimetric analysis (TGA) technique and concluded that biomass of low lignin content produces a lighter pyrolysis liquid product, which may be considered a better bio-oil for use as a fuel. Therefore, the lignin in biomass can affect the bio-oil quality in terms of viscosity, molecular weight, stability, and combustion behaviour, and a lower lignin content of biomass appears to be beneficial to the bio-oil properties.

1.3.1.3 Extractives

Extractives in biomass refer to the nonstructural components that can be extracted by solvents (e.g. water, ethanol, acetone, benzene, hexane, dichloromethane, and toluene) such as fats, waxes, proteins, terpenes, gums, resins, simple sugars, starches, phenolics, essential oils, pectins, mucilages, glycosides and saponins, fatty acids, sterols, and flavonoid [53,54]. The role of wood extractives was investigated in vacuum pyrolysis of wood by Roy et al. [54], and it was found that wood extractives showed a significant inhibiting effect on bio-oil yield and levoglucosan formation. Removal of wood extractives led to 34% and 64% loss in hydrogen and oxygen content of solid char, respectively.

1.3.1.4 Inorganics

Inorganics or mineral matter in biomass becomes ash after combustion. The influence of mineral matter present in biomass during pyrolysis has previously been studied [55–63] and found that the inorganic species catalyse biomass decomposition and char-forming reactions, resulting in the reduction of liquid yields and the formation of char, noncondensable gases, and reaction water. The most active element is potassium, followed by sodium [61–63]. Small quantity of inorganics such as a few parts per million of sodium salts can greatly accelerate char formation as it was found that the char yield from the pyrolysis of pure crystalline cellulose was increased by a factor of >10 [64]. Alkali metals also influence the thermal decomposition mechanism during fast pyrolysis by enhancing the fragmentation (ring scission) of the monomers making up the macropolymer chains [60,65]. These suggest that the content and composition of ash are important parameters that affect the yield and chemical composition of the pyrolysis products.

1.3.2 Feedstock pretreatment

Feedstock pretreatment can affect pyrolysis processing and products. The purpose of the pretreatment is to change or remove biomass components so as to improve the pyrolysis products. The pretreatment can be performed physically, chemically, thermally, and biologically.

1.3.2.1 Physical pretreatment

The most basic pretreatment of biomass feedstock prior to feeding into a fast pyrolysis reactor is the reduction of biomass particle size by milling or grinding.

Because biomass has a very low thermal conductivity of around 0.1 W/(m K) [66], the temperature gradient across the particles can influence the biomass pyrolysis mechanism [46,67]. Therefore, quick heating to a desired pyrolysis temperature is very difficult and can only be achieved by reduction of particle size to a few millimetres [66]. Typical reactors such as fluidised bed require 5 mm or less biomass particle size. However, in some systems such as ablative reactors, larger biomass can also be utilised. It is important to note that too large biomass can lead to lower bio-oil yields and higher char yield because the internal biomass may not be completely pyrolysed. Too small biomass particles can escape the pyrolysis reactor before completely pyrolysed in a system like fluidised-bed reactor due to the entrainment of carrier gas. Therefore, appropriate size of biomass particle should be carefully considered by taking into account the product yields and the cost of biomass size reduction.

1.3.2.2 Chemical pretreatment

Chemical pretreatment of biomass can be done by washing with liquid solvents such as water, acid, and basic solutions with an aim to remove minerals or inorganics in biomass. Demineralisation of the biomass to remove alkali cations can significantly increase the organic liquid yields and enhance the formation of anhydrosugars [68]. Blasi et al. [69] examined the effect of water washing on pyrolysis characteristics of straw and found that water washing increases the bio-oil yields, while char formation is decreased. Washing of biomass by acids, such as HNO_3 and HF, was also found to reduce the mineral content of biomass [70].

1.3.2.3 Thermal pretreatment

Basic thermal pretreatment of biomass is drying, which can be performed by natural sunlight or additional process heat. As mentioned earlier, that water in biomass can affect the bio-oil production by affecting the pyrolysis efficiency and the bio-oil water content and the related properties. Drying of biomass can reduce the heat load for evaporation of the water in the pyrolysis reactor and the cooling load for condensing the water into bio-oil portion. For commercial or large-scale pyrolysis plants, water evaporated from drying of biomass can be used or sold as steam to increase the economic feasibility of the pyrolysis plants.

Another thermal technique that has been used for biomass pretreatment prior to fast pyrolysis reaction is called torrefaction, which can be regarded as mild pyrolysis at a temperature of below 300°C. In torrefaction, moisture is removed together with part of hemicellulose or volatile. The main purpose of torrefaction is to remove water and to enhance grindability, energy density, bacteria resistance, and hydrophobicity. The treatment of biomass by torrefaction was found to affect the bio-oil yield and properties. The bio-oil yields decreased after torrefaction compared with torrefied and biomass feeds [71–73]. However, the bio-oil quality was improved by the torrefaction pretreatment as it was found when torrefied biomass was pyrolysed that the bio-oils produced had lower acidity and higher energy density [71].

1.3.2.4 Biological pretreatment

Biological pretreatment by white-rot fungus of biomass prior to pyrolysis has been proved to enhance the pyrolysis performance by reducing the pyrolysis temperature and decomposing the lignin component [74–76]. Yang et al. [76] employed three white-rot fungi *Pleurotus ostreatus BP2*, *Echinodontium taxodii 2538*, and *Irpex lacteus CD2* to biologically pretreat corn stover and then studied its thermal characteristics during pyrolysis in a TGA instrument. The results showed that the fungal pretreatment could decrease activation energy and the pyrolysis temperature of hemicellulose and cellulose up to 36°C.

1.3.3 Pyrolysis temperature

In fast pyrolysis processing, temperature can be either temperature of reaction or reactor temperature. The former is the temperature at which biomass particles pyrolyse and are difficult to measure. The reactor temperature is higher than the reaction temperature due to the heat loss or temperature gradient from the reactor wall to heat transfer medium and to biomass particles. Increasing the reactor temperature would increase the temperature of reaction. It is therefore important to define where the temperature is measured.

Generally, increasing temperature would lead to higher gas yield and lower char yield. Although the overall gas yield is increased with temperature, the yields of individual gases are not increased uniformly as it is known that increasing temperatures can enhance secondary cracking of pyrolysis vapour that ultimately produces CO and CH_4 rather than CO_2 [77]. The effect of temperature on bio-oil yield is more complex as bio-oil yield is the sum of organics and water yields. Therefore, it is easier to consider the yields of organics and water separately. In pyrolysis temperature range of 400–600°C, organic yields tend to reach a maximum at a specific temperature depending on feedstocks (500°C is a typical maximum temperature for wood feedstock). Temperatures below or above this would give lower organic yield. The effect of temperature on water yield is relatively small, but the yield appears to increase marginally with increasing temperature.

The effect of temperature on bio-oil chemical properties has been reported by Elliott in 1988 [78]. He concluded that for short-residence-time processing, there is a direct correlation between chemical composition and operating temperature. As the temperature increases, the oxygen content decreases, and the hydrogen to carbon ratio decreases. In addition, he also reported that the biological activity of pyrolysis oils, as measured by mutagenic and tumour-initiating activity, correlates with pyrolysis temperature. The activity appears only in the high-temperature (750°C or more) tars that contain high levels of polycyclic aromatic hydrocarbons (PAH). Similar finding has also been found by Horne and Williams in 1996 [79] that increasing pyrolysis temperature (from 400 to 550°C) led to an increase in PAH concentration of bio-oils produced from fast pyrolysis of mixed wood waste, but the overall PAH concentration in all bio-oils was low (<120 ppm).

1.3.4 Vapour residence time

Vapour residence time is measured as how long the pyrolysis vapour spends in the hot environment prior to condensation. It is mentioned earlier that one of the essential features of fast pyrolysis is the short vapour residence time of typically <2 s. This is to reduce secondary reactions such as thermal cracking, recondensation, repolymerisation, and char formation, which can cause a reduction of organic yields and an increase in the yields of permanent gases and char. It has been shown by Piskorz et al. [80] when pyrolysing whole sweet sorghum and sweet sorghum bagasse that increasing vapour residence time from ~0.2 to ~1.0 s led to a reduction of organic yields and an increase in the yields of char and gas.

1.3.5 Carrier gas

Carrier gas or environment in fast pyrolysis is usually nonoxidative gas such as nitrogen. However, other gases such as H_2, CO_2, CO, and CH_4, steam can also be used [81]. Zhang et al. [82] studied the effects of N_2, CO_2, CO, CH_4, and H_2 atmospheres on biomass pyrolysis in a fluidised-bed reactor and found that CH_4 atmosphere gave the highest bio-oil yields of 59% and CO gave the lowest liquid yield of 50%. According to GC/MS analysis of bio-oil, less methoxy-containing compounds and more monofunctional phenols in the bio-oils were found when applying CO and CO_2 as the carrier gas. Under CO and H_2 atmospheres, the HHV of the bio-oils could reach the highest value of around 24 MJ/kg. The char from biomass pyrolysis under CO_2 atmosphere had increased surface area and different chemical composition [83] compared with the chars produced under N_2 atmosphere.

1.3.6 Char separation

In a fast pyrolysis process, solid char is normally separated by a cyclone or a series of cyclones. Separation of char out of the reactor zone as soon as possible is recommended as the char formed during pyrolysis serves as a vapour-cracking catalyst, which can decrease bio-oil yield and increase the yield of noncondensable gases. Generally, the efficiency of the char separation by cyclones is not 100%. As a consequence, some char is inevitably carried over from cyclones and ends up in the bio-oil as suspended submicron particles, thus increasing the bio-oil solid content and viscosity. The presence of high concentrations of submicron char particles in bio-oils will make them problematic for combustion in steam boilers, diesel engines, and turbine operations because of the potential release of the ash and alkali metals during combustion [84]. In addition, agglomeration of the alkali metal-laden submicron char particles in the bio-oil taking place during cold room storage may cause problems for its long-term storage, transportation, and use [84]. The presence of char can also worsen the stability of bio-oil by catalysing reactions that lead to the increase in bio-oil viscosity [19]. Since the inefficient char separation during pyrolysis can deteriorate the pyrolysis liquid product, improving the char removal system would be useful. This can be achieved by, for example, applying hot vapour filtration technique whose detail is provided in Section 1.4.1.

1.4 Bio-oil upgrading options

Depending upon the desired application, bio-oils can be used without modification, or they may be chemically changed to meet the requirements of a specific application, for example, storable boiler fuels or high-octane gasoline [20]. A selection of upgrading methods, namely, hot vapour filtration, solvent addition, and hydrotreating, is outlined, whereas catalytic vapour cracking or catalytic pyrolysis is discussed in Chapter 2.

1.4.1 Hot vapour filtration

As discussed earlier, char fines that are entrained out of cyclones and collect in bio-oil can affect the viscosity, stability, and applications of bio-oil. Therefore, hot vapour filtration has been proposed as a more effective technique for capturing these fine particles. Depending on the feedstock particle size and ash content as well as the design of the hot filter, the efficiency of hot vapour filtration varies considerably.

Diebold [19] suggested that although cyclonic separation is the easiest way to remove char, even the best cyclones begin to lose their efficiency with char particles smaller than about 10 μm in diameter. This statement was supported by Lee et al. [58] as they found when pyrolysing rice straw samples in a fluidised-bed reactor that by using a cyclone char, particle size 10–170 μm was captured, whereas a hot vapour filter could remove solids of particle size around 0.3 μm, leading to the bio-oil alkali metals (Na + K) content of 25 ppm and alkaline earth metals (Mg + Ca) content of 63 ppm. It has been reported by Diebold [19] that the best job of hot gas filtration to date at the National Renewable Energy Laboratory (NREL) resulted in <2 ppm alkali and 2 ppm alkaline earth metals in bio-oil. The main difference between the NREL and Lee et al. [58] works could be due to the biomass raw materials used and the design of the filtration units.

Although the main purpose of a hot vapour filtration unit is to capture fine particles in order to improve the char removal efficiency that leads to a reduction of bio-oil solid content, there are some side effects due to the existence of the unit in the pyrolysis system. When the char accumulates in the filter unit as a filter cake, it can function as a vapour-cracking catalyst that can cause a reduction of bio-oil yield. This problem is projected to be more severe in large-scale operations because of the cracking of the reactive molecules or polymerisation of some of the bio-oil components to form char. A solution to this problem could be the use of very compact filters that can remove the char particles and maintain a very low residence time of 1 s or less in the hot vapour filter unit [84]. Moreover, the char cake may crack heavy molecules in pyrolysis vapour, particularly those derived from lignin. This would lead to the production of bio-oil of better quality in terms of solid content, viscosity, and stability.

1.4.2 Solvent addition

The addition of solvents even with small quantities can dramatically improve bio-oil properties. Various solvents have been tested including alcohols (ethanol, methanol, and isopropanol), acetone, ethyl acetate, tetralin, and mixtures of two different solvents (methyl isobutyl ketone + methanol and acetone + methanol) [18,85–89].

General beneficial effects of solvent addition on bio-oil properties are reduction of viscosity, improvement of stability measured by the rates of increase in viscosity and in average molecular weight during storage or ageing, decrease of density that would allow the bio-oil to be filtered more easily, increase in heating value or added heating value, improved homogeneity and burning characteristics, and delay of the bio-oil phase separation process. However, a disadvantage of the addition of solvents, especially alcohols, is the decrease of bio-oil flash point. To overcome this drawback, Oasmaa et al. [90] proposed a simple technique called 'concentration method' by which most of the water and part of the light reactive volatiles of fast pyrolysis liquid are replaced by alcohol. The water and light volatiles are removed from the main bio-oil collection container by increasing the condensation temperature to 50 ± 4°C. This method results in an improvement of bio-oil quality in terms of homogeneity, viscosity, stability, odour, and acidity without significantly decreasing the flash point of the bio-oil product.

The function of alcohol, an organic solvent, in bio-oil/alcohol blend appears to involve molecular dilution to slow the bio-oil ageing reactions and the formation of intermediate products by the reactions of alcohol with compounds present in bio-oil [18]. It is also believed that the addition of alcohols can enhance the solubility of the hydrophobic compounds such as high-molecular-mass lignin derivatives and extractives, leading to an improvement of bio-oil homogeneity [89]. Since an analysis of the mixture of bio-oil and ethanol shows the presence of acetals, hemiacetals, and ester, the reactions probably include esterification of carboxylic acid groups and acetalisation and hemiacetal formation of aldehyde groups [86]. These chemical reactions are summarised in Fig. 1.4. The esters, acetals, and hemiacetals produced can terminate the chains of the hydroxyl group leading to more stable liquid products.

1.4.3 Catalytic upgrading

Catalytic upgrading in biomass fast pyrolysis can produce transportation biofuel. The possible routes are shown in Fig. 1.5. It can be divided into two main routes—bio-oil upgrading and pyrolysis vapour upgrading. The *first* one is the catalytic upgrading of bio-oil after production from fast pyrolysis process. Based on this route, bio-oil can be

$$R\text{-}OH + R'COOH \leftrightarrow R'COOR + H_2O$$

Esterification

$$2ROH + R'COH \leftrightarrow R'CH\text{-}(OR)_2 + H_2O$$

Acetalisation

$$ROH + R'COH \leftrightarrow R'CH\text{-}OR\text{-}OH$$

Hemiacetal formation

Fig. 1.4 Reactions probable to occur in bio-oil with the addition of alcohol [19].

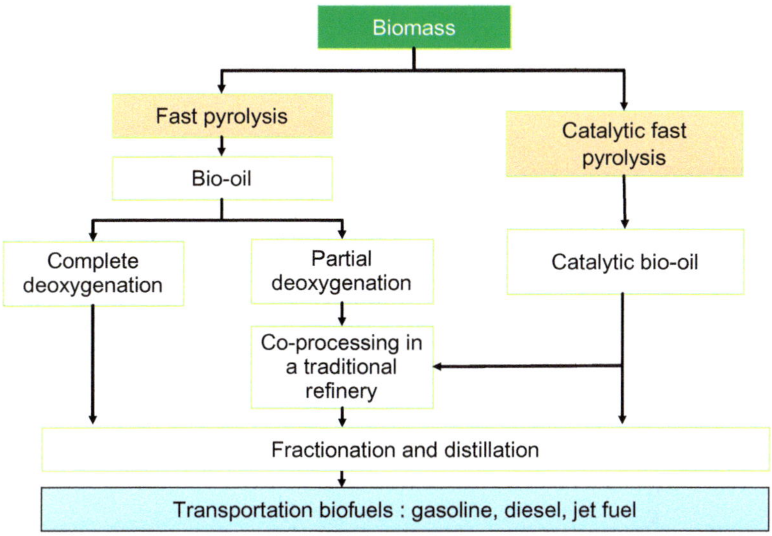

Fig. 1.5 Possible routes for transportation biofuel from fast pyrolysis of nonfood biomass.

either completely or partially deoxygenated. If partial deoxygenation is performed, the liquid product may be able to coprocess in a traditional refinery. This bio-oil deoxygenation is often known as hydrotreating. The purpose of hydrotreating is to reject the oxygen in the form of water. Depending on the severity of the hydrotreating, >95% of oxygen in bio-oil can be removed, thus producing nearly pure hydrocarbons including aromatics and naphthenes (cycloalkanes). Hydrotreating can be conceptually characterised as follows [41,91]:

$$C_6H_8O_4\left(\text{bio-oil}\right)+6H_2 \rightarrow 6CH_2\left(\text{hydrocarbon equivalent to naptha}\right)+4H_2O$$

To achieve hydrotreating, high pressures of 70–200 bars to provide a high hydrogen partial pressure and a two-stage process operating at about 250–400°C are normally required [41]. The first stage of hydrotreating is an initial stabilisation where a temperature of 250–275°C is applied. The second stage is a more conventional hydrotreating process operating at 350–400°C. The initial step consumes little hydrogen but is essential to avoid polymerisation of the bio-oil at the higher temperature of the second stage. Another factor that is important for hydrotreating is the presence of solid catalysts. Typical hydrotreating catalysts include sulphided CoMo- and NiMo-based catalysts. During hydrotreating at high hydrogen pressure in the presence of catalyst, bio-oil reacts with H_2 to form water and saturate carbon-carbon bonds. To minimise the hydrogen consumption without decreasing the octane number of the liquid product, it is suggested that hydrogenation of aromatics in bio-oil should be avoided [38].

In addition to the benefit of hydrotreating in terms of bio-oil properties improvement, a useful side effect of full hydrotreating (<2 wt% oxygen in the final product) is that the product water readily separates from the hydrocarbon product and is relatively clean, and the water may also strip out the alkali metal derived from the wood ash and resolve one of the potential problems in using biomass-derived fuels in turbines [41]. Although hydrotreating is a promising upgrading option, the requirements of high pressure and high hydrogen consumption (~300–700 L/L of bio-oil) [92] may reduce the process feasibility, especially in economic aspects.

Another promising approach to upgrading bio-oil is the direct catalytic conversion in the vapour phase at near-atmospheric pressure and without hydrogen by catalytic vapour cracking and reforming. This can avoid the problem of re-evaporation of liquid bio-oil and allows pyrolysis and catalytic upgrading to occur in one thermal cycle. This method is therefore called 'catalytic fast pyrolysis' with more detail in Chapter 2.

1.5 Examples of plants/implementation

Three reactor types, including bubbling and circulating fluidised beds and rotating cone, have been commercialised in the market, while others still remain at the stage of demonstration, pilot, or even laboratory research [93]. Currently, there are several industrial fast pyrolysis plants around the world. The country that has developed this technology extensively is Canada. As can be seen from Table 1.2, the world's largest fast pyrolysis plants are located in Canada by Dynamotive Energy Systems with the maximum biomass input of 200 tonne/day. The plant is based on a fluidised-bed process configuration. It is located at the city of Guelph in Ontario, Canada (Fig. 1.6). Although the Dynamotive Energy Systems' plants met technical performance specifications and demonstrated continuous operation over sustained periods of time (days), the lack of a bio-oil market in North America led to their temporary closure [66]. The second largest is also in Canada by Ensyn with the capacity of 100 tonne/day. The Ensyn reactor is based on transported bed or circulating fluidised-bed reactor. It is located at the city of Renfrew in Ontario, Canada (Fig. 1.7). In addition, the well-known Canadian companies that are involved in industrialization of fast pyrolysis technology include ABRI-Tech (Figs. 1.8 and 1.9) and Agri-Therm (Figs. 1.10 and 1.11), which have auger and fluidised-bed reactors, respectively. Both companies also support the concept of mobile pyrolysis plant.

Table 1.2 Examples of industrial-scale fast pyrolysis plants

Organisation	Process type	Maximum capacity (tonne/day)
Dynamotive, Canada	Fluid bed	200
Ensyn, Canada	Transported bed/CFB	100
ABRI-Tech, Canada	Augur/screw	50
Biomass Technology Group (BTG), the Netherlands	Rotating cone	50
Agri-Therm, Canada	Fluid bed	10

Fig. 1.6 World's largest industrial-scale fast pyrolysis plant with a biomass input of 200 tonne/day by Dynamotive Energy Systems located at Guelph, Ontario, Canada [94].

Fig. 1.7 Ensyn 100 tonne/day industrial fast pyrolysis plant located in Renfrew, Ontario, Canada [95].

1.6 Conclusions and recommendation

Fast pyrolysis of biomass is a promising technology for biofuel production. Although the process is simple, it requires careful considerations in design, construction, commissioning, and operation. The need for in-depth technical and practical knowledge and know-how is crucial in the development of this technology. It has been proved in the laboratory scale that biofuel either as a fuel oil or a transportation fuel can be produced by fast pyrolysis incorporation with upgrading processes. One of the high

Fig. 1.8 ABRI-Tech 50 tonne/day industrial fast pyrolysis plant located in Namur, Quebec, Canada [96].

Fig. 1.9 ABRI-Tech 1 tonne/day mobile fast pyrolysis plant located in Namur, Quebec, Canada [96].

Fig. 1.10 Agri-Therm 10 tonne/day mobile fast pyrolysis plant [97,98].

Fig. 1.11 Agri-Therm mobile fast pyrolysis plant concept design [98,99].

potential upgrading processes is the catalytic processing. Nevertheless, the upscaling of this catalytic processing requires careful catalyst development and engineering process design so as to overcome catalyst premature deactivation. The main barrier preventing the industrialisation of fast pyrolysis technology can be related to the biomass collection since biomass tends to spread around the agricultural lands. It is therefore proposed that the development of mobile fast pyrolysis plants should be supported in order to produce an energy carrier like bio-oil in the biomass field. The bio-oil is then transported to a large-scale upgrading plant for production of transportation fuel. Moreover, to enhance the economic feasibility of biofuel production from biomass by fast pyrolysis process, the coproduction of biochar and other value-added specialty chemicals should be considered based on a biorefinery concept.

References

[1] Balat M, et al. Main routes for the thermo-conversion of biomass into fuels and chemicals. Part 1: pyrolysis systems. Energ Convers Manage 2009;50(12):3147–57.
[2] Park HJ, et al. Catalytic vapor cracking for improvement of bio-oil quality. Catal Surv Asia 2011;15(3):161–80.
[3] Mohan D, Pittman CU, Steele PH. Pyrolysis of wood/biomass for bio-oil: a critical review. Energy Fuel 2006;20(3):848–89.
[4] Zhang Q, et al. Review of biomass pyrolysis oil properties and upgrading research. Energ Convers Manage 2007;48(1):87–92.
[5] Bridgwater AV, Peacocke GVC. Fast pyrolysis processes for biomass. Renew Sustain Energy Rev 2000;4:1–73.
[6] Theodore D, Juan S. Catalytic fast pyrolysis: a review. Energies 2013;6(1):514–38.
[7] Bridgwater AV. Renewable fuels and chemicals by thermal processing of biomass. Chem Eng J 2003;91(2–3):87–102.
[8] Scott DS, Piskorz J. The continuous flash pyrolysis of biomass. Can J Chem Eng 1984;62:404–12.
[9] Mohan D, Pittman CU, Steele PH. Pyrolysis of wood/biomass for bio-oil: a critical review. Energy Fuel 2006;20(3):848–89.

[10] Daugaard DE, Brown RC. Enthalpy for pyrolysis for several types of biomass. Energy Fuel 2003;17(4):934–9.

[11] Piskorz J, Scott DS, Radlein D. Composition of oils obtained by fast pyrolysis of different woods. In: Soltes EJ, Milne TA, editors. Pyrolysis oils from biomass: producing, analyzing, and upgrading. Washington, DC: American Chemical Society; 1988. pp. 167–77.

[12] Duanguppama K, Suwapaet N, Pattiya A. Fast pyrolysis of contaminated sawdust in a circulating fluidised bed reactor. J Anal Appl Pyrolysis 2016;118:63–74.

[13] Sirijanusorn S, Sriprateep K, Pattiya A. Pyrolysis of cassava rhizome in a counter-rotating twin screw reactor unit. Bioresour Technol 2013;139:343–8.

[14] Janse AMC, et al. A novel interconnected fluidised bed for the combined flash pyrolysis of biomass and combustion of char. Chem Eng J 1999;76(1):77–86.

[15] Pattiya A, Sukkasi S, Goodwin V. Fast pyrolysis of sugarcane and cassava residues in a free-fall reactor. Energy 2012;44(1):1067–77.

[16] Oasmaa A, et al. Fast pyrolysis of forestry residue. 1. Effect of extractives on phase separation of pyrolysis liquids. Energy Fuel 2003;17(1):1–12.

[17] Oasmaa A, et al. Norms and standards for pyrolysis liquids. End-user requirements and specifications. Energy Fuel 2005;19(5):2155–63.

[18] Diebold JP, Czernik S. Additives to lower and stabilize the viscosity of pyrolysis oils during storage. Energy Fuel 1997;11(5):1081–91.

[19] Diebold JP. A Review of the Chemical and Physical Mechanisms of the Storage Stability of Fast Pyrolysis Bio-Oils. Golden, CO: National Renewable Energy Laboratory (NREL); 2000. p. 1–59.

[20] Diebold JP, et al. In: Klass DL, editor. Low-Pressure Upgrading of Primary Pyrolysis Oils from Biomass and Organic Wastes, in Energy from biomass and wastes X. London: Elsevier Applied Science; 1987. p. 801–30.

[21] Sipila K, et al. Characterization of biomass-based flash pyrolysis oils. Biomass Bioenergy 1998;14(2):103–13.

[22] Scholze B, Hanser C, Meier D. Characterization of the water-insoluble fraction from fast pyrolysis liquids (pyrolytic lignin) Part II. GPC, carbonyl goups, and C-13-NMR. J Anal Appl Pyrolysis 2001;58:387–400.

[23] Piskorz J, Majerski P, Radlein D. Pyrolysis of biomass—aerosol generation: properties, applications, and significance for process engineers. In: Overend RP, Chornet E, editors. Biomass: a growth opportunity in green energy and value-added products, Vols. 1 and 2. Oxford: Pergamon-Elsevier Science Ltd; 1999. p. 1153–9.

[24] Fratini E, et al. SANS analysis of the microstructural evolution during the aging of pyrolysis oils from biomass. Langmuir 2006;22(1):306–12.

[25] Oasmaa A, Kuoppala E. Fast pyrolysis of forestry residue. 3. Storage stability of liquid fuel. Energy Fuel 2003;17(4):1075–84.

[26] Stoikos T. Upgrading of Biomass Pyrolysis Liquids to High-Value Chemicals and Fuel Additives. In: Bridgwater AV, Grassi G, editors. Biomass pyrolysis liquids upgrading and utilisation. Elsevier Applied Science; 1991. p. ix, 377 p..

[27] Mortensen PM, et al. A review of catalytic upgrading of bio-oil to engine fuels. Appl Catal A Gen 2011;407(1):1–19.

[28] Gust S. Combustion experiences of flash pyrolysis fuel in intermediate size boilers. In: Bridgwater AV, Boocock DGB, editors. Developments in thermochemical biomass conversion. Springer-Science+Business Media, B.V.; 1997. p. 481–8.

[29] Solantausta Y, et al. Wood-pyrolysis oil as fuel in a diesel-power plant. Bioresour Technol 1993;46:177–88.

[30] Chiaramonti D, et al. Development of emulsions from biomass pyrolysis liquid and diesel and their use in engines—part 2: tests in diesel engines. Biomass Bioenergy 2003;25(1):101–11.
[31] Crayford, A.P., P.J. Bowen, and P.J. Kay, Comparison of gas-oil and bio-oil spray performance for use in a gas turbine. Proceedings of the ASME Turbo Expo, vol. 1 2010. : p. 659–667.
[32] Strenziok, R., U. Hansen, and H. Kiinstner, Combustion of Bio-oil in a Gas Turbine. Progress in thermochemical biomass conversion, 2001(Oxford: Blackwell Science): p. 1452–1458.
[33] Goyal HB, Seal D, Saxena RC. Bio-fuels from thermochemical conversion of renewable resources: a review. Renew Sustain Energy Rev 2008;12(2):504–17.
[34] Bridgwater AV. In: Fast pyrolysis. 15th European Biomass Conference and Exhibition, Berlin; 2007.
[35] Czernik S, Bridgwater AV. Overview of applications of biomass fast pyrolysis oil. Energy Fuel 2004;18(2):590–8.
[36] Balat M. An overview of the properties and applications of biomass pyrolysis oils. Energy Source A Recover Util Environ Eff 2011;33(7):674–89.
[37] Bridgwater AV, Meier D, Radlein D. An overview of fast pyrolysis of biomass. Org Geochem 1999;30(12):1479–93.
[38] Huber GW, Iborra S, Corma A. Synthesis of transportation fuels from biomass: chemistry, catalysts, and engineering. Chem Rev 2006;106(9):4044–98.
[39] Ahmad M, et al. Biochar as a sorbent for contaminant management in soil and water: a review. Chemosphere 2014;99:19–33.
[40] Mohan D, et al. Organic and inorganic contaminants removal from water with biochar, a renewable, low cost and sustainable adsorbent—a critical review. Bioresour Technol 2014;160:191–202.
[41] Bridgwater AV. Catalysis in thermal biomass conversion. Appl Catal A Gen 1994; 116(1–2):5–47.
[42] Stefanidis SD, et al. A study of lignocellulosic biomass pyrolysis via the pyrolysis of cellulose, hemicellulose and lignin. J Anal Appl Pyrolysis 2014;105:143–50.
[43] Richards GN. Chemistry of pyrolysis of polysaccharides and lignocellulosics. In: Bridgwater AV, editor. Advances in thermochemical biomass conversion. London: Blackie Academic & Professional; 1994. p. 727–45.
[44] Blanco A, Chejne F. Modeling and simulation of biomass fast pyrolysis in a fluidized bed reactor. J Anal Appl Pyrolysis 2016;118:105–14.
[45] Yang H, et al. Characteristics of hemicellulose, cellulose and lignin pyrolysis. Fuel 2007;86(12–13):1781–8.
[46] Williams PT, Besler S. The in fluence of temperature and heating rate on the slow pyrolysis of biomass. Renew Energy 1996;7(3):233–50.
[47] Akhtar J, Saidina Amin N. A review on operating parameters for optimum liquid oil yield in biomass pyrolysis. Renew Sustain Energy Rev 2012;16(7):5101–9.
[48] Gable P, Brown RC. Effect of biomass heating time on bio-oil yields in a free fall fast pyrolysis reactor. Fuel 2016;166:361–6.
[49] Zhou G, et al. Direct upgrading of fast pyrolysis lignin vapor over HZSM-5 catalyst. Green Chem 2016;18(7):1965–75.
[50] Freda C, et al. In: Pyrolysis oils obtained from different biomasses. 14th European biomass conference, Paris, France; 2005.
[51] Fahmi R, et al. The effect of lignin and inorganic species in biomass on pyrolysis oil yields, quality and stability. Fuel 2008;87(7):1230–40.

[52] Ghetti P, Ricca L, Angelini L. Thermal analysis of biomass and corresponding pyrolysis products. Fuel 1996;75(5):565–73.
[53] Kallioinen A, et al. Effects of bacterial treatments on wood extractives. J Biotechnol 2003;103(1):67–76.
[54] Roy C, Pakdel H, Brouillard D. The role of extractives during vacuum pyrolysis of wood. J Appl Polym Sci 1990;41(1–2):337–48.
[55] Nik-Azar M, et al. Mineral matter effects in rapid pyrolysis of beech wood. Fuel Process Technol 1997;51:7–17.
[56] Raveendran K, Ganesh A, Khilart KC. Influence of mineral matter on biomass pyrolysis characteristics. Fuel 1995;74(12):1812–22.
[57] Muller-Hagedorn M, et al. A comparative kinetic study on the pyrolysis of three different wood species. J Anal Appl Pyrolysis 2003;68-69:231–49.
[58] Lee KH, et al. Influence of reaction temperature, pretreatment, and a char removal system on the production of bio-oil from rice straw by fast pyrolysis, using a fluidized bed. Energy Fuel 2005;19(5):2179–84.
[59] Das P, Ganesh A, Wangikar P. Influence of pretreatment for deashing of sugarcane bagasse on pyrolysis products. Biomass Bioenergy 2004;27(5):445–57.
[60] Fahmi R, et al. The effect of alkali metals on combustion and pyrolysis of Lolium and Festuca grasses, switchgrass and willow. Fuel 2007;86(10–11):1560–9.
[61] Bridgwater AV. Review of fast pyrolysis of biomass and product upgrading. Biomass Bioenergy 2012;38:68–94.
[62] Talmadge MS, et al. A perspective on oxygenated species in the refinery integration of pyrolysis oil. Green Chem 2014;16(2):407–53.
[63] Venderbosch RH, Prins W. Fast pyrolysis technology development. Biofuels Bioprod Biorefin 2010;4(2):178–208.
[64] Radlein D, Piskorz J, Scott DS. Control of selectivity in the fast pyrolysis of cellulose. Elsevier; 1992643–9.
[65] Scott DS, et al. Pretreatment of poplar wood for fast pyrolysis: rate of cation removal. J Anal Appl Pyrolysis 2001;57(2):169–76.
[66] Radlein D, Quignard A. A short historical review of fast pyrolysis of biomass. Oil Gas Sci Technol—Revue d'IFP Energies nouvelles 2013;68(4):765–83.
[67] Shuping Z, et al. Pyrolysis characteristics and kinetics of the marine microalgae Dunaliella Tertiolecta using thermogravimetric analyzer. Bioresour Technol 2010;101(1):359–65.
[68] Scott DS, et al. Pretreatment of poplar wood for fast pyrolysis: rate of cation removal. J Anal Appl Pyrolysis 2000;57(2):169–76.
[69] Blasi CD, Branca C, D'Errico G. Degradation characteristics of straw and washed straw. Thermochimica Acta 2000;364:133–42.
[70] Carrier M, et al. Influence of chemical pretreatment on the internal structure and reactivity of pyrolysis chars produced from sugar cane bagasse. Energy Fuel 2012;26(7):4497–506.
[71] Boateng AA, Mullen CA. Fast pyrolysis of biomass thermally pretreated by torrefaction. J Anal Appl Pyrolysis 2013;100:95–102.
[72] Zheng A, et al. Effect of torrefaction on structure and fast pyrolysis behavior of corncobs. Bioresour Technol 2013;128:370–7.
[73] Meng J, et al. The effect of torrefaction on the chemistry of fast-pyrolysis bio-oil. Bioresour Technol 2012;111:439–46.
[74] Pandey KK, Pitman AJ. FTIR studies of the changes in wood chemistry following decay by brown-rot and white-rot fungi. Int Biodeter Biodegr 2003;52(3):151–60.
[75] Singh D, et al. Investigation of wheat straw biodegradation by Phanerochaete chrysosporium. Biomass Bioenergy 2011;35(3):1030–40.

[76] Yang X, et al. Effect of biopretreatment on thermogravimetric and chemical characteristics of corn stover by different white-rot fungi. Bioresour Technol 2010;101 (14):5475–9.
[77] Boroson ML, et al. Product yields and kinetics from the vapor phase cracking of wood pyrolysis tars. AIChE J 1989;35(1):120–8.
[78] Elliott DC. Relation of Reaction Time and Temperature to Chemical Composition of pyrolysis Oils. In: Soltes EJ, Milne TA, editors. Pyrolysis oils from biomass : producing, analyzing, and upgrading. Washington, DC: American Chemical Society; 1988. p. 55–65.
[79] Horne PA, Williams PT. Influence of temperature on the products from the flash pyrolysis of biomass. Fuel 1996;75(9):1051–9.
[80] Piskorz J, et al. Fast pyrolysis of sweet sorghum and sweet sorghum bagasse. J Anal Appl Pyrolysis 1998;46(1):15–29.
[81] Liu C, et al. Catalytic fast pyrolysis of lignocellulosic biomass. Chem Soc Rev 2014; 43(22):7594–623.
[82] Zhang H, et al. Biomass fast pyrolysis in a fluidized bed reactor under N_2, CO_2, CO, CH_4 and H_2 atmospheres. Bioresour Technol 2011;102(5):4258–64.
[83] Guizani C, Escudero Sanz FJ, Salvador S. Effects of CO_2 on biomass fast pyrolysis: reaction rate, gas yields and char reactive properties. Fuel 2014;116:310–20.
[84] Agblevor FA, Besler S. Inorganic compounds in biomass feedstocks. 1. Effect on the quality of fast pyrolysis oils. Energy Fuel 1996;10(2):293–8.
[85] Adjaye JD, Bakhshi NN. Catalytic conversion of a biomass-derived oil to fuels and chemicals I: model compound studies and reaction pathways. Biomass Bioenergy 1995;8(3):131–49.
[86] Radlein D, Piskorz J, Majerski P. Method of Upgrading Biomass Pyrolysis Liquids for Use as Fuels and as a Source of Chemicals by Reaction with Alcohols. Canadian Patents Database, U.K.; 1996.
[87] Boucher ME, et al. Bio-oils obtained by vacuum pyrolysis of softwood bark as a liquid fuel for gas turbines. Part II: stability and ageing of bio-oil and its blends with methanol and a pyrolytic aqueous phase. Biomass Bioenergy 2000;19(5):351–61.
[88] Boucher ME, Chaala A, Roy C. Bio-oils obtained by vacuum pyrolysis of softwood bark as a liquid fuel for gas turbines. Part I: properties of bio-oil and its blends with methanol and a pyrolytic aqueous phase. Biomass Bioenergy 2000;19(5):337–50.
[89] Oasmaa A, et al. Fast pyrolysis of forestry residue and pine. 4. Improvement of the product quality by solvent addition. Energy Fuel 2004;18(5):1578–83.
[90] Oasmaa A, et al. Quality improvement of pyrolysis liquid: effect of light volatiles on the stability of pyrolysis liquids. Energy Fuel 2005;19(6):2556–61.
[91] Bridgwater AV, Cottam ML. Opportunities for biomass pyrolysis liquids production and upgrading. Energy Fuel 1992;6(2):113–20.
[92] Elliott DC, et al. Catalytic Hydroprocessing of fast pyrolysis bio-oil from pine sawdust. Energy Fuel 2012;26(6):3891–6.
[93] Kan T, Strezov V, Evans TJ. Lignocellulosic biomass pyrolysis: a review of product properties and effects of pyrolysis parameters. Renew Sustain Energy Rev 2016; 57:1126–40.
[94] Stucley C. In: Fast pyrolysis, for renewable liquid fuels from wood. Biofuels Conference, Wellington, New Zealand; 2010.
[95] ENSYN. The RTP pyrolysis pathway maximizing value of biomass residues; 2011.
[96] ABRI-Tech. ABRI-Tech Inc. 2014 ; Available from: http://www.advbiorefineryinc.ca/ (1 January).

[97] Berruti F, et al. A mobile pyrolyzer for converting agricultural and forestry residues into liquid bio-oil and bio-char. In: The International Conference on Thermochemical Conversion Science; Chicago, IL: Agri-Therm Inc.; 2009.
[98] Berruti FM. ICFAR, mobile pyrolysis system (MPS200), LPS100, MFR & feeding technology, pyrolysis future outlook. Ontario: Institute for Chemicals and Fuels from Alternative Resources. University of Western Ontario; 2013.
[99] Marshall AJ. Commercial application of pyrolysis technology in agriculture. Guelph, ON: Ontario Federation of Agriculture (OFA); 2013.

Catalytic pyrolysis

A. Pattiya
Mahasarakham University, Kamriang, Maha Sarakham, Thailand

2

2.1 Introduction

2.1.1 Background

Fast pyrolysis is a thermochemical liquefaction of lignocellulosic biomass for bio-oil production. Although the as-produced bio-oils have versatile potential applications, they can further be upgraded with the use of catalysts. As already mentioned in Chapter 1, bio-oil can be hydrotreated with catalysts like sulphided Co–Mo or Ni–Mo using high hydrogen pressure in order to remove oxygen from liquid bio-oil, giving hydrocarbon-rich liquid biofuel as product. Another possible upgrading route for bio-oil is the use zeolite catalysts such as ZSM-5 to deoxygenate the liquid bio-oil at atmospheric pressure without hydrogen input. With this technique, bio-oil properties are improved by cracking and aromatisation reactions, giving lower molecular weight and oxygen content together with higher proportion of hydrocarbons.

Although technically possible, the liquid bio-oil upgrading technique is not attractive because the bio-oil needs to be revapourised prior to reactions over the catalyst. In fast pyrolysis process, pyrolysis vapour is condensed into bio-oil; the condensation followed by evaporation for upgrading is not thermally efficient. Therefore, it would be more efficient to introduce catalysts into the pyrolysis process prior to condensation of the vapour so as to induce vapour-phase catalytic reactions. Since the catalytic vapour-phase upgrading is carried out immediately after or at the same time as pyrolysis in one thermal cycle, the combination of the upgrading and the pyrolysis process is referred to as 'catalytic pyrolysis of biomass', which is the main focus of this chapter. It is important to note that the main purpose of the catalyst in catalytic pyrolysis of biomass is the removal of the oxygenated compounds simultaneously with catalytically cracking the high-molecular-weight compounds into products of desired carbon chain length [1].

2.1.2 Mode of catalysis

There are four possible configurations (see Fig. 2.1) that can be used to incorporate a catalyst into a fast pyrolysis system: (i) biomass and catalyst cofeeding, (ii) use of catalyst as the whole or part of heat-transfer medium in fast pyrolysis reactor, (iii) close-coupled in-bed catalysis, and (iv) close-coupled catalyst in a secondary reactor [2].

In cofeeding biomass and catalyst, the catalyst is added to the biomass before experiments. Depending on the nature of the catalyst, the addition can be by either wet impregnation or dry mixing. Possible drawbacks of this mode of catalyst addition

Direct Thermochemical Liquefaction for Energy Applications. https://doi.org/10.1016/B978-0-08-101029-7.00002-3

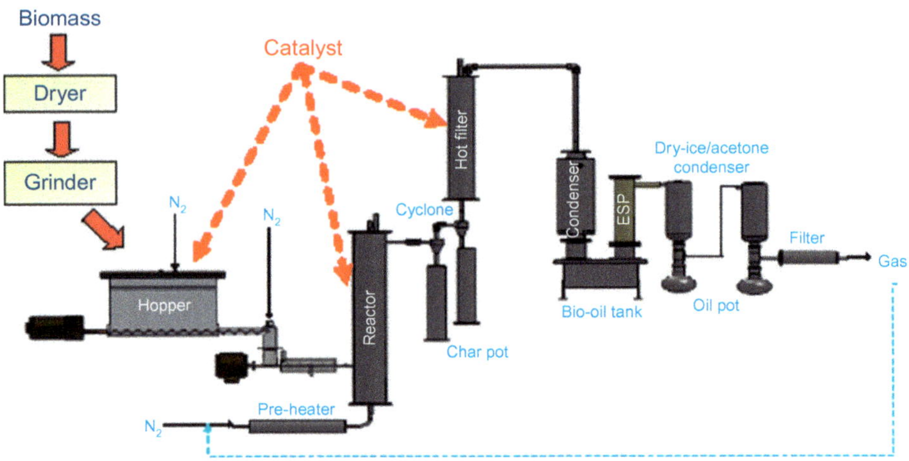

Fig. 2.1 Possible positions for incorporating catalysts in a fast pyrolysis system.

include difficulty in recovering the catalyst and a high chance of catalyst particles entraining out of the cyclone and ending up in the liquid product.

The use of catalyst as the whole or part of the heat-transfer medium allows the catalyst to be mixed directly with the feedstock in the pyrolysis reactor. This mode is referred to as in situ catalytic fast pyrolysis (in situ CFP). If a fixed- or fluidised-bed reactor is used for fast pyrolysis, this mode can be called in-bed catalytic fast pyrolysis. It is considered as a simple and economical way to incorporate catalyst into the pyrolysis process. In this mode, the pyrolysis and catalytic reactions take place under the same operating temperature. It also allows an immediate contact of pyrolysis vapour with the catalyst. Therefore, the catalyst is involved in the cracking and deoxygenation reactions at a very early stage of the evolved pyrolysis vapour, allowing the reactive and large molecules to be immediately reacted before seconday thermal reactions take place. The catalyst surface can also aid the thermal decomposition of the biomass. However, depending on the thermal conductivity of the catalytic material, the heating rate or heat transfer for pyrolysis reaction can be lower than when using typical sand as heat-transfer medium. Since the contact time between the pyrolysis vapour and catalyst in this mode is very short, significant proportion of catalyst is required in order to modify the quality of the bio-oil product or to produce high yield of liquid hydrocarbons. Otherwise, only partial upgrading is achieved. With this catalytic mode, the catalyst may be rapidly deactivated by coking or deposition of the char fines and condensed pyrolysis vapour. The char and catalyst are mixed in the process, which make it difficult to separate the catalyst after the reaction. These problems can be solved by burning char and coke in a separate combustor to generate heat for keeping the temperature of the catalyst. The catalyst is then recycled back to the pyrolysis reactor in a similar way as in the circulating fluidised-bed reactor. This facilitates continuous operation of the in situ catalytic fast pyrolysis system.

The third mode of catalyst addition is the close-coupled in-bed catalysis where catalyst is placed as fixed or fluidised bed at the reactor freeboard.

In this configuration, pyrolysis vapour together with char particles passes over the catalyst bed, leading to the accumulation of the char that could ultimately increase the pressure drop across the bed. Therefore, the char particles should be separated from the pyrolysis vapour stream by a cyclone prior to being passed over a catalyst bed. This idea leads to an occurrence of the fourth incorporation mode, which is close-coupled catalysis in a secondary reactor. This catalytic position is referred to as ex situ or ex-bed catalytic fast pyrolysis. This mode allows the catalyst to be operated under different operating temperatures and residence time from the primary pyrolysis reactor. Therefore, pyrolysis and catalysis conditions can be separately optimised [3]. It can eliminate the problems of char accumulation and catalyst poisoning by metals in biomass, thus preventing the premature catalyst deactivation. The ex-bed reactor can be fixed, fluidised, moving, or circulating beds, and only pyrolysis vapour after removing char by cyclone and hot filter units passes through the catalyst bed. Although this catalytic pyrolysis mode is flexible and promising, the economics related to the capital and operating costs due to the requirement of a secondary reactor is the main disadvantage for this mode.

2.1.3 Catalytic pyrolysis reactor

Catalytic pyrolysis of biomass has been investigated with several types of pyrolysis reactors. The most widely used reactor for catalyst testing or screening is a microscale fixed-bed reactor coupled with gas chromatography/mass spectrometry (GC/MS) and/or gas chromatography/flame ionisation detector (GC/FID). This reactor system is known as pyroprobe, pyrolysis-GC/MS, py-GC/MS, or analytical pyrolysis. Several research studies have been reported using this system [4–10]. The benefit of py-GC/MS is the rapid analysis of the catalyst activities, biomass types, and process parameters during catalytic pyrolysis of biomass. However, the py-GC/MS applied milligram quantity of biomass and catalyst to produce pyrolysis vapour. The vapour is then analysed by GC/MS without being condensed into liquid product. It is likely that a number of the chemicals identifed in the pyrolysis vapour by the GC/MS are the same as those found in liquid bio-oil. However, it is also possible that some chemicals in the pyrolysis vapour can further react during condensation to become other chemicals in bio-oils. In addition, bio-oil is composed of large molecules such as oligomers or polymers derived from the degradation of cellulose, hemicellulose, and lignin, which cannot be detected by GC/MS system. Therefore, the analysis of the py-GC/MS may not entirely represent the catalytic bio-oil.

Second type of reactor that is easy to set up for catalyst, feedstock, and process parameter testing is a typical fixed-bed reactor. Previous studies have been performed using rice husk [11–13], cotton seed [14,15], and *Euphorbia rigida* [16–19] in fixed-bed reactors with selected studies discussed below.

Ates et al. [18,19] studied the catalytic pyrolysis of *E. rigida* using a commercial Co–Mo catalyst (Criterion-534) in a bench-scale fixed-bed reactor in order to investigate the effects of catalyst ratios (0%, 5%, 10%, and 25% by weight) and water vapour velocities (0.0, 0.6, 1.3, and 2.7 cm/s) on bio-oil yields. Ten grams of the biomass sample having average particle size of 0.55 mm was mixed with the catalyst and placed in the reactor. The pyrolysis experiments were conducted at a slow heating rate of 7°C/min to a final temperature of 550°C. The results showed that by using 20 wt% of the Criterion-534 catalyst in static atmosphere, bio-oil yield on dry, ash-free basis was increased from 21.65 (without any catalyst) to 30.98% [18]. The yield was increased further when applying water vapour and reached the maximum of 42.56% with 1.3 cm/s water vapour velocity [19]. In addition, the bio-oil analysis results suggested that the presence of steam enhanced the formation of aromatic compounds [19].

The same group of researchers, Ates and coworkers, also tested two commercial catalysts from petrochemical industry, namely, DHC-32 (Ni-W/Al_2O_3) and HC-K 1.3Q (Ni-Mo/Al_2O_3), in catalytic pyrolysis of biomass so as to explore the influence of catalysts on bio-oil yields and properties [17]. The former contains mainly (wt%) aluminium oxide (40.0%–60.0%), nickel oxide (1.0%–6.0%), silicon oxide (10.0%–30.0%), and tungsten oxide (15.0%–25.0%), while the latter is composed mainly of (wt%) aluminium oxide (35.0%–65.0%), molybdenum trioxide (0.1%–28.0%), silica (amorphous) (0.1%–20.0%), molybdenum metal/powder (1.0%–15.0%), nickel oxide (0.1%–6.0%), and aluminium phosphate (0.1%–3.0%). Two biomass samples (*E. rigida* and sesame stalk) were pyrolysed at a slow heating rate of 7°C/min using a bench-scale fixed-bed reactor. The results showed that at a final pyrolysis temperature of 500°C, the use of catalysts led to the increase of bio-oil yields and the decrease of gas yields for both biomass samples. The increases in the yields of bio-oils were more pronounced when using HC-K 1.3Q catalyst. It is important to note that typically the bio-oil yield is decreased with the use of catalyst in catalytic fast pyrolysis, but in this case, the bio-oil yields were increased after catalysis. This could be due to the lower heating rate applied in the process, and catalyst may aid the heat transfer, thus increasing the heating rate of the biomass pyrolysis reactions. With regard to the bio-oil composition, different catalysts have generally been used with an aim to decrease the amount of oxygenated compounds in the bio-oils. However, it was found that the oxygenated compounds increased in the catalytic *E. rigida* bio-oils compared with the noncatalytic one. It is thought that this phenomenon is a result of the terpenoid structure of the *E. rigida* biomass. Nevertheless, for the sesame stalk, this situation is vice versa as adding a catalyst could decrease the amount of these compounds. These results are consistent with the works carried out by Adam et al. [20] and Antonakou et al. [21] as mentioned earlier that the biomass types and/or composition also play an important role during the catalytic pyrolysis.

Another type of reactor that has been tested in catalytic pyrolysis is microwave reactors. Catalytic microwave-induced pyrolysis of biomass (hybrid poplar, loblolly pine, and Douglas fir) was performed to produce liquid organics rich in aromatics followed by hydrogenation in n-heptane medium for production of renewable high-density cycloalkanes for jet fuels. In microwave reactor, activated carbon powder was used as an absorber for the microwave heating. The pyrolysis vapour passed through a fixed-bed catalytic reactor filled with ZSM-5 catalyst [22].

In addition, several other reactors have been used to study catalytic pyrolysis of biomass including in situ [23–33] and ex situ [34–37] bubbling fluidised bed, in situ conical spouted bed [38,39], in situ fixed beds with constant catalysts feed [40–43], and in situ circulating fluidised-bed [44–46] reactors.

2.2 Mechanism of catalytic reactions

Bio-oils contain significant amount of oxygenated compounds including lignin derivatives that results in the relatively low heating values, low stability, high viscosity, low volatility, and low pH. To overcome these problems, a catalyst can be incorporated into a pyrolysis unit to induce catalytic reactions. Conceptual catalytic pyrolysis of biomass is shown in Fig. 2.2.

A combination of several reactions is expected to take place during catalytic pyrolysis of biomass including cracking, protolytic cracking (cleavage of C–C bonds), hydrocracking, aromatic side-chain scission, hydrogen transfer, hydrogenation, dehydrogenation, isomerisation, oligomerisation, polymerisation, dehydration, decarboxylation, decarbonylation, ketonisation, condensation, alkylation, cyclisation, and aromatisation reactions [1,4,6,21,23,47–57]. These reactions lead to the removal of oxygen and the reduction of molecular weight of the produced catalytic bio-oils. Oxygen is rejected from the pyrolysis vapour or bio-oil as CO (decarbonylation), CO_2 (decarboxylation), and H_2O (dehydration). The CO_2 route is most desirable since two moles of oxygen are removed per mole of carbon, whereas hydrogen, which is deficient in biomass, is preserved.

Regarding the cracking reactions, a desired catalyst would selectively crack the heavy molecules present in pyrolysis vapour, particularly lignin oligomers expelled from the lignin macropolymer of the original biomass during pyrolysis [58] to produce lower-molecular-weight compounds, while minimising the cracking of low-molecular-weight molecules to permanent gases. In cracking lignin derivatives, deoxygenation can also occur; for example, when ether bonds (C–O–C), both ether bridge joining subunits and diaryl ether are cleaved, and carbon monoxide is generated [59]. Moreover, during cracking of pyrolysis vapours, some olefinic gases such as ethylene and propylene may occur as intermediates that can undergo further

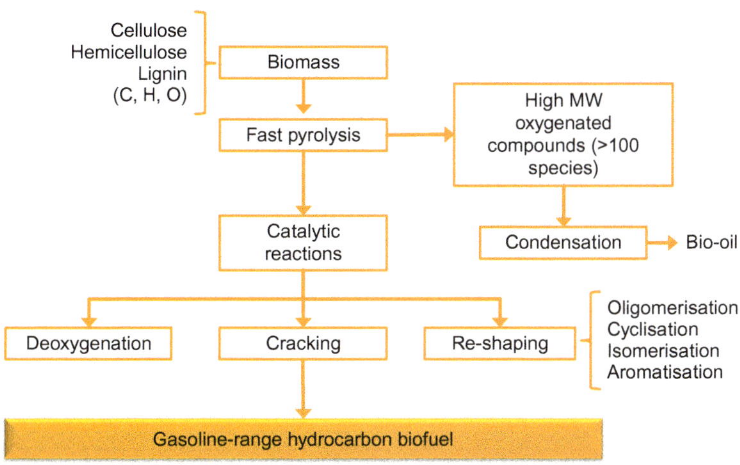

Fig. 2.2 Catalytic pyrolysis concept.

reforming reactions such as oligomerisation, cyclisation, isomerisation, or aromatisation to produce high-value liquid hydrocarbons.

The mechanisms of cracking and reforming of biomass pyrolysis vapours on ZSM-5 catalysts are not well understood. However, Bridgwater [60] suggested a global mechanism, which is a combination of cracking on the catalyst surface followed by synthesis of aromatics in the catalyst pores.

Williams and Horne [61,62] suggested that the formation of aromatic and polycyclic aromatic compounds during the zeolite ZSM-5 catalytic upgrading of biomass pyrolysis oils occurs by a dual mechanistic route. First, low-molecular-weight hydrocarbons such as alkene gases are formed on the catalyst, which then undergo aromatisation reactions (possibly Diels-Alder reactions) to produce aromatic and polycyclic aromatic hydrocarbons. Additionally, deoxygenation of oxygenated compounds found in the nonphenolic fraction of the pyrolysis oils may directly form aromatic compounds without going through the alkene route.

In addition, a hydrocarbon pool mechanism widely accepted for conversion of methanol and ethanol to hydrocarbons over HZSM-5 [63–66] can be used to explain the conversion of biomass-derived compounds into aromatics over zeolites [67,68]. According to this theory, oxygenates, such as furans and aldehydes, diffuse into the zeolite pores and undergo a series of reactions, eventually leading to aromatics and olefins [5,63,67,69–71].

The severity of reactions such as cracking, deoxygenating, and reforming influences the liquid product distribution and depends largely on feedstock composition, types and properties of catalyst, and process parameters. The effects of feedstock composition, catalyst types, catalyst properties, and process parameters are reviewed in the next sections.

2.3 Effect of feedstock composition

2.3.1 Cellulose, hemicellulose and lignin

The woody biomass typically contains about 40–47 wt% cellulose, 25–35 wt% hemicellulose, and 16–31 wt% lignin [72,73] with small content of inorganics. The proportion of these components in the feedstock affects the yield of and properties of liquid product in catalytic pyrolysis. Pure cellulose with empirical formula $C_6H_{10}O_5$ has been investigated in catalytic fast pyrolysis in a bubbling fluidised bed [33] or in a micropyrolyser coupled with GC–MS/FID/TCD [73–75]. Carlson et al. found that cellulose gave approximately 30% carbon yield of aromatics when using ZSM-5 with catalyst-to-feed ratio of 19 in a pyroprobe reactor [74]. By comparing the three major biomass components in the in situ catalytic fast pyrolysis using a pyroprobe reactor with H-ZSM-5 catalyst, it has been shown that cellulose gave highest yield of aromatic hydrocarbons followed by hemicellulose and lignin [75]. Similar findings were reported by Huang et al. [76] when testing 6.0 wt% La/HZSM-5 catalyst in a fixed-bed reactor using rice husk, sawdust, sugarcane bagasse, and cellulose, hemicellulose, and lignin as feedstocks. Feedstocks with higher cellulose or hemicellulose contents gave higher olefin yields compared with those having higher lignin content. This finding is also confirmed by another study [22] that catalytically pyrolysed three lignocellulosic biomass feedstocks (hybrid poplar, loblolly pine, and Douglas fir) with different contents of cellulose, hemicellulose, and lignin in a microwave reactor with ZSM-5 as a catalyst. They found that a maximum carbon yield of aromatics (25%) was obtained from hybrid poplar, which was higher than those for loblolly pine (23%) and Douglas fir (22%). This is possibly due to the higher cellulose content of hybrid poplar [22].

One of the reasons why lignin contributes to low hydrocarbon yields in catalytic pyrolysis of biomass is due to the coke formation on catalyst surface and pores by lignin derivatives [76]. Nevertheless, lignin itself can give a reasonable yield of hydrocarbon products suitable process conditions are applied. For example, it has been found when testing lignin from a wheat straw-based bioethanol plant in a continuous fast pyrolysis reactor with a close-coupled fixed bed of HZSM-5 catalyst (SiO_2/Al_2O_3 ratio of 23) that optimum catalyst temperature for obtaining high liquid yield was 600°C [77]. Without the catalyst, 27.6 wt% daf (dry, ash-free) of organic liquid was obtained. With the catalyst at 600°C, the yield of organics decreased to 5.7 wt% daf containing 70 wt% aromatic hydrocarbons (mainly benzene and toluene). Thus, 4 wt% daf (or 6% carbon yield) of aromatics was produced from lignin catalytic pyrolysis at 600°C compared with 14% carbon yield when wood was catalytically pyrolysed in a fluidised-bed reactor with HZSM-5 catalyst [78]. The lignin organic yield corresponds to 8.7% energy recovery in liquid product compared with 33% for the noncatalytic counterpart [77]. Deoxygenation of lignin catalytic pyrolysis occurred mainly by dehydration and decarbonylation, thus rejecting oxygen as H_2O and CO, respectively. This leads to a very low oxygen content of 4% for catalytic lignin organic liquid. Compared with wood or cellulose fast pyrolysis, lignin gave higher proportion of char product [77].

2.3.2 Inorganics

Although woody biomass typically contains small proportion of inorganics, other types of biomass such as agricultural residues have a relatively high ash or inorganic content. The inorganics can deposit onto catalyst surface or internal pore structure during catalytic pyrolysis processing, resulting in catalyst deactivation [79]. The in situ or in-bed mode of catalysis seems to be affected more than the ex situ mode because in the ex situ mode, most of the inorganics are separated by cyclone separators and hot gas filters prior to catalytic reactions. Mullen and Boateng [80] studied the deposition of inorganic metals during catalytic pyrolysis of switchgrass using H-ZSM-5 catalyst and found that in addition to the deposition of coke species, inorganics like transition metals such as Fe and Cu and alkali and alkaline earth metals such as K, Ca, and Mg could be deposited onto the surface and within the catalyst's pores that ultimately slow down the catalytic activity.

2.3.3 Feedstock pretreatment

Pretreatment of biomass feedstock prior to catalytic pyrolysis is also known to affect the product yields and properties. Three pretreatment techniques can be applied to biomass prior to catalytic pyrolysis processing including biomass washing/demineralisation, fungal treatment, and torrefaction. Washing or leaching of biomass with water or acid aqueous solution either at room temperature or at elevated temperature can remove inorganic species from biomass. As mentioned earlier that inorganics can be harmful to catalyst as they are involved in blocking the active sites of catalyst. By removing these species, coke formation and catalyst life can be extended to a certain level. Yildiz et al. [81] investigated the effect of biomass ash in catalytic fast pyrolysis of pine wood using ZSM-5-based catalyst in a continuously operated mechanically stirred-bed reactor at 500°C and found that products. The addition of ash to the catalytic fast pyrolysis reduced the yields of both the organics and coke by 2 wt% (on feed basis), while increasing 1 wt% water and 4 wt% permanent gas. The total yield of CO plus CO_2 was also found to be increased by >10%, while the CO_2 production (decarboxylation reactions) was clearly favoured with the addition of ash [81]. This can be regarded as the benefit of ash in biomass as the preferred pathway is usually decarbonylation (de-CO_2).

The effect of biological pretreatment of corn stover by white-rot fungus *Irpex lacteus* CD2 was investigated in catalytic fast pyrolysis using ZSM-5 catalyst in a pyroprobe reactor [82]. The results show that the fungal pretreatment could help increase the aromatic hydrocarbon yield from 10.03 wt% without the treatment to 11.49 wt% and the coke yield decreased from 14.29 to 11.93 wt%. This indicated that fungal pretreatment has a beneficial effect on biomass catalytic fast pyrolysis as it can improve aromatic production together with reduction of the undesirable coke.

Biomass can also be thermally pretreated prior to catalytic pyrolysis by torrefaction technique. Torrefaction is a mild pyrolysis process operated at a lower temperature than fast pyrolysis, typically around 250–280°C with the aim to remove moisture and some volatiles, while maintaining most carbon in the biomass. The effect of biomass torrefaction on catalytic pyrolysis has been investigated by several researchers [83–87]. Depending on the feedstock type, the torrefaction pretreatment can lead to the increase or decrease of aromatic hydrocarbon yields. When using wood and corncob, the aromatic hydrocarbon yields from catalytic pyrolysis of torrefied biomass in pyroprobe reactors with ZSM-5 catalyst were increased [84,85,87]. For torrified hemicellulose and lignin, the aromatic hydrocarbon yields were reduced due to the structural transformation of the biomass during pretreatment [83,86]. Nevertheless, selectivity of benzene, toluene, and xylenes (BTX) from catalytic fast pyrolysis of torrefied hemicellulose and lignin was enhanced [86]. The severity of torrefaction condition also affected the product yield in catalytic pyrolysis of torrefied biomass. Severe torrefaction (torrefaction at 270°C or 300°C) can lead to the sharp increase of coke yield and reduction of aromatic yield [87].

2.4 Types of catalysts

In catalytic pyrolysis of biomass, selection of catalyst type is crucial. Ideally, the catalysts should be highly active, selective to particular products, resistant to deactivation, readily recycled, and cheap [88]. Various types of catalysts can be applied for catalytic pyrolysis of biomass. The most widely investigated catalysts for this purpose include zeolites, especially ZSM-5 type, and zeolite-like mesoporous catalytic materials. In addition, metal oxides and certain natural catalysts have also been studied in the field of catalytic pyrolysis. The review of catalytic pyrolysis of biomass with these different types of catalyst is given below.

2.4.1 Zeolite

Zeolites are microporous three-dimensional crystalline solids with well-defined channels or pores of uniform diameter. They are aluminosilicate materials that consist of SiO_4 and AlO_4^- tetrahedra, which are interlinked by the sharing of oxygen atoms to give a three-dimensional network through which long channels run [89]. Different types of zeolites have different structures or frameworks, which result in their different properties and applications. The molecular size dimensions of pores and channels of zeolites make them suitable for shape selective catalysis in which the reaction selectivity depends on the size and architecture of the pores [90]. Zeolite such as chabazite and mordenite can be found naturally in rocks; however, the synthetic zeolites such as H-MFI (i.e. H-ZSM-5), H-BEA (i.e. H-Beta), and H-FAU (i.e. H–X and H–Y) and their modified forms are widely used for industrial chemical reactions [79]. Zeolites can be used as catalysts in many applications such as cracking of heavy oil in petroleum refinery industries. The catalytic activities of zeolites are related to their shape selectivity and acidity. Additional details on these catalyst properties are in Sections 2.5.1 and 2.5.2.

2.4.2 ZSM-5 based catalysts

ZSM-5 is a zeolite that possesses the three-dimensional system of intersecting channels that is made up of elliptical straight channels (0.51 × 0.55 nm) and near-circular zigzag channels (0.54 × 0.5 nm) [91]. It is initially developed by Mobil and is used in methanol-to-gasoline (MTG) process. The use of a monofunctional zeolite ZSM-5 catalyst in converting biomass pyrolysis liquid into gasoline-range hydrocarbons was suggested by Bridgwater and Cottam [92] to conceptually follow the equation:

$$C_6H_8O_4 \rightarrow 4.6CH_{1.2} + 1.4CO_2 + 1.2H_2O$$

where $C_6H_8O_4$ and $CH_{1.2}$ represent pyrolysis liquid and hydrocarbon products (mainly aromatics), respectively. This gives a maximum stoichiometric yield of 42 wt% on liquid or a maximum energetic yield of about 50 wt%. If a bifunctional or multifunctional catalyst that can generate in situ hydrogen through shifting the carbon monoxide in the product gas is used, the process would be expected to operate more in a carbon-limited environment rather than hydrogen limited, as depicted in the following equation with a maximum stoichiometric yield of 55 wt% on liquid [92]:

$$C_6H_8O_4 + 3.6H_2 \rightarrow 6CH_{1.2} + 4H_2O$$

The processing conditions for the upgrading are atmospheric pressure, temperature from 350°C to 600°C, and hourly space velocities of around two [60]. There are a number of beneficial process parameters including low-pressure (atmospheric) processing; no hydrogen addition; temperature similar to those preferred for optimum yields of bio-oil; and close-coupled process, which offer significant processing and economical advantages over hydrotreating [60]. In addition, a higher grade of gasoline (high-octane gasoline) is produced than from hydrotreating as the high yields of aromatics (BTX) are produced from the zeolite [92].

Catalytic pyrolysis of biomass in a bench-scale reactor using ZSM-5 catalysts has been carried out extensively at the University of Leeds [61,62,93–99]. The experimental system used was a combined fluidised-bed pyrolysis reactor unit with the freeboard of the reactor packed with the zeolite to form a fixed catalyst bed. The reactor was 10cm diameter × 100cm high, constructed of stainless steel with full gas flow and temperature control. The reactor was heated externally, and the fluidised bed and freeboard could be separately temperature controlled. Biomass feedstocks tested included pine wood [93], mixed wood waste [61,62,94–97,99,100], and rice husk [98]. The primary pyrolysis and catalyst bed temperatures investigated ranged from 400°C to 600°C, and the weight hourly space velocity (WHSV) was around $1\,h^{-1}$. It was found from these studies that ZSM-5 catalysts affected not only the product yields but also the bio-oil composition. Generally, the catalysts decreased the yield of liquid bio-oils along with increasing the yields of permanent gases, which included carbon dioxide (CO_2), carbon monoxide (CO), hydrogen (H_2), methane (CH_4), ethylene (C_2H_4), and propylene (C_3H_6) with minor concentrations of other hydrocarbon gases. The gas yields also increased with increasing catalytic bed temperature.

The liquid bio-oil before catalysis was homogeneous and single phase. However, after the pyrolysis vapours were passed over the catalyst bed, the bio-oil products became two phases (aqueous and oil phases), which can be easily separated. This is because the oil phase contained significant amounts of hydrocarbons, mainly aromatics in gasoline boiling range such as benzene, toluene, and alkylated benzenes. By raising the catalyst bed temperature, the concentration of these single-ring aromatic hydrocarbons was also increased. Apart from the increase of the aromatics, phenol and methyl phenol were also reported to increase in concentration after the zeolite catalysis [94].

The ZSM-5 catalyst was also found to be effective in deoxygenation reactions as could be seen from the significantly lowered oxygen content of bio-oils. The oxygen was removed in the form of water at low catalytic temperature and carbon oxides (CO and CO_2) at high catalytic temperatures [96,98]. When analyzing bio-oils by size exclusion chromatography (SEC) (also known as gel permeation chromatography (GPC)) technique, it was found that the molecular weight distribution of the catalytic bio-oils was shifted to lower range and with increasing catalyst bed temperatures, the molecular weights were further decreased [62,96,98].

In addition, Williams and Horne [62] investigated the influence of different zeolite catalysts (H-ZSM-5, Na-ZSM-5, and Y-zeolite) and activated alumina in comparison with an inert material (stainless steel ball bearings). The results showed that there were only small differences in the product yields and compositions from the catalysis of biomass-derived pyrolysis oils for the Na-ZSM-5 and H-ZSM-5 catalysts. The ZSM-5 catalysts gave the highest yields of hydrocarbon products when compared with the Y-zeolite and activated alumina catalysts.

In general, the catalytic bio-oils produced at the University of Leeds appeared to be better for usage than the noncatalytic ones due to the lower concentrations of oxygenated compounds, the higher concentrations of hydrocarbons and the lower molecular weight range. However, the higher concentrations of polycyclic aromatic hydrocarbon (PAH) in the catalytic bio-oils are their drawbacks since some of these PAH are known to be carcinogenic and/or mutagenic. To overcome this problem, the PAH may need to be further processed by for example hydrocracking in a typical refinery; this could give gasoline-range hydrocarbons as products.

Another notable drawback of the catalytic pyrolysis process reported was the formation of coke on the catalyst, which was in the range of 11–13 wt% on biomass fed basis for the ZSM-5 catalyst [62,96], but the coke yield increased to 19.1 wt% for the Y-zeolite and 18.4 wt% for the activated alumina [62]. This was believed to be due to the larger pore size of Y-zeolite (7.4Å) and activated alumina (20Å), which allows larger coke precursors to enter the pore structure of the catalyst, initiating and contributing to the increased formation of coke [62]. The deactivation of zeolite ZSM-5 is discussed in more details in Section 2.5.5.

ZSM-5 catalyst can be improved by adding acid/base materials or doping with certain active elements. With the addition of acid or base materials, such as Al_2O_3, CaO, and MgO into ZSM-5, the yield of aromatics and olefin from catalytic fast pyrolysis of rice stalk and pine could be increased. The added acid or base was speculated to enhance the cracking of large oxygenates into smaller molecules, which could access micropores in the ZSM-5, thus increasing hydrocarbon yield [101,102]. ZSM-5 catalyst may be doped with Ni and Co to allow steam reforming to take place during catalytic pyrolysis reactions as it was found that Ni- and Co-doped ZSM-5 catalysts gave H_2-rich gas as products [1]. Therefore, hydrogen transfer and hydrogenation reactions can be enhanced at the metallic Ni and Co active sites [103]. Adding Pt into HZSM-5 catalyst was also found to be beneficial to aromatics production from miscanthus catalytic pyrolysis as it is expected that Pt participates in cracking, hydrogenolysis, hydrocracking, and dehydrogenation reactions [104].

One active element that has been used to dope the ZSM-5 catalyst is gallium (Ga). Park et al. [105] compared HZSM-5 and Ga-modified HZSM-5 in ex situ catalytic fast pyrolysis of pine sawdust at 400°C and found that Ga/HZSM-5 gave higher bio-oil yield with higher selectivity towards aromatics. The yields of aromatics and benzene, toluene, and xylenes (BTX) over Ga/HZSM-5 were found to be twice that over the HZSM-5 catalyst. In another study that modifying ZSM-5 catalyst with only 1% Ga during catalytic fast pyrolysis of Radiata pine sawdust, high yields of bio-oil and hydrocarbons were obtained [106]. The use of Ga/ZSM-5 in catalytic pyrolysis of glucose was also found to increase the aromatic carbon yield from 30 (without Ga) to 40% on Ga/ZSM-5 [107,108]. Ga/HZSM-5 appears to be a bifunctional catalyst, where Ga species increase the rate of decarbonylation and olefin aromatisation and the ZSM-5 catalyses other reactions such as oligomerisation [32].

2.4.3 Zeolite-like catalysts

According to Baerlocher et al. [109], zeolites and zeolite-like materials do not comprise an easily definable family of crystalline solids; a simple criterion for distinguishing zeolites and zeolite-like materials from denser tectosilicates is based on the framework density (FD), the number of tetrahedrally coordinated atoms (T-atoms) per 1000 Å^3. Basically, zeolite-like catalysts are crystalline micro- or mesoporous materials having zeolite-like structure or layer structure.

Zeolite-like catalysts that have been previously tested in catalytic pyrolysis of biomass include MCM-41 [20,21,42,110,111], SBA-15 [20,112] and MSU-S [41] mesoporous materials using either a microscale batch reactor [110,112] or a bench-scale fixed-bed reactor [20,21,41,42,111]. Mobile crystalline material-41 (MCM-41) is an ordered mesoporous silicate molecular sieve of the M41S type, which possesses a porous system consisting of hexagonally arranged channels with diameters varying from 15 to 100 Å and has a high surface area of about 1000 m^2/g [113]. SBA-15 is also an ordered mesoporous silicate material, which possesses hexagonal structure and has long-range order, large monodispersed mesopores (up to 500 Å) and thick walls (typically between 3 and 9 nm) that make them more thermally and hydrothermally stable than MCM-41-type materials [20]. Another mesoporous molecular sieve tested in catalytic pyrolysis of biomass is MSU-S type, which exhibits high hydrothermal stability and increased acidity compared with Al-MCM-41 catalysts [41]. The great interest in attempting these mesoporous catalytic materials during catalytic pyrolysis of biomass arises from the fact that the main targeted compounds that need to undergo reactions (cracking, deoxygenating, and reforming to produce a liquid product with lighter and less oxygenated compounds and rich in high-value products) are bulky molecules, particularly those derived from lignin. This therefore requires catalysts structures with channel diameters at mesopore scale.

Pyrolysis of bark-free spruce wood was conducted at 500°C in a microscale batch reactor in order to test Al-MCM-41 [110] and SBA-15 [112] catalysts. In the reactor system, also known as pyrolysis-gas chromatography/mass spectrometry (Py-GC/MS), approximately 1.5–2.0 mg of biomass samples were pyrolysed generating pyrolysis vapours, which were then passed over catalyst beds where the catalytic reactions took place. The catalysed pyrolysis vapours were subsequently entrained to the GC/MS equipment connected online to the pyrolysis unit in order to analyse the pyrolysis products. However, the Py-GC/MS is a rapid method and is extremely useful for catalyst screening as various catalytic materials can be comparatively evaluated with different process parameters such as biomass/catalyst ratio, pyrolysis temperature, and heating rate. The Al-MCM-41 catalysts tested by Adam et al. [110] included an unmodified, a transition metal (Cu) modified, a pore enlarged with a spacer, and a pore enlarged by altering the template chain length. The biomass/catalyst ratio applied was 1:1. The purpose of the work was to find out whether aluminosilicate mesopore catalysis can improve the bio-oils quality, which was monitored by aldehyde yield (for stability indicator), organic acid yield (for pH or corrosiveness indicator), and phenol yield (because phenol is one of the high-value chemicals). Results showed that the presence of the catalysts led to a complete elimination of levoglucosan, a decrease in aldehyde yields, an increase in acetic acid yields, an increase in the yield of phenol, and a slight increase in hydrocarbon yield [110]. In addition, the catalysts were found to promote the scission of the aliphatic side chain of phenol derivatives, leading to a decrease in the yield of high-molecular-mass (≥138) phenols. When SBA-15 and Al-SBA-15 were tested in comparison with a commercial fluid cracking catalyst (FCC) by the same group of researchers [112], similar results were found; the presence of catalysts led to an increase in light phonols, a decrease in heavy phenolics compounds, and an increase in organic acids. The amounts of the furan and furan derivatives were also increased, whereas the aldehyde (hydroxyacetaldehyde, propanal, and 5-hydroxymethyl furfural) yields were decreased. Char and water formation was enhanced by the use of catalysts. Among the three catalysts, FCC was found to be the most effective catalyst for the changes of the pyrolysis products. Moreover, aluminium incorporation into SBA-15 structure enhanced the effectiveness of the compounds in the upgrading of pyrolysis vapours.

As mentioned earlier, the zeolite-like mesoporous catalysts (MCM-41, SBA-15, and MSU-S types) were tested in a bench-scale fixed-bed reactor [20,21,41,42,111]. In all experiments, the reactor was filled with 0.7 g of catalyst or glass beads for the noncatalytic tests, and the piston was filled with 1.5 g of biomass. As soon as the reactor reached the desired reaction temperature (500°C), the biomass was transferred into the pyrolysis zone by means of a specially designed piston system, and the pyrolysis products were in contact with the catalyst for 15 min. During the run, a constant stream of N_2 (100 cm^3/min) was fed from the top of the reactor for the continuous withdrawal of the products and the maintenance of the inert atmosphere. By using this experimental setup, Adam et al. [20] investigated the catalytic activities of Al-MCM-41 materials with a Si/Al ratio of 20 and with three different mean pore sizes (24, 28, and 30 Å) in comparison with Cu-Al-MCM-41, SBA-15, and Al-SBA-15. They found that after catalysis, gas yields increased, coke/char yields remained the same or slightly decreased, and aqueous part in the liquid phase increased. The organic yields increased with increasing the pore size of Al-MCM-41 catalysts. The analysis of the organic liquid showed that the hydrocarbon and phenol yields increased, while the carbonyls and acids yields decreased due to the catalytic activities. All the catalysts studied reduced the amounts of the undesirable compounds (defined as carbonyls, acids, and PAHs) in the bio-oil, while more desirable compounds (hydrocarbons, phenols and alcohols) were obtained. However, one of the catalysts drawbacks is the PAH formation. The pore enhancement of the Al-MCM-41 appears to worsen the bio-oil quality because by increasing pore size, the yield of undesirable products increased at a higher rate than the desirable products. The incorporation of Cu into the Al-MCM-41 was found to increase the amounts of desirable products more than undesirable ones. In addition, the incorporation of Al into SBA-15 framework led to very high content of desirable products in the bio-oils; however, the undesirable products also increase. This confirms the Py-GC/MS results as mentioned above that aluminium incorporation into SBA-15 structure enhanced the effectiveness of the compounds in upgrading pyrolysis vapours.

Apart from the effect of pore sizes of Al-MCM-41 materials that have been studied using the bench-scale fixed-bed reactor, the effect of Si/Al ratios [21,42], the effect of metal incorporation [21,111], and the effect of different species of biomass raw materials [20,21] on pyrolysis products were recently investigated. The results can be summarised as follows.

The lowest Si/Al ratio (20) enhanced the production of phenols, while the intermediate Si/Al ratios (34 or 51) promoted the reduction of undesirable fractions (carbonyls, acids, and heavy compounds) [21]. High Si/Al ratios (50 or 51) of the Al-MCM-41 samples enhanced the production of the organic phase of the bio-oil, while low Si/Al ratios (20, 30, or 34) favoured the production of noncondensable gases [21,42].

The metals that were incorporated into Al-MCM-41 and tested for catalytic pyrolysis of biomass included copper (Cu), iron (Fe), and zinc (Zn). Without the metal incorporation, Al-MCM-41 enhanced the production of hydrocarbons; however, in the presence of the metals, the amounts of hydrocarbons were not increased [21]. Incorporation of Cu led to a significant increase of the concentration of hydrogen in the gas stream [21]. Furthermore, the quality of the bio-oil based on the yield of phenols is increased for Cu and Fe incorporated Al-MCM-41 catalysts. However, for Zn–Al-MCM-41, the yield of phenols was lower than the unmodified Al-MCM-41 material, indicating that zinc is a less favourable metal for conversion of biomass to phenols than iron and copper [111].

Comparison of different biomass feedstocks during the catalytic pyrolysis was carried out by Adam et al. [20] and Antonakou et al. [21]. The former studied miscanthus and spruce, whereas the latter investigated miscanthus and Lignocel beech wood. The results from both works provide a general conclusion that the effect of catalysts on pyrolysis products was influenced by the type of biomass feedstocks, and in some cases, the Al-MCM-41 performed no better than the conventional zeolites. In other words, a catalyst that can be used to give desired pyrolysis products for a specific type of biomass may not provide the same results when other types of biomass are catalytically pyrolysed. Therefore, this needs to be taken into consideration when studying catalytic pyrolysis of biomass.

Hydrothermally stable mesoporous aluminosilicates (MSU-S) assembled from zeolite beta (BEA) seeds were tested by Triantafyllidis et al. [41] in catalytic pyrolysis of biomass using the same bench-scale fixed-bed reactor and pyrolysis conditions as described above. MSU-S led to significantly lower organic yields and also to higher coke and char yields, compared with noncatalytic pyrolysis. With regard to the quality of the organic phase of the bio-oil, the MSU-S catalytic materials were found to increase the formation of polycyclic aromatic hydrocarbons (PAHs) and heavy fractions, while they dramatically decreased the yields of acids, alcohols and carbonyls, and phenols. These effects were more pronounced in the wormhole-like MSU-S/W_{BEA} sample with the high textural porosity possibly because diffusion of large reactants and products was facilitated compared with the hexagonal pore structure of MSU-S/H_{BEA}. Since the bio-oils produced with the two MSU-S catalysts contained mainly hydrocarbons, PAH and heavy compounds together with a relatively low content of oxygenated compounds (phenols, acids, carbonyls, and alcohols), it is suggested that the bio-oils may be of high value for refinery upgrading processes such as fluid catalytic cracking (FCC) units for 'green gasoline' production or hydrotreating/hydrocracking processes for 'green diesel' production [41].

2.4.4 Metal oxides

Metal oxides possess either redox properties due to their multivalent nature or certain acid-base properties that could potentially catalyse the thermal decomposition of biomass and/or the reaction of pyrolysis intermediates to form more stable products [114–116].

Acidic metal oxides such as Al_2O_3, SiO_2, and SiO_2/Al_2O_3 and the sulphated metal oxides such as SO_4^{2-}/TiO_2, SO_4^{2-}/ZrO_2, and SO_4^{2-}/SnO_2 were tested in catalytic fast pyrolysis. Acidic metal oxides could deoxygenate pyrolysis vapour via decarbonylation reactions, giving off CO [43].

Basic metal oxides such as MgO and CaO can help improve the quality of bio-oil in terms of oxygen content. These basic metal oxides are active catalysts for ketonisation and aldol condensation of carboxylic acid and carbonyl compounds [116]. The oxide ions behave as bases and the metal cations can function as Lewis acids [116]. CaO was tested in catalytic pyrolysis of white pine in a fluidised-bed reactor by Lin et al. [117]. The results showed that CaO reduced the bio-oil yield from 39.4 to 34.1 wt%, but the oxygen content of bio-oil was also reduced from 39 to 31 wt% compared with the noncatalytic one. It was speculated that CaO catalyst could catalyse the deoxygenation reaction by CO_2 absorption and dehydration, leading to the formation of calcium carbonate on the spent catalyst [73,117]. MgO catalyst was found to improve bio-oil properties in terms of oxygen content, heating value, and hydrocarbon distribution [14,43]. The oxygen content of cotton seed bio-oil decreased from 9.56 to 4.9 wt% when using MgO [14]. MgO is speculated to catalyse the ketonisation of carboxylic acids giving off CO_2 as products [43].

Transition metal oxides such as ZnO were also investigated in catalytic pyrolysis of biomass. A zinc oxide catalyst from Haldor Topsøe containing 99% ZnO and 0.5% Mg was investigated by Nokkosmaki et al. [118] in a bench-scaled fluidised-bed pyrolysis reactor close-coupled with a secondary catalyst bed reactor. Pine sawdust was pyrolysed at 525°C, and nitrogen was used as carrier gas. After the pyrolysis reactor, the pyrolysis vapours were passed through a hot gas filter kept at 475°C to remove char particles. Subsequently, the vapours were led through a bed of catalyst maintained at 400°C, followed by a condensation unit to obtain bio-oil products. The results showed that the ZnO was a mild catalyst as the liquid yield did not reduce substantially. According to the bio-oil analysis, the catalyst had no effect on water-insoluble fraction (lignin-derived compounds), but it decomposed the diethyl ether-insoluble fraction (water-soluble anhydrosugars and polysaccharides). The presence of the catalyst lowered the initial viscosity of bio-oil (measured at 50°C) from 9.5 (without catalyst) to 5.7 cSt (with the ZnO catalyst). According to an accelerated ageing test of bio-oil, the increase in viscosity was significantly lower for the ZnO-treated bio-oil (55%) than for the reference bio-oil without any catalyst (129%), thus indicating an improvement in the stability.

2.4.5 Natural minerals

Natural catalysts that may be promising for catalytic pyrolysis of biomass include alkali and alkaline earth metal-based catalysts and slate. The inorganic compounds (containing alkali and alkaline earth metals) present in biomass have catalytic effects on pyrolysis reactions such as cracking and char forming, thus increasing gas yields and decreasing liquid yields. Since cracking reactions are involved, the bio-oil produced would expect to have lower molecular weight and possibly lower initial viscosity. It was found from a study by Nik-Azar et al. [119] that alkali metals could decrease the molecular weight of bio-oil from 300 amu when raw beech wood was used as feedstock to about 190 amu when the wood samples were impregnated with potassium (K) or sodium (Na) cations, although the use of these metals also decreased the bio-oil yield. The alkali and alkaline earth metal-based catalysts are readily available in the forms of ash and char. Therefore, these two materials may be used for catalytic pyrolysis of biomass with an aim of cracking heavy molecules in pyrolysis vapours to produce bio-oils of lower molecular weight and viscosity. Another type of natural catalysts that could be used for bio-oil upgrading is slate since it was found to improve the bio-oil quality in terms of stability, initial viscosity, and heating value with no significant loss in the liquid yield [2,120].

In addition, calcite, a natural carbonate mineral, was found to be active for deoxygenation reactions in catalytic pyrolysis of rice husk as it could help reduce the bio-oil oxygen content from 48.3 to 41.1 wt%, leading to 20% increase in the HHV, although bio-oil yield was decreased from 34 to 12 wt% [121]. Enhanced cracking and ring-opening reactions were proposed as reaction pathway for calcite catalyst since more light organics were produced.

Acid-leached olivine was reported to be active for deoxygenation performance as it could decrease the oxygen content of the bio-oil from 37 to 25 wt%. The char yield also decreased from 20 to 10 wt% with the use of olivine. The key to the successful deoxygenation activity of the olive is due to the acid-leaching treatment because little deoxygenation was observed for the untreated catalysts [122,123].

2.5 Catalyst properties

2.5.1 Shape selectivity

The product distribution obtained from catalytic pyrolysis of biomass is a strong function of shape selectivity and acidity of catalyst. Shape selectivity effects are mainly imposed by pore-opening size, pore shape, and crystallite size of catalyst [47]. There are three types of shape selectivity, namely, reactant selectivity, product selectivity, and restricted transition-state selectivity [89].

Reactant selectivity means that only starting materials of a certain size and shape can enter the interior of the zeolite pores and undergo reactions at the catalytically active sites. If the molecular size of a reactant is larger than the pore apertures, it cannot react. Therefore, zeolites are also called molecular sieves. In spite of the considerable molecular sieve effect, 100% selectivity is often not attained because the reactants can also react to a certain degree on the outer surface of the zeolite crystals [89].

Product selectivity occurs when only products of a certain size and shape that can leave the pore system are formed. This type of shape selectivity can also have drawbacks. Large molecules that are unable to exit the pores can be converted to undesired by-products or undergo coking, thus leading to catalyst deactivation.

The third form of shape selectivity depends on the fact that chemical reactions often proceed via intermediates. Owing to the pore system, only those intermediates that geometrically fit to the zeolite cavities can be formed during catalysis. Therefore, only products produced through intermediates that are suitable for a specific type of zeolites can be formed.

With an attempt to elucidate the functions of shape selectivity and acidity, the catalytic activities of ZSM-5 and silicalite were comparatively investigated. Silicalite is a molecular sieve catalyst consisting of all silica and very low alumina possessing a uniform pore diameter with structure similar to ZSM-5, the only exception being the absence of acid sites [53,91]. The results show that it was the combination of shape selectivity and acidity that is responsible for alkylation, isomerisation, cyclisation, and aromatisation reactions that ultimately resulted in the formation of aromatic hydrocarbons [124].

2.5.2 Acidity

Apart from the shape selectivity functionality of zeolites, acidity is an important property in catalysis application, and it has been found to play the most critical role in upgrading reactions consisting of cracking, deoxygenation, oligomerisation, cyclisation, aromatisation, alkylation, isomerisation, and polymerisation [125,126]. These reactions are expected to take place via a carbonium ion mechanism [65,126].

There are two types of acid sites, namely, Brønsted (proton donor) and Lewis (electron-pair acceptor) sites that exist in zeolites. Although both sites are involved in catalytic activities, the former appear to be stronger. The strength of zeolite acidity relates not only to the number of the acid sites but also to the natures of the acid sites (Brønsted or Lewis). The zeolite acidity can be measured experimentally by, for example, temperature-programmed desorption of ammonia (NH_3-TPD), FTIR measurements of pyridine adsorption, and ^1H MAS solid-state NMR for the determination of the total acidity, nature of acid sites, and number of hydroxyl groups present on zeolite catalysts, respectively. The number of Brønsted acid sites can be calculated based on the TCD signal for NH_3 and propylene from IPA-TPD; the products of isopropylamine decomposition. Then, the number of Lewis acid sites can be calculated as the difference between total acid sites and Brønsted acid sites [33]. For ZSM-5-based catalysts, the concentration of the total acid sites is 0.1–0.4 mmol/g depending of the catalyst composition and synthesis, whereas the ratio of the Brønsted and Lewis acid sites is between 1:1 and 5:1 [1,33].

The acidity of a zeolite catalyst can normally be modified by changing the synthesis conditions, varying the Si/Al ratio, incorporating metals, or using other alternative methods to tailor the characteristics for the reaction of interest [127–129]. Zeolite acidity can be estimated from the Si/Al ratio; the high value of this ratio (>10) indicates the low concentration of acid sites of high strength, whereas the low Si/Al ratio (1–1.5) results in the high concentration of acid sites of medium strength [89]. However, low Si/Al ratio can result in instability of the catalyst framework due to dealuminisation under hydrothermal conditions [1]. Therefore, the acidity of the catalyst needs to be fine-tuned to achieve the desirable acidity with good stability.

Acidity of catalyst has been varied and investigated in catalytic pyrolysis of biomass. General finding is that the high acidity of catalyst by lowering the Si/Al ratio or modification with certain metals such as Ga results in lower bio-oil yield, higher gas yield, higher aromatic hydrocarbon yield, higher deoxygenation (dehydration, decarbonylation, and decarboxylation), cracking and reshaping (oligomerisation, alkylation, cyclisation, and aromatisation) activities, and higher coke formation [4,79,130–133]. Since the high acidity has both advantages in terms of product yields and disadvantages in terms of catalyst stability and coke formation, optimisation of catalyst acidity is of great importance for maximising the yields of desired products with high quality.

2.5.3 Pore size

Catalyst pore size greatly influences reaction activity by affecting mass transfer of reactants and products into and out of catalyst; smaller pore size decreases mass transfer and restricts the diffusion of large molecules [47]. The pore size of catalyst such as zeolite can be modified by treatment with steam [134] or alkaline substance such as NaOH [135–137] to obtain a hierarchical zeolite, which is also called mesoporous zeolite. The advantage of the hierarchical zeolite is its high accessibility of reactants to active sites. It is known that pyrolysis vapour contains large molecules derived from degradation of cellulose, hemicellulose, and lignin components. Some molecules are too large to access the active sites of catalysts, thus condensing on their surface, blocking the pores, eventually deactivating the catalysts. The enlargement of pore size would allow these molecules to enter the pore structure towards active sites, such as Brønsted acid sites, react, and exit the catalyst. For catalysts with small pores, it is likely that small oxygenated compounds from pyrolysis vapour can enter the pores; react to become larger molecules, for example, PAH; and cannot exit the pore structure, thus blocking the active sites.

Agblevor et al. [30] observed when studying the fractional catalytic pyrolysis of hybrid poplar wood using an ExxonMobil HZSM-5 zeolite as a fluidised-bed material that the catalyst pore size played a critical role during catalytic pyrolysis. The lignin decomposition intermediates larger than the pore size of the catalyst were selectively converted into phenolic compounds such as phenols, cresols, and catechols, while the carbohydrate fraction smaller than the catalyst pore size was converted into gas products.

Studies [134,136] on the comparison of H-ZSM-5 catalysts and their modified mesoporous or hierarchical forms reveal that the yield of aromatics can be improved with the increased accessibility to Brønsted acid sites due to the increase in mesoporosity. Catalytic pyrolysis of beech wood at 550°C with a 10:1 catalyst/biomass ratio showed that the aromatic yields increased from 24% for untreated catalyst to 26%–30% for mesoporous catalyst [136]. Carbon deposits could also be reduced from 44% for untreated catalyst to 40%–42% for mesoporous catalyst [136]. When lignin was upgraded, the aromatic yield was increased from 12% to 20% when catalysts were modified by steam treatment to increase their active site accessibility [134].

Therefore, it is logical to state that the pore size of catalyst is crucial in catalytic pyrolysis of biomass for hydrocarbon productions or bio-oil upgrading.

Mordenites were also modified by alkaline treatment to obtain mesoporous structure and were tested in catalytic pyrolysis of beech wood [137]. The results show that the combined effects of higher surface area and larger mesopore volume, resulting from alkaline treatment of the mordenite, could improve the deoxygenation activity due to enhanced active site accessibility. The use of the modified mordenite caused a reduction of organics yield from 37% to 11% by weight, and the oxygen content was reduced from 40 to 22 wt%.

2.5.4 Catalyst pretreatment and calcination

Prior to catalytic fast pyrolysis processing of biomass, most catalysts require calcination or pretreatment. Typically, catalysts such as ZSM-5 are calcined by heating in a furnace at 500–600°C with a flow of air or nitrogen for 2–5 h. Calcination temperature is normally slightly higher than the operating temperature of the catalyst. Calcination can change the ammonium cation form of ZSM-5 into its proton form (H-ZSM-5). ZSM-5 catalyst can also be pretreated by hydrothermal method, which is done by stirring the catalyst powder in deionised water at 60°C for 2 h followed by drying the slurry at 105°C and calcination in a furnace [22]. Properties of catalyst can be improved by proper pretreatment and calcination. As mentioned earlier, pretreatment methods such as by steam and alkaline solution can affect the pore size of catalyst, thus influencing the product yields and properties. Not only is calcination applied for fresh catalysts, but also spent or used catalysts may also be calcined in order to restore their catalytic activity.

2.5.5 Catalyst deactivation

According to previous studies [61,94,138–140] on catalytic upgrading of pyrolysis liquids, vapours, or model compounds using zeolite ZSM-5, deactivation of the catalyst can be either reversible or irreversible. The reversible deactivation occurs by coking. Coke deposition leads to blockage of pores and active sites of catalysts, thus reducing the catalyst lifetime and deteriorating the product yields and properties.

In catalytic fast pyrolysis, coke formation is a major undesired competing reaction with the production of aromatics and olefins [47]. Gayubo et al. [140] suggested that the coke deposits can be divided into two fractions based on the two peaks of DTG curves obtained from the spent catalyst. The first fraction, which burns at lower temperature, is the thermal coke, which is probably the residue deposited on the matrix of the catalyst particles. The second fraction corresponds to the internal catalytic coke (possibly has a graphitic carbon structure similar to that generated in the transformation of methanol and ethanol) located in the internal channels of the catalyst, which burns at higher temperature. Since the thermal coke deposits contain significant quantities of oxygen [61,94], it is believed that these deposits that condensed on the catalyst surface are derived from the oxygenated components of pyrolysis vapours such as high-molecular-weight lignin derivatives (phenolic-based compounds), high-molecular-weight carbohydrate-derived compounds, and small but reactive molecules such as aldehydes (especially acetaldehyde) and furfural [94,138,140,141]. The catalytic coke is formed in the internal pore structure by the transformation of oxygenated compounds over catalyst acid sites through reactions of oligomerisation, cyclisation, aromatisation, and condensation [142,143].

The irreversible deactivation of the catalyst was suggested to take place via the dealuminating effect of steam at high temperature (≥450°C), which result in a sharp decrease of the catalyst acidity when it is used in successive reaction-regeneration cycles [138]. Deactivation by dealumination is irreversible due to a significant loss of acidity or acid sites, particularly the stronger one [144]. This deactivation may involve the formation of polycyclic aromatic hydrocarbons that are too large to leave the pores, thus blocking the active acid sites because the presence of steam enhanced the formation of PAH [99]. It is therefore suggested that to avoid irreversible loss of catalyst activity, the operation should be performed at temperature below 450°C and without addition of steam, although it is known that the presence of steam can have beneficial effects such as the increase in the formation of monocyclic aromatic hydrocarbons [99] and the attenuation of the overall coke content [139].

Apart from the coke deposition and dealumination by steam, deactivation of catalyst can occur by deposition of inorganic metals derived from biomass feedstock such as alkali and alkaline earth metals and transition metals such as Fe and Cu on the surface, pores, and active sites of the catalysts [79,80]. This deactivation cannot be recovered by simple combustion. Therefore, it may be called irreversible. However, depending on the binding mechanism of the metals and the catalyst, the catalyst may be washed by water to dissolve the inorganic metals followed by drying and calcination of the catalyst in order to restore its properties and catalytic activity.

The catalyst deactivation also influences the pyrolysis products. It is known that ZSM-5 catalysts can increase the concentrations of hydrocarbons (both monocyclic and polycyclic aromatic hydrocarbons), reduce oxygenated compounds, and decrease molecular weight range of bio-oils. These changes are reported to reduce in their severity when the catalyst deactivated as observed from either increasing the catalysis run duration [94] or increasing the number of regeneration cycles [61]. The former occurred by the combination of reversible and irreversible deactivation, whereas the latter indicates merely irreversible deactivation.

Factors influencing the catalyst deactivation are feedstock composition, catalyst properties, and operating parameters. As mentioned earlier in the mechanism of coking, feedstock containing high proportion of lignin tended to cause high coke. Huang et al. investigated catalytic pyrolysis of several biomass feedstocks with different contents of cellulose, hemicellulose, and lignin over La/HZSM-5 and reported that lignin or feedstocks with high lignin content caused higher coke formation [76]. The major element in the feed that is known to affect coke formation is hydrogen. Feedstock with low hydrogen content or low hydrogen-to-carbon effective ratio (H/C_{eff}) can induce more coke as the hydrogen is deficient in the hydrocarbon pool inside the catalyst [47]. In addition, the presence of water either from biomass, from pyrolysis reaction product or from added steam can inhibit coke formation due to the production of hydrogen in the feed by partial steam reforming [145–147].

Pore size of catalyst is known to affect the formation of coke [47]. In a catalytic pyrolysis study using pine wood in a fluidised-bed reactor, large cavities in zeolite structure were found to result in the high yield of coke [23]. Jae et al. [148] compared zeolites with different pore sizes and revealed that low pore size resulted in less coke formation as ZSM-5 and ZSM-11 with medium pore size resulted in the least coke formation, whereas Beta zeolite, SSZ-55, and Y-zeolite having the largest pores gave highest coke.

Process parameters that can influence the coke formation include reaction temperatue, heating rate for in situ process, and catalyst-to-feed ratio. High reaction temperature, lower heating rate, and lower catalyst-to-feed ratio resulted in high coke formation [149,150].

2.5.6 Catalyst regeneration

A spent catalyst in catalytic pyrolysis of biomass is normally covered with thermal and catalytic cokes as can be seen from its black surface. Coke can be burned off by introducing air or a mixture of air and nitrogen at 500–600°C. Although the catalytic activity of the regenerated catalysts is restored, their efficiency in catalytic upgrading of biomass fast pyrolysis vapours was found to drop to a certain extent [133]. It is also possible that if the regeneration temperature is too high, loss of Brønsted acid sites through dehydroxylation may occur, thus lowering the catalyst performance [47].

2.6 Effect of process parameters

Several process parameters are known to affect the catalytic pyrolysis process and products including catalyst position, catalyst temperature, weight hourly space velocity (WHSV), biomass-to-catalyst ratio, carrier gas, and liquid collection. The effects of these parameters are discussed below.

2.6.1 Efeect of catalyst position

In situ and ex situ configurations of catalytic pyrolysis of biomass using ZSM-5 are compared in Table 2.1. This table shows that in situ mode gives higher yields of bio-oil, aromatics, phenols, gas, and CO than the ex situ ones when using bubbling fluidised-bed (BFB), auger, or pyroprobe reactors. Overall performance seems to be better with the in situ mode. However, according to oxygen content of bio-oil, the BFB and auger reactors gave different results. In the case of auger reactor, in situ configuration is better than ex situ one for oxygen removal. However, opposite trend was reported when the BFB reactor was used. When testing faujasite [151] and ReUS-Y [152] catalysts in a fixed-bed catalytic pyrolysis reactor, ex situ mode appeared to give better results for deoxygenation of bio-oils. Based on catalyst deactivation by either coking or accumulation of metals on catalyst's surface, ex situ configuration is

Table 2.1 Comparison of in situ and ex situ catalytic pyrolysis of biomass using ZSM-5 catalysts

Catalyst	Reactor	Comparison criteria	In situ[a]	Ex situ[a]	Reference
HZSM-5	BFB and Auger	Bio-oil yield	Higher	Lower	[153,154]
	Auger	Yield of aromatics	Higher	Lower	[154]
	Pyroprobe		Higher	Lower	[3,155–157]
	BFB	Oxygen content of	Higher	Lower	[153]
	Auger	bio-oil	Lower	Higher	[154]
	BFB	Acid content in bio-oil	Higher	Lower	[153]
	Auger	Char yield	Lower	Higher	[154]
		Gas yield	Higher	Lower	[154]
		Yield of phenols	Higher	Lower	[154]
		Overal performance	Higher	Lower	[154]
		CO yield	Higher	Lower	[154]
	BFB	Catalyst deactivation	Faster	Slower	[153]
		Accumulation of metals on spent catalyst	Higher	Lower	[153]
Faujasite	Fixed bed	Degree of oxygen	Lower	Higher	[151]
ReUS-Y		removal			[152]

[a] The words higher/lower/faster/slower mean the value/degree of the comparison criteria is higher/lower/faster/slower than the ex situ or in situ.

better. Therefore, it is not possible to conclude that mode is better because the results are not conclusive. This implies that the effect of reactor configuration may depend on the reactor types, biomass feedstocks, catalyst types and properties, and process parameters.

2.6.2 Catalyst temperature

The temperature of catalyst in catalytic pyrolysis of biomass affects the rates of thermal and catalytic reactions such as cracking, dehydration, decarbonylation, decarboxylation, and aromatisation [22], thus affecting the bio-oil and gas yields, the productivity and selectivity of aromatic hydrocarbons and coke, and the deoxygenation pathways. Increasing catalyst temperature leads to high selectivity and yields of aromatic hydrocarbons [22,77]. For lignin catalytic pyrolysis, the highest aromatic yield of 4.0 wt%daf was obtained at the highest catalyst temperature investigated of 600°C [77]. The yield of aromatics with two methyl substituents decreases, and the aromatics selectivity is mainly towards benzene and toluene when applying a high catalyst temperature [77]. High catalyst temperatures increased gas yield due to the cracking of the pyrolysis vapour into carbon oxides and light hydrocarbons [77]. The catalyst temperature also affects catalyst deactivation or the formation of coke. The coke yield decreases with increasing catalyst temperature [77].

Williams et al. [158] investigated the effect of catalyst temperature in catalytic pyrolysis of rice husk in a fluidised-bed reactor with downstream fixed bed of HZSM-5 catalyst and found that coke yield decreased from 12.0 wt% at 400°C to 7.8 wt% at 600°C catalyst temperature. At a too low temperature of 350°C, pyrolysis vapour may condense on the surface of catalyst or inside the catalyst pores becoming coke, which can reduce organic liquid yield [77]. Catalyst temperature affects the deoxygenation pathway in the way that at high catalyst temperature, oxygen is removed from bio-oil or pyrolysis vapour by dehydration giving off H_2O, whereas at high catalyst temperature, oxygen is rejected by decarbonylation or decarboxylation giving off CO_x as product [158].

2.6.3 WHSV or catalyst contact time

Weight hourly space velocity (WHSV, h^{-1}) is defined as the ratio of the biomass feed rate (kg/h) to the catalyst weight used (kg). The inverse of WHSV (1/WHSV) is catalyst contact time (h or s). WHSV or catalyst contact time is known to affect not only the catalytic product yields and composition but also the catalyst deactivation.

Yang et al. [33] found from catalytic fast pyrolysis of cellulose with ZSM-5 catalyst in a bubbling fluidised-bed reactor that aromatics and CO_2 yields are favoured at high catalyst contact times (> 10,000 s), whereas at medium catalyst contact times of around 1000–10,000 s, oxygenated and unidentified compounds are favoured. In addition, carbon monoxide and methane are favoured at catalyst contact times below 1000 s. Based on these results, if aromatics are targeted products, WHSV of the pyrolysis system should be below $0.36\,h^{-1}$. Carlson et al. [5] also reported that by lowering the WHSV, the yields of monocyclic aromatics can be increased with decreasing the selectivity towards polycyclic aromatic hydrocarbon (PAH).

Apart from the catalyst contact time, the effect of vapour residence over catalytic bed was also investigated by Horne and Williams [95]. In their study, biomass in the form of waste wood chips was pyrolysed in a fluidised-bed reactor at 550°C to maximise the formation of pyrolytic vapours. The pyrolysis vapours were then passed over a fixed bed of ZSM-5 zeolite catalyst located at the reactor freeboard at 500°C. The catalyst bed was diluted with stainless steel ball bearings to increase the residence time of the pyrolysis vapours in the bed and to provide additional hot surface for thermal cracking of the vapours. Bio-oils produced without the secondary bed appeared to contain low quantities of aromatic hydrocarbons. The presence of the steel alone did not increase the amount of aromatic hydrocarbons significantly. The presence of the undiluted catalyst did increase the formation of aromatic hydrocarbons, and the combination of the catalyst and the steel together gave a further increase. From the analysis of the aromatic hydrocarbon fractions, the 1:2 catalyst/steel experiment gave three times the yield of monocyclic aromatic hydrocarbons observed in the undiluted catalyst experiment. Therefore, the presence of steel in the catalyst bed appears to be very beneficial for the formation of monocyclic aromatic hydrocarbons since these compounds are considered as target chemicals for the upgrading of biomass pyrolysis vapours for use as either premium grade fuel or high-value chemicals. In addition, the presence of steel in the catalyst bed also increased the formation of polycyclic aromatic hydrocarbons.

2.6.4 Biomass-to-catalyst ratio

Biomass-to-catalyst ratio, also called weight turnover, is the ratio of biomass or feed weight to catalyst weight. This parameter also represents time-on-stream of a catalytic pyrolysis system with continuous biomass feeding. For a constant biomass feeding system and a constant catalyst weight, increasing time-on-stream leads to higher biomass-to-catalyst ratio. It is known in catalytic pyrolysis field that catalyst can be deactivated upon exposure to long-duration reactions or high amount of feed materials due to coke accumulation on external surface and micropores of the catalyst, thus blocking the active sites. For catalytic pyrolysis of cellulose in a bubbling fluidised bed with ZSM-5, the aromatic-producing activity was significantly reduced after a weight turnover of 2.0 since more than half of the Brønsted acidity and microporosity was lost [33]. The inverse of this parameter is sometimes used, which is catalyst-to-feed ratio because often weight of catalyst used is much higher than that of biomass feedstock. For example, Li et al. [130] reported that increasing the catalyst-to-feed ratio could increase the yield of aromatic hydrocarbons.

2.6.5 Environment or carrier gas

In catalytic pyrolysis of biomass, reducing or nonoxidising environment is required in order to prevent oxidation of biomass. Gases like nitrogen, helium, and carbon dioxide can be used to remove oxygen in the system. In a laboratory scale, nitrogen is usually applied due to its availability and low cost. In a larger or commercial scale, nitrogen gas can be initially used to remove oxygen in the system. After that, product gas from pyrolysis reactions can be recycled back into the system and used as carrier gas. The benefit of the recycle gas is not only the cost of the gas but also the advantage in terms of the yields of liquid products. It has been found by Mante et al. [25] when investigating the influence of recycling the noncondensable gases such as CO/N_2, CO_2/N_2, $CO/CO_2/N_2$, and H_2/N_2 in catalytic pyrolysis of hybrid poplar wood that the bio-oil yield can be increased and the char/coke yield can be decreased when noncondensable gas is recycled [25]. Another important parameter related to the carrier gas is the temperature of the carrier gas. In a system like fluidised-bed reactors either bubbling or circulating, significant amount of carrier gas is required to lift or fluidise the heat-transfer medium. If the carrier gas is at room temperature, it is likely that the temperature of the bed in the reaction may be dropped. This would affect the pyrolysis reaction and heating rate, thus producing less liquid and high char yield. To prevent this, a preheater must be installed or incorporated in the catalytic pyrolysis system. The carrier gas can be preheated at the same pyrolysis temperature or slightly lower.

Actively reducing gas such as hydrogen can also be used in catalytic pyrolysis process. The pyrolysis process using hydrogen can be called hydropyrolysis. With enough proportion of hydrogen in the catalytic pyrolysis system, hydrodeoxygenation or hydrogenolysis can take place. Oxygen from pyrolysis vapour can then be selectively rejected via dehydration giving off water as product, thus maintaining most of the carbon in the product [1,88,159]. In the presence of hydrogen, the increase in hydrocarbon compositions was also found [159].

Catalytic pyrolysis of biomass has also been explored under steam and alcohol cofeeding environments. In the presence of steam, small oxygenated species in pyrolysis vapour can undergo steam-reforming reactions, producing in situ hydrogen, which can be involved in catalytic reactions for hydrocarbon production [160–165]. It has been found when biomass was pyrolysed in a fluidised-bed reactor at 550°C and the pyrolysis vapours in the presence of steam were passed over a zeolite catalyst bed maintained at either 500°C or 550°C that the formation of hydrocarbons including PAH was enhanced by the presence of steam [99]. In addition, the presence of steam can also retard the coke formation reaction due to the increased ratio of hydrogen to carbon in the catalytic pyrolysis environment [145–147].

Cofeeding of alcohol in catalytic pyrolysis system results in the increase of hydrogen proportion or effective H/C ratio. Horne et al. [166] injected methanol into the top section of a fluidised-bed reactor to mix with the pyrolysis vapour. The mixture was then passed over a fixed bed of zeolite HZSM-5 catalyst at 500°C. They found that the key effects of cofeeding methanol were the increase in the amount of oxygen removed from the pyrolysis vapours as water at the expense of CO and CO_2 formation and the significant increase in the formation of alkylated phenolic and aromatic hydrocarbon products. Additionally, the presence of methanol was found to suppress the formation of the PAH.

2.6.6 Liquid collection

Liquid collection is very important for catalytic pyrolysis since some aromatic hydrocarbons, especially those with low molecular weight, are difficult to condense at room temperature or even with the aid of an electrostatic precipitator. This is due to the low partial pressure of the volatile in the carrier gas. In most lab-scale units, fresh carrier gas is continuously used throughout the experiments without recycling the noncondensable gas to the unit. This leads to dilution of liquid hydrocarbon products in the flow of carrier gas. As a consequence, cooling of the catalytic vapour stream to below room temperature is crucial for liquid product recovery. Yang et al. [33] used four gas-washing bottles containing isopropanol at 0°C as primary liquid collectors where most organic liquid was dissolved in isopropanol (see Fig. 2.3). Then, the residual organics were collected by additional four condensers maintained at −78°C by a dry-ice/acetone bath.

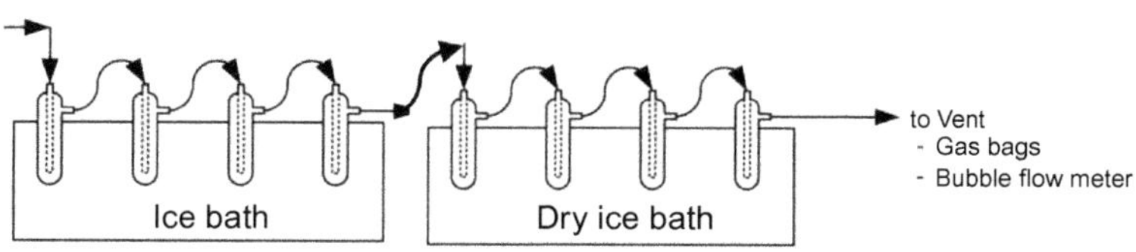

Fig. 2.3 Catalytic bio-oil collection [33].

2.7 Catalytic pyrolysis products

Products from catalytic pyrolysis of biomass include solid, liquid, and gas. There are two fractions of solid produced in the process. The first one is char, which is a residue from biomass thermal decomposition. Char from ex situ catalytic pyrolysis is similar to noncatalytic fast pyrolysis, but the char from in situ catalytic pyrolysis may be slightly different in composition from noncatalytic ones. Secondary solid produced in the catalytic pyrolysis reactions is called coke, either thermal or catalytic coke. It is on the external or internal surface of catalysts. Coke is typically burned after the process to recover the catalyst, and the heat from the combustion is used to reheat the catalyst before recycling back into the process.

Catalytic pyrolysis liquid products are varied depending on several factors including biomass feedstock composition and the severity and pathway of the catalytic reactions. Typical noncatalytic bio-oil is homogeneous single-phase liquid, even though it can be fractionated into organic-rich and aqueous phases depending on the liquid collection system. Noncatalytic organic-rich bio-oils contain 15%–30% water without phase separation. For mild catalytic pyrolysis reaction, the appearance of the liquid product can be almost the same as the noncatalytic ones. However, increasing the severity of the catalytic reactions leads to bio-oil phase separation, which can be two or three phases. In catalytic pyrolysis of pine in a bench-scale fluidised-bed reactor system [153], three phases of biomass were obtained: top organic phase, middle aqueous phase, and bottom heavy organic phase. The top oils had lower organic oxygen contents than the bottom oils (7%–8% vs. 16%–18%) and had higher hydrogen/carbon molar ratio (1.15 vs. 1.06–1.08). One- to two-ring aromatic hydrocarbons were predominant in the top oils, whereas 2–4-ring aromatics were predominant in the bottom oils. The bottom and top oils contained similar groups of chemicals with heavier fractions found in the bottom oil. Chemical analysis of the aromatic hydrocarbons found in catalytic pyrolysis bio-oil included mainly benzene, toluene, ethylbenzene, xylenes, styrene, indene, and naphthalene, whereas the oxygenates include benzofuran, phenol, acetaldehyde, acetic acid, hydroxyacetone, and 5-hydroxymethylfurfural (HMF) [33].

An example of bio-oil produced from the catalytic pyrolysis is shown by Fig. 2.4 in comparison with noncatalytic bio-oil. As can be seen, the appearance of the catalytic bio-oil is different from the typical one. According to the GC/MS analysis of the catalytic fast pyrolysis bio-oil, approximately 93% of the total chromatographic peak area are hydrocarbons including mainly xylene, toluene, styrene, 1,2,4-trimethylbenzene, ethylbenzene, naphthalene, and 1-methylnaphthalene. These compounds are typically found in gasoline. Therefore, the catalytic bio-oil may be used as transportation biofuel to replace petrol.

Product gas stream in catalytic pyrolysis of biomass is composed of CO, CO_2, H_2, and C_xH_y. It is also likely that water and some volatile liquid that cannot be condensed or collected in the liquid collection may be in the gas stream. The heating value of the gas varied depending on the severity and pathway of the catalytic reactions. The catalytic gas products can be recycled into the fast pyrolysis system as carrier gas or be combusted for providing pyrolysis process heat together with the char.

Fig. 2.4 Pyrolysis products.

2.8 Conclusions

Catalytic pyrolysis has demonstrated high potential for direct thermochemical liquefaction of biomass for energy applications. All liquid, solid, and gaseous products can be used as biofuels. Catalysts can be incorporated into a fast pyrolysis system as in situ (mixed with biomass feedstock or as heat-transfer medium) or ex situ (close-coupled in the reactor above the bed or as a secondary reactor). Several parameters affect catalytic pyrolysis reactions leading to varied product yields, composition, and properties. These parameters include biomass feedstock composition, catalyst types and properties, and process parameters. Although significant research has been carried out in the past, there are still barriers in developing commercial-scale catalytic pyrolysis of biomass to be overcome. The most important is the low liquid hydrocarbon product yield and the high coke formation. Catalyst development and engineering process design are on the way to identify sustainable and economical routes for energy and chemical applications of products from catalytic pyrolysis of biomass.

References

[1] Liu C, Wang H, Karim AM, Sun J, Wang Y. Catalytic fast pyrolysis of lignocellulosic biomass. Chem Soc Rev 2014;43(22):7594–623.

[2] Salter EH. Catalytic pyrolysis of biomass for improved liquid fuel quality [thesis (Ph.D.)-University of Aston in Birmingham, 2001]. Birmingham: University of Aston in Birmingham; 2001.

[3] Wan S, Wang Y. A review on ex situ catalytic fast pyrolysis of biomass. Front Chem Sci Eng 2014;8(3):280–94.
[4] Mihalcik DJ, Mullen CA, Boateng AA. Screening acidic zeolites for catalytic fast pyrolysis of biomass and its components. J Anal Appl Pyrolysis 2011;92(1):224–32.
[5] Carlson TR, Cheng Y-T, Jae J, Huber GW. Production of green aromatics and olefins by catalytic fast pyrolysis of wood sawdust. Energy Environ Sci 2011;4(1):145–61.
[6] Pattiya A, Titiloye JO, Bridgwater AV. Fast pyrolysis of cassava rhizome in the presence of catalysts. J Anal Appl Pyrolysis 2008;81(1):72–9.
[7] Torri C, Reinikainen M, Lindfors C, Fabbri D, Oasmaa A, Kuoppala E. Investigation on catalytic pyrolysis of pine sawdust: catalyst screening by Py-GC-MIP-AED. J Anal Appl Pyrolysis 2010;88(1):7–13.
[8] Zhu X, Lu Q, Li W, Zhang D. Fast and catalytic pyrolysis of xylan: effects of temperature and M/HZSM-5 (M=Fe, Zn) catalysts on pyrolytic products. Front Energy Power Eng China 2010;4(3):424–9.
[9] Qiang L, XiFeng Z, WenZhi L, Ying Z, DengYu C. On-line catalytic upgrading of biomass fast pyrolysis products. Chin Sci Bull 2009;54:1941–8.
[10] Pattiya A, Titiloye JO, Bridgwater AV. Evaluation of catalytic pyrolysis of cassava rhizome by principal component analysis. Fuel 2010;89(1):244–53.
[11] Naqvi SR, Uemura Y, Yusup SB. Catalytic pyrolysis of paddy husk in a drop type pyrolyzer for bio-oil production: the role of temperature and catalyst. J Anal Appl Pyrolysis 2014;106:57–62.
[12] Zhou L, Yang H, Wu H, Wang M, Cheng D. Catalytic pyrolysis of rice husk by mixing with zinc oxide: characterization of bio-oil and its rheological behavior. Fuel Process Technol 2013;106:385–91.
[13] Abu Bakar MS, Titiloye JO. Catalytic pyrolysis of rice husk for bio-oil production. J Anal Appl Pyrolysis 2013;103:362–8.
[14] Pütün E. Catalytic pyrolysis of biomass: effects of pyrolysis temperature, sweeping gas flow rate and MgO catalyst. Energy 2010;35(7):2761–6.
[15] Pütün E, Uzun BB, Pütün AE. Production of bio-fuels from cottonseed cake by catalytic pyrolysis under steam atmosphere. Biomass Bioenergy 2006;30(6):592–8.
[16] Pütün E, Ateş F, Pütün AE. Catalytic pyrolysis of biomass in inert and steam atmospheres. Fuel 2008;87(6):815–24.
[17] Ates F, Putun A, Putun E. Pyrolysis of two different biomass samples in a fixed-bed reactor combined with two different catalysts. Fuel 2006;85(12–13):1851–9.
[18] Ates F, Putun AE, Putun E. Fixed bed pyrolysis of *Euphorbia rigida* with different catalysts. Energ Convers Manage 2005;46(3):421–32.
[19] Ates F, Putun AE, Putun E. Catalytic pyrolysis of perennial shrub, *Euphorbia rigida* in the water vapour atmosphere J Anal Appl Pyrol 2005; vol.73 2: 299–304.
[20] Adam J, Antonakou E, Lappas A, Stocker M, Nilsen MH, Bouzga A, et al. In situ catalytic upgrading of biomass derived fast pyrolysis vapours in a fixed bed reactor using mesoporous materials. Microporous Mesoporous Mater 2006;96(1–3):93–101.
[21] Antonakou E, Lappas A, Nilsen MH, Bouzga A, Stöcker M. Evaluation of various types of Al-MCM-41 materials as catalysts in biomass pyrolysis for the production of biofuels and chemicals. Fuel 2006;85(14–15):2202–12.
[22] Zhang X, Lei H, Zhu L, Qian M, Chan JC, Zhu X, et al. Development of a catalytically green route from diverse lignocellulosic biomasses to high-density cycloalkanes for jet fuels. Catal Sci Technol 2016;6(12):4210–20.
[23] Aho A, Kumar N, Eränen K, Salmi T, Hupa M, Murzin DY. Catalytic pyrolysis of woody biomass in a fluidized bed reactor: influence of the zeolite structure. Fuel 2008;87(12):2493–501.

[24] Agblevor FA, Mante O, Abdoulmoumine N, McClung R. roduction of stable biomass pyrolysis oils using fractional catalytic pyrolysis. Energy Fuel 2010;24(7):4087–9.
[25] Mante OD, Agblevor FA, Oyama ST, McClung R. The influence of recycling non-condensable gases in the fractional catalytic pyrolysis of biomass. Bioresour Technol 2012;111:482–90.
[26] Mullen CA, Boateng AA, Mihalcik DJ, Goldberg NM. Catalytic fast pyrolysis of white oak wood in a bubbling fluidized Bed. Energy Fuel 2011;25(11):5444–51.
[27] Zhang H, Xiao R, Huang H, Xiao G. Comparison of non-catalytic and catalytic fast pyrolysis of corncob in a fluidized bed reactor. Bioresour Technol 2009;100(3):1428–34.
[28] Zhang H, Xiao R, Wang D, Zhong Z, Song M, Pan Q, et al. Catalytic fast pyrolysis of biomass in a fluidized bed with fresh and spent fluidized catalytic cracking (FCC) catalysts. Energy Fuel 2009;23(12):6199–206.
[29] Jae J, Coolman R, Mountziaris TJ, Huber GW. Catalytic fast pyrolysis of lignocellulosic biomass in a process development unit with continual catalyst addition and removal. Chem Eng Sci 2014;108:33–46.
[30] Agblevor FA, Beis S, Mante O, Abdoulmoumine N. Fractional catalytic pyrolysis of hybrid poplar wood. Ind Eng Chem Res 2010;29:3533–8.
[31] Karanjkar PU, Coolman RJ, Huber GW, Blatnik MT, Almalkie S, de Bruyn Kops SM, et al. Production of aromatics by catalytic fast pyrolysis of cellulose in a bubbling fluidized bed reactor. AIChE J 2014;60(4):1320–35.
[32] Cheng YT, Jae J, Shi J, Fan W, Huber GW. Production of renewable aromatic compounds by catalytic fast pyrolysis of lignocellulosic biomass with bifunctional Ga/ZSM-5 catalysts. Angew Chem 2012;124:1416–9.
[33] Yang H, Coolman R, Karanjkar P, Wang H, Dornath P, Chen H, et al. The effects of contact time and coking on the catalytic fast pyrolysis of cellulose. Green Chem 2017;19(1):286–97.
[34] Aho A, Kumar N, Lashkul AV, Eränen K, Ziolek M, Decyk P, et al. Catalytic upgrading of woody biomass derived pyrolysis vapours over iron modified zeolites in a dual-fluidized bed reactor. Fuel 2010;89(8):1992–2000.
[35] Aho A, Tokarev A, Backman P, Kumar N, Eränen K, Hupa M, et al. Catalytic pyrolysis of pine biomass over H-Beta zeolite in a dual-fluidized bed reactor: effect of space velocity on the yield and composition of pyrolysis products. Top Catal 2011;54(13–15):941–8.
[36] Mante OD, Agblevor FA. Catalytic conversion of biomass to bio-syncrude oil. Biomass Convers Biorefin 2011;1(4):203–215.
[37] Mante OD, Agblevor FA, McClung R. Fluid catalytic cracking of biomass pyrolysis vapors. Biomass Convers Biorefin 2011;1(4):189–201.
[38] Atutxa A, Aguado R, Gayubo AG, Olazar M, Bilbao J. Kinetic description of the catalytic pyrolysis of biomass in a conical spouted bed reactor. Energy Fuel 2005;19(3):765–74.
[39] Du S, Sun Y, Gamliel DP, Valla JA, Bollas GM. Catalytic pyrolysis of miscanthus x giganteus in a spouted bed reactor. Bioresour Technol 2014;169:188–97.
[40] Stephanidis S, Nitsos C, Kalogiannis K, Iliopoulou EF, Lappas AA, Triantafyllidis KS. Catalytic upgrading of lignocellulosic biomass pyrolysis vapours: effect of hydrothermal pre-treatment of biomass. Catal Today 2011;167(1):37–45.
[41] Triantafyllidis KS, Iliopoulou EF, Antonakou EV, Lappas AA, Wang H, Pinnavaia TJ. Hydrothermally stable mesoporous aluminosilicates (MSU-S) assembled from zeolite seeds as catalysts for biomass pyrolysis. Microporous Mesoporous Mater 2007;99(1–2):132–9.
[42] Iliopoulou EF, Antonakou EV, Karakoulia SA, Vasalos IA, Lappas AA, Triantafyllidis KS. Catalytic conversion of biomass pyrolysis products by mesoporous materials: effect of steam stability and acidity of Al-MCM-41 catalysts. Chem Eng J 2007;134(1–3):51–7.
[43] Stefanidis SD, Kalogiannis KG, Iliopoulou EF, Lappas AA, Pilavachi PA. In-situ upgrading of biomass pyrolysis vapors: catalyst screening on a fixed bed reactor. Bioresour Technol 2011;102(17):8261–7.

[44] Leibold H, Seifert H. Energetische Aspekte bei der Heibetagasreinigung von biomassebasierten Synthesegasen. Erneuerbare Energien 2010;4:95–107.
[45] Iliopoulou EF, Stefanidis S, Kalogiannis K, Psarras AC, Delimitis A, Triantafyllidis KS, et al. Pilot-scale validation of Co-ZSM-5 catalyst performance in the catalytic upgrading of biomass pyrolysis vapours. Green Chem 2014;16(2):662–74.
[46] Lappas AA, Samolada MC, Iatridis DK, Voutetakis SS, Vasalos IA. Biomass pyrolysis in a circulating fluid bed reactor for the production of fuels and chemicals. Fuel 2002;81(16):2087–95.
[47] Rezaei PS, Shafaghat H, Daud WMAW. Production of green aromatics and olefins by catalytic cracking of oxygenate compounds derived from biomass pyrolysis: a review. Appl Catal A Gen 2014;469:490–511.
[48] Chang CD, Silvestri AJ. The conversion of methanol and other O-compounds to hydrocarbons over zeolite catalysts. J Catal 1977;47:249–59.
[49] Adjaye JD, Bakhshi NN. Production of hydrocarbons by catalytic upgrading of a fast pyrolysis bio-oil. Part I: conversion over various catalysts. Fuel Process Technol 1995;45:161–83.
[50] Corma A, Huber G, Sauvanaud L, Oconnor P. Processing biomass-derived oxygenates in the oil refinery: catalytic cracking (FCC) reaction pathways and role of catalyst. J Catal 2007;247(2):307–27.
[51] Seddegi ZS, Budrthumal U, Al-Arfaj A, Al-Amerc AM, Barri SAI. Catalytic cracking of polyethylene over all-silica MCM-41 molecular sieve. Appl Catal A 2002;225:167–76.
[52] Shafizadeh F, Lai YZ. Thermal degradation of 1,6-Anhydro-B-D-glucopyranose. J Org Chem 1972;37:278–84.
[53] Adjaye JD, Bakhshi NN. Production of hydrocarbons by catalytic upgrading of a fast pyrolysis bio-oil. Part I: conversion over various catalysts. Fuel Process Technol 12//1995;45(3):161–83.
[54] French R, Czernik S. Catalytic pyrolysis of biomass for biofuels production. Fuel Process Technol 2010;91(1):25–32.
[55] Sharma RK, Bakhshi NN. Catalytic upgrading of fast pyrolysis oil over HZSM-5. Can J Chem Eng 1993;71:383–91.
[56] Mebold J, Scahill J. Production of Primary Pyrolysis Oils i n a Vortex Reactor. Div Fuel Chem 1987;32:297–307.
[57] Mortensen PM, Grunwaldt JD, Jensen PA, Knudsen KG, Jensen AD. A review of catalytic upgrading of bio-oil to engine fuels. Appl Catal A Gen 11/4/2011;407(1–2):1–19.
[58] Piskorz J, Majerski P, Radlein D. Pyrolysis of biomass—aerosol generation: properties, applications, and significance for process engineers. In: Overend RP, Chornet E, editors. Biomass: a growth opportunity in green energy and value-added products. Vols. 1 and 2. Oxford: Pergamon-Elsevier Science Ltd; 1999. p. 1153–9.
[59] Ferdous D, Dalai AK, Bej SK, Thring RW. Pyrolysis of lignins: experimental and kinetics studies. Energy Fuel 2002;16(6):1405–12.
[60] Bridgwater AV. Catalysis in thermal biomass conversion. Appl Catal A Gen 1994;116(1–2):5–47.
[61] Williams PT, Horne PA. The influence of catalyst regeneration on the composition of zeolite-upgraded biomass pyrolysis oils. Fuel 1995;74(12):1839–51.
[62] Williams PT, Horne PA. The influence of catalyst type on the composition of upgraded biomass pyrolysis oils. J Anal Appl Pyrolysis 1995;31:39–61.
[63] Haw JF, Song W, Marcus DM, Nicholas JB. The mechanism of methanol to hydrocarbon catalysis. Acc Chem Res 2003;36:317–26.
[64] White JL. Methanol-to-hydrocarbon chemistry: the carbon pool (r)evolution. Catal Sci Technol 2011;1(9):1630–5.

[65] Sun J, Wang Y. Recent advances in catalytic conversion of ethanol to chemicals. ACS Catal 2014;4(4):1078–90.
[66] Johansson R, Hruby SL, Rass-Hansen J, Christensen CH. The hydrocarbon pool in ethanol-to-gasoline over HZSM-5 catalysts. Catal Lett 2008;127(1–2):1–6.
[67] Carlson TR, Jae J, Huber GW. Mechanistic insights from isotopic studies of glucose conversion to aromatics over ZSM-5. ChemCatChem 2009;1(1):107–10.
[68] To AT, Resasco DE. Role of a phenolic pool in the conversion of m-cresol to aromatics over HY and HZSM-5 zeolites. Appl Catal A Gen 2014;487:62–71.
[69] Stocker M. Methanol-to-hydrocarbons: catalytic materials and their behavior. Microporous Mesoporous Mater 1999;29:3–48.
[70] Bjorgen M, Svelle S, Joensen F, Nerlov J, Kolboe S, Bonino F, et al. Conversion of methanol to hydrocarbons over zeolite H-ZSM-5: On the origin of the olefinic species. J Catal 2007;249(2):195–207.
[71] Zhang J, Choi YS, Shanks BH. Catalytic deoxygenation during cellulose fast pyrolysis using acid–base bifunctional catalysis. Catal Sci Technol 2016;6(20):7468–76.
[72] Mohan D, Pittman CU, Steele PH. Pyrolysis of wood/biomass for bio-oil: a critical review. Energy Fuel 2006;20(3):848–89.
[73] ASTM, Standard Specification for Pyrolysis Liquid Biofuel (ASTM D7544 - 12). 2012, American Society for Testing and Materials.
[74] Carlson TR, Vispute TP, Huber GW. Green gasoline by catalytic fast pyrolysis of solid biomass derived compounds. ChemSusChem 2008;1(5):397–400.
[75] Wang K, Kim KH, Brown RC. Catalytic pyrolysis of individual components of lignocellulosic biomass. Green Chem 2014;16(2):727–35.
[76] Huang W, Gong F, Fan M, Zhai Q, Hong C, Li Q. Production of light olefins by catalytic conversion of lignocellulosic biomass with HZSM-5 zeolite impregnated with 6wt.% lanthanum. Bioresour Technol Oct2012;121:248–55.
[77] Zhou G, Jensen PA, Le DM, Knudsenb NO, Jensen AD. Direct upgrading of fast pyrolysis lignin vapor over HZSM-5 catalyst. Green Chem 2016;18(7):1965–75.
[78] Zhang H, Carlson TR, Xiao R, Huber GW. Catalytic fast pyrolysis of wood and alcohol mixtures in a fluidized bed reactor. Green Chem 2012;14(1):98–110.
[79] Galadima A, Muraza O. In situ fast pyrolysis of biomass with zeolite catalysts for bioaromatics/gasoline production: a review. Energy Convers Manag 11/15/2015;105:338–54.
[80] Mullen CA, Boateng AA. Accumulation of inorganic impurities on HZSM-5 zeolites during catalytic fast pyrolysis of switchgrass. Ind Eng Chem Res 2013;52(48):17156–61.
[81] Yildiz G, Ronsse F, Venderbosch R, Rv D, SRA K, Prins W. Effect of biomass ash in catalytic fast pyrolysis of pine wood. Appl Catal B Environ 2015;168–169:203–11.
[82] Yu Y, Zeng Y, Zuo J, Ma F, Yang X, Zhang X, et al. Improving the conversion of biomass in catalytic fast pyrolysis via white-rot fungal pretreatment. Bioresour Technol Apr2013;134:198–203.
[83] Mahadevan R, Adhikari S, Shakya R, Wang K, Dayton DC, Li M, et al. Effect of torrefaction temperature on lignin macromolecule and product distribution from HZSM-5 catalytic pyrolysis. J Anal Appl Pyrolysis 2016;122:95–105.
[84] Hassan EB, Elsayed I, Eseyin A. Production high yields of aromatic hydrocarbons through catalytic fast pyrolysis of torrefied wood and polystyrene. Fuel 2016;174:317–24.
[85] Chen D, Li Y, Deng M, Wang J, Chen M, Yan B, et al. Effect of torrefaction pretreatment and catalytic pyrolysis on the pyrolysis poly-generation of pine wood. Bioresour Technol 2016;214:615–22.
[86] Zheng A, Jiang L, Zhao Z, Huang Z, Zhao K, Wei G, et al. Impact of torrefaction on the chemical structure and catalytic fast pyrolysis behavior of hemicellulose, lignin, and cellulose. Energy Fuel 2015;29(12):8027–34.

[87] Zheng A, Zhao Z, Huang Z, Zhao K, Wei G, Wang X, et al. Catalytic fast pyrolysis of biomass pretreated by torrefaction with varying severity Energy Fuel 2014; vol.28 9: 5804–11.

[88] Theodore D, Juan S. Catalytic fast pyrolysis: a review. Energies 2013;6(1):514–38.

[89] Hagen J. 2006 Industrial catalysis: a practical approach, 9783527311446 (hbk.): ^{1}115.00 3527311440 (hbk.): ^{1}115.00, Journal.

[90] Weitkamp J. Zeolites and catalysis. Solid State Ionics 2000;131:175–88.

[91] Adjaye JD, Bakhshi NN. Production of hydrocarbons by catalytic upgrading of a fast pyrolysis bio-oil. Part II: comparative catalyst performance and reaction pathways. Fuel Process Technol 1995;45(3):185–202.

[92] Bridgwater AV, Cottam ML. Opportunities for biomass pyrolysis liquids production and upgrading Energy Fuel 1992; vol.6 2: 113–20.

[93] Horne PA, Williams PT. Premium quality fuels and chemicals from the fluidized-bed pyrolysis of biomass with zeolite catalyst. Upgrading Renewable Energy 1994;5(5–8):810–2.

[94] Horne PA, Williams PT. The effect of zeolite ZSM-5 catalyst deactivation during the upgrading of biomass-derived pyrolysis vapours. J Anal Appl Pyrolysis 1995;34(1): 65–85.

[95] Horne PA, Williams PT. Upgrading of biomass-derived pyrolytic vapours over zeolite ZSM-5 catalyst: effect of catalyst dilution on product yields. Fuel 1996;75(9): 1043–50.

[96] Williams PT, Horne PA. Characterization of oils from the fluidized-bed pyrolysis of biomass with zeolite catalyst upgrading. Biomass Bioenergy 1994;7(1–6):223–36.

[97] Williams PT, Horne PA. Analysis of aromatic-hydrocarbons in pyrolytic oil derived from biomass. J Anal Appl Pyrolysis Feb1995;31:15–37.

[98] Williams PT, Nugranad N. Comparison of products from the pyrolysis and catalytic pyrolysis of rice husks. Energy Jun2000;25(6):493–513.

[99] Williams PT, Nugranad N, Horne PA. Influence of steam on the formation of polycyclic aromatic hydrocarbons in the zeolite catalytic upgrading of biomass pyrolysis oils. In: Kaltschmitt M, Bridgwater AV, editors. Biomass gasification and pyrolysis: state of the art and future prospects. Newbury: Cpl Press; 1997. p. 422–30.

[100] Chen NY, Walsh DE, Koenig LR. Fluidized-bed upgrading of wood pyrolysis liquids and related compounds. In: Soltes EJ, Milne TA, editors. Pyrolysis oils from biomass: producing, analyzing, and upgrading. Washington, D.C.: American Chemical Society; 1988. p. 277–89.

[101] Zhang H, Xiao R, Jin B, Xiao G, Chen R. Biomass catalytic pyrolysis to produce olefins and aromatics with a physically mixed catalyst. Bioresour Technol Jul2013;140:256–62.

[102] Zhang H, Zheng J, Xiao R, Jia Y, Shen D, Jin B, et al. Study on pyrolysis of pine sawdust with solid base and acid mixed catalysts by thermogravimetry–fourier transform infrared spectroscopy and pyrolysis–gas chromatography/mass spectrometry Energy Fuel 2014; vol.28 7: 4294–9.

[103] Iliopoulou EF, Stefanidis SD, Kalogiannis KG, Delimitis A, Lappas AA, Triantafyllidis KS. Catalytic upgrading of biomass pyrolysis vapors using transition metal-modified ZSM-5 zeolite. Appl Catal B Environ 2012;127:281–90.

[104] Park HJ, Park K-H, Jeon J-K, Kim J, Ryoo R, Jeong K-E, et al. Production of phenolics and aromatics by pyrolysis of miscanthus. Fuel 2012;97:379–84.

[105] Park HJ, Park Y-K, Kim J-S, Jeon J-K, Yoo K-S, Yim J-H, et al. Bio-oii upgrading over Ga modified zeolites in a bubbling fluidized bed reactor. Stud Surf Sci Catal 2006;159:553–6.

[106] Park HJ, Heo HS, Jeon J-K, Kim J, Ryoo R, Jeong K-E, et al. Highly valuable chemicals production from catalytic upgrading of radiata pine sawdust-derived pyrolytic vapors over mesoporous MFI zeolites. Appl Catal B Environ 2010;95(3–4):365–73.
[107] Carlson TR, Tompsett GA, Conner WC, Huber GW. Aromatic production from catalytic fast pyrolysis of biomass-derived feedstocks. Top Catal 2009;52(3):241–52.
[108] Duanguppama K, Suwapaet N, Pattiya A. Fast pyrolysis of contaminated sawdust in a circulating fluidised bed reactor. J Anal Appl Pyrolysis 2016;118:63–74.
[109] Baerlocher C, Meier WMAozst, Meier WM, Olson D. 2001 Atlas of zeolite framework types, 0444507019 (pbk.): No price, Journal.
[110] Adam J, Blazso M, Meszaros E, Stocker M, Nilsen MH, Bouzga A, et al. Pyrolysis of biomass in the presence of Al-MCM-41 type catalysts. Fuel 2005;84(12–13):1494–502.
[111] Nilsen MH, Antonakou E, Bouzga A, Lappas A, Mathisen K, Stocker M. Investigation of the effect of metal sites in Me-Al-MCM-41 (Me = Fe, Cu or Zn) on the catalytic behavior during the pyrolysis of wooden based biomass. Microporous Mesoporous Mater 2007;105(1–2):189–203.
[112] Adam J, Blazsó M, Mészáros E, Stöcker M, Nilsen MH, Bouzga AM, et al. Vapour phase upgrading of biomass pyrolysis products with mesoporous SBA-15 and FCC catalysts Conference Vapour phase upgrading of biomass pyrolysis products with mesoporous SBA-15 and FCC catalysts, Paris, France 857–861.
[113] Melo RAA, Giotto MV, Rocha J, Urquieta-Gonzulez EA. MCM-41 ordered mesoporous molecular sieves synthesis and characterization. Mater Res 1999;2:173–9.
[114] Wachs IE. Recent conceptual advances in the catalysis science of mixed metal oxide catalytic materials. Catal Today 2005;100(1–2):79–94.
[115] Fierro JLG. Metal oxides: chemistry and applications. Boca Raton, FL: CRC Press/ Taylor & Francis; 2006.
[116] Cavani F, Centi G, Marion P. Catalytic Ammoxidation of Hydrocarbons on Mixed Oxides, in Metal Oxide Catalysis. Wiley-VCH Verlag GmbH & Co. KGaA; 2009; p. 771–818.
[117] Lin Y, Zhang C, Zhang M, Zhang J. Deoxygenation of bio-oil during pyrolysis of biomass in the presence of CaO in a fluidized-bed reactor. Energy Fuel 2010; vol.24 10: 5686–95.
[118] Nokkosmäki MI, Kuoppala ET, Leppämäki EA, Krause AOI. Catalytic conversion of biomass pyrolysis vapours with zinc oxide. J Anal Appl Pyrolysis 2000;55(1):119–31.
[119] Nik-Azar M, Hajaligol MR, Sohrabi M, Dabir B. Mineral matter effects in rapid pyrolysis of beech wood. Fuel Process Technol 1997;51:7–17.
[120] Scholze B. Long-term stability, catalytic upgrading, and application of pyrolysis oils—improving the properties of a potential substitue for fossil fuels. Hamburg: University of Hamburg; 2002.
[121] Li R, Zhong ZP, Jin BS, Zheng AJ. Application of mineral bed materials during fast pyrolysis of rice husk to improve water-soluble organics production. Bioresour Technol 2012;119(0):324–30.
[122] Sanna A, Li S, Linforth R, Smart KA, Andresen JM. Bio-oil and bio-char from low temperature pyrolysis of spent grains using activated alumina. Bioresour Technol 2011;102(22):10695–703.
[123] Sanna A, Andrésen JM. Bio-oil deoxygenation by catalytic pyrolysis: new catalysts for the conversion of biomass into densified and deoxygenated bio-oil. ChemSusChem 2012;5(10):1944–57.
[124] Katikaneni SPR, Idem RO, Bakhshi NN. Catalytic conversion of a biomass-derived oil using various catalysts. In: Kaltschmitt M, Bridgwater AV, editors. Biomass gasification and pyrolysis: state of the art and future prospects. Newbury: Cpl Press; 1997. p. 411–21.

[125] Murata K, Liu Y, Inaba M, Takahara I. Catalytic fast pyrolysis of jatropha wastes. J Anal Appl Pyrolysis 2012;94:75–82.
[126] Vitolo S, Bresci B, Seggiani M, Gallo MG. Catalytic upgrading of pyrolytic oils over HZSM-5 zeolite: behaviour of the catalyst when used in repeated upgrading±regenerating cycles. Fuel 2001;80:17–26.
[127] Calsavara V, Machado Jr. NRCF, JLB SAEF. On the acidity and/or basicity of usy zeolites after basic and acid treatment. Braz J Chem Eng 2000;17:91–8.
[128] Kantarelis E, Yang W, Blasiak W. Effect of zeolite to binder ratio on product yields and composition during catalytic steam pyrolysis of biomass over transition metal modified HZSM5. Fuel 2014;122:119–25.
[129] Qiao K, Li X, He L, Liu X, Yan Z, Xing W, et al. An efficient modification of ultra-stable Y zeolites using citric acid and ammonium fluosilicate. Appl Petrochem Res 2014;4(4):373–8.
[130] Li X, Su L, Wang Y, Yu Y, Wang C, Li X, et al. Catalytic fast pyrolysis of Kraft lignin with HZSM-5 zeolite for producing aromatic hydrocarbons. Front Environ Sci Eng 2012;6(3):295–303.
[131] Wang S, Cai Q, Chen J, Zhang L, Wang X, Yu C. Green aromatic hydrocarbon production from cocracking of a bio-oil model compound mixture and ethanol over Ga2O3/HZSM-5. Ind Eng Chem Res 2014;53(36):13935–44.
[132] Kelkar S, Saffron CM, Andreassi K, Li Z, Murkute A, Miller DJ, et al. A survey of catalysts for aromatics from fast pyrolysis of biomass. Appl Catal B Environ 2015;174–175:85–95.
[133] Zhang Y, Chen P, Lou H. In situ catalytic conversion of biomass fast pyrolysis vapors on HZSM-5. J Energy Chem 2016;25(3):427–33.
[134] Neumann GT, Pimentel BR, Rensel DJ, Hicks JC. Correlating lignin structure to aromatic products in the catalytic fast pyrolysis of lignin model compounds containing β–O–4 linkages. Catal Sci Technol 2014;4(11):3953–63.
[135] Johan C, Groen ZW, Brouwer S, Huynink SJ, Kapteijn F, Moulijn JA, et al. Direct demonstration of enhanced diffusion in mesoporous ZSM-5 zeolite obtained via controlled desilication. J Am Chem Soc 2007;129:355–60.
[136] Li J, Li X, Zhou G, Wang W, Wang C, Komarneni S, et al. Catalytic fast pyrolysis of biomass with mesoporous ZSM-5 zeolites prepared by desilication with NaOH solutions. Appl Catal A Gen 2014;470:115–22.
[137] Stefanidis S, Kalogiannis K, Iliopoulou EF, Lappas AA, Triguero JM, Navarro MT, et al. Mesopore-modified mordenites as catalysts for catalytic pyrolysis of biomass and cracking of vacuum gasoil processes. Green Chem 2013;15(6):1647–58.
[138] Gayubo AG, Aguayo AT, Atutxa A, Prieto R, Bilbao J. Deactivation of a HZSM-5 zeolite catalyst in the transformation of the aqueous fraction of biomass pyrolysis oil into hydrocarbons. Energy Fuel 2004;18(6):1640–7.
[139] Gayubo AG, Aguayo AT, Atutxa A, Aguado R, Bilbao J. Transformation of oxygenate components of biomass pyrolysis oil on a HZSM-5 zeolite. I. Alcohols and phenols. Ind Eng Chem Res 2004;43(11):2610–8.
[140] Gayubo AG, Aguayo AT, Atutxa A, Valle B, Bilbao J. Undesired components in the transformation of biomass pyrolysis oil into hydrocarbons on an HZSM-5 zeolite catalyst. J Chem Technol Biotechnol 2005;80(11):1244–51.
[141] Gayubo AG, AsT A, Atutxa A, Aguado R, Olazar M, Bilbao J. Transformation of oxygenate components of biomass pyrolysis oil on a HZSM-5 zeolite. II. aldehydes, ketones, and acids. Ind Eng Chem Res 2004;43:2619–26.

[142] Gayubo AG, Valle B, Aguayo AT, Olazar M, Bilbao J. Pyrolytic lignin removal for the valorization of biomass pyrolysis crude bio-oil by catalytic transformation. J Chem Technol Biotechnol 2010;85(1):132–44.
[143] Repellin V, Govin A, Rolland M, Guyonnet R. Modelling anhydrous weight loss of wood chips during torrefaction in a pilot kiln. Biomass Bioenergy 2010; 34(5):602–9.
[144] Gayubo AG, AsT A, Atutxa A, Aguado R, Bilbao J. Role of reaction-medium water on the acidity deterioration of a HZSM-5 zeolite. Ind Eng Chem Res 2004;43:5042–8.
[145] Valle B, Gayubo AG, AsT A, Olazar M, Bilbao J. Selective production of aromatics by crude bio-oil valorization with a nickel-modified HZSM-5 zeolite catalyst. Energy Fuel 2010;24(3):2060–70.
[146] Sharma RK, Bakhshi NN. Catalytic conversion of fast pyrolysis oil to hydrocarbon fuels over HZSM-5 in a dual reactor system. Biomass Bioenergy 1993;5:445–55.
[147] Adjaye JD, Sharma RK, Bakhshi NN. Catalytic conversion of wood derived bio-oil to fuels and chemicals. Stud Surf Sci Catal 1992;73:301–8.
[148] Jae J, Tompsett GA, Foster AJ, Hammond KD, Auerbach SM, Lobo RF, et al. Investigation into the shape selectivity of zeolite catalysts for biomass conversion J Catal 2011; vol.279 2: 257–68.
[149] Methods and apparatus for producing syngas and alcohols 2009; WO 2009079127.
[150] Sharma RK, Bakhshi NN. Conversion of non-phenolic fraction of biomass derived pyrolysis oil to hydrocarbon fuels over HZSM-5 using a dual reactor system. Bioresour Technol 1993;45:195–203.
[151] Nguyen TS, Zabeti M, Lefferts L, Brem G, Seshan K. Catalytic upgrading of biomass pyrolysis vapours using faujasite zeolite catalysts. Biomass Bioenergy 2013;48:100–10.
[152] Güngör A, Önenç S, Uçar S, Yanik J. Comparison between the "one-step" and "two-step" catalytic pyrolysis of pine bark. J Anal Appl Pyrolysis 2012;97:39–48.
[153] Iisa K, French RJ, Orton KA, Yung MM, Johnson DK, ten Dam J, et al. In situ and ex situ catalytic pyrolysis of pine in a bench-scale fluidized bed reactor system. Energy Fuel 2016;30(3):2144–57.
[154] Yildiz G, Pronk M, Djokic M, Geem KMv, Ronsse F, Duren Rv, et al. Validation of a new set-up for continuous catalytic fast pyrolysis of biomass coupled with vapour phase upgrading J Anal Appl Pyrolysis 9//2013; vol.103: 343–51.
[155] Gamliel DP, Du S, Bollas GM, Valla JA. Investigation of in situ and ex situ catalytic pyrolysis of miscanthus x giganteus using a PyGC-MS microsystem and comparison with a bench-scale spouted-bed reactor. Bioresour Technol 2015;191:187–96.
[156] Wang K, Johnston PA, Brown RC. Comparison of in-situ and ex-situ catalytic pyrolysis in a micro-reactor system. Bioresour Technol 2014;173:124–31.
[157] Wan S, Waters C, Stevens A, Gumidyala A, Jentoft R, Lobban L, et al. Decoupling HZSM-5 catalyst activity from deactivation during upgrading of pyrolysis oil vapors. ChemSusChem 2015;8(3):552–9.
[158] Williams PT, Nugranad N. Comparison of products from the pyrolysis and catalytic pyrolysis of rice husks. Energy 2000;25:493–513.
[159] Melligan F, Hayes MHB, Kwapinski W, Leahy JJ. hydro-pyrolysis of biomass and online catalytic vapor upgrading with Ni-ZSM-5 and Ni-MCM-41. Energy Fuel 2012;26(10):6080–90.
[160] Trane R, Dahl S, Skjøth-Rasmussen MS, Jensen AD. Catalytic steam reforming of bio-oil. Int J Hydrog Energy 2012;37(8):6447–72.

[161] Wang D, Czernik S, Chornet E. Production of hydrogen from biomass by catalytic steam reforming of fast pyrolysis oils. Energy Fuel 1998; vol.12 1: 19–24.
[162] Wang D, Czernik S, Montane D, Mann M, Chornet E. Biomass to hydrogen via fast pyrolysis and catalytic steam reforming of the pyrolysis pil or its fractions. Ind Eng Chem Res 1997;36:1507–18.
[163] Wang D, Montane D, Chornet E. Catalytic steam reforming of biomass-derived oxygenates: acetic acid and hydroxyacetaldehyde. Appl Catal A 1996;143:245–70.
[164] Wang Z, Pan Y, Dong T, Zhu X, Kan T, Yuan L, et al. Production of hydrogen from catalytic steam reforming of bio-oil using C12A7-O− based catalysts. Appl Catal A Gen 2007;320:24–34.
[165] Marquevich M, Czernik S, Chornet E, Montane D. Hydrogen from biomass: steam reforming of model compounds of fast-pyrolysis oil. Energy Fuel 1999;13:1160–6.
[166] Horne PA, Nugranad N, Williams PT. Catalytic coprocessing of biomass-derived pyrolysis vapours and methanol. J Anal Appl Pyrolysis 1995;34(1):87–108.

Part Two

Hydrothermal Liquefaction

Recent advances in energy recovery from wastewater sludge

3

L. Nazari, S. Sarathy*,†, D. Santoro*,†, D. Ho*,†, M.B. Ray*, C. (Charles) Xu**
*Western University, London, ON, Canada, †Trojan Technologies, London, ON, Canada

3.1 Introduction

Renewable energy has attracted increasing interest due to the worldwide rising demand for primary energy and the shrinking reserves of fossil fuels as well as the growing concerns on climate change and greenhouse gas emission. Biofuels, broadly defined as solid, gaseous, or liquid fuels produced from bio-renewable resources, that is, biomass, are sustainable alternative to fossil fuels [1].

Biomass resources mainly include wood and forestry residues/by-products (e.g. sawdust, bark, tree tops, and lignin), agricultural crops and crop residues (e.g. wheat/rice straws and corn stover), marine products, municipal solid wastes (MSW), wastewater sludge, and waste streams from animal farms (manure) and food processing [2–4]. Biomass has a lower energy density based on either volume or mass, for example, its higher heating value (HHV) is only 15–20 MJ/kg (dry basis), compared with ~30 and 40 MJ/kg for coal and petroleum, respectively. Thus, a proper conversion method is required to densify biomass into gaseous or liquid biofuels. Based on the production technologies, the produced biofuels can be classified into three generations. First-generation biofuels are produced by conversion of food crops (e.g. sugar, starch, and vegetable oils), such as bioalcohols and biodiesel. Second-generation biofuels are produced from non-food biomass such as agroforestry residues. Some examples of second-generation biofuels are bio-oil, bio-dimethylfuran (bio-DMF), biohydrogen, and bio-Fischer-Tropsch diesel. Third-generation biofuels are derived mainly from algae [5–7].

The two main types of technologies for conversion of biomass to biofuels and chemicals are biochemical and thermochemical processes. Biological conversion refers mainly to fermentation of carbohydrate materials into bioethanol and biogas [4,8,9]. Thermochemical conversions mainly include combustion, gasification, pyrolysis, and hydrothermal liquefaction. The above-mentioned biomass conversion methods and their products are illustrated in Fig. 3.1.

Biogas and bio-oil are the two typical biofuels produced from biochemical and thermochemical conversions, respectively. Biogas is a mixture of mostly methane and carbon dioxide with trace amounts of other gases and is considered as a valuable renewable energy source. Biogas can more or less be used in all applications that have been developed for natural gas [10]. For example, biogas can be used for production of heat and steam, electricity generation/cogeneration, as vehicle fuel, and production of chemicals [10]. Biogas is most commonly produced by anaerobic digestion (AD) of

Direct Thermochemical Liquefaction for Energy Applications. https://doi.org/10.1016/B978-0-08-101029-7.00011-4

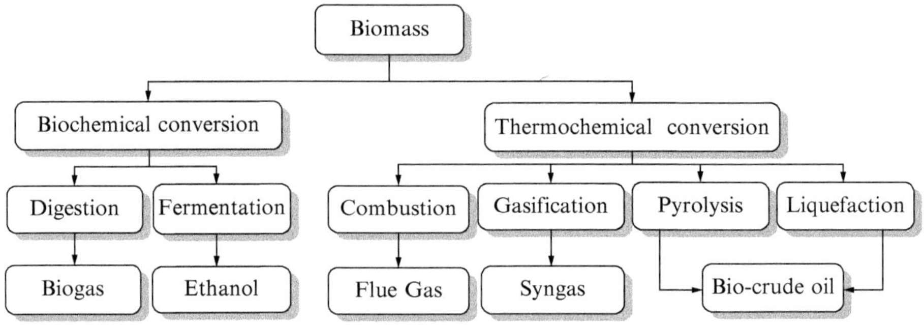

Fig. 3.1 Biomass conversion methods [8,9].

organic materials, a slow biological process (requiring a residence time of 10–20 days) consisting of four stages, that is, hydrolysis, acidogenesis, acetogenesis, and methanogenesis [11]. Bio-crude oils or simply biocrude or bio-oils are typical biofuels from biomass thermochemical conversion. Bio-oils have been considered as the promising alternatives to petroleum fuels for power or heat generation and extraction of valuable chemicals [12]. Bio-oils can be produced by two main routes: pyrolysis or hydrothermal liquefaction (HTL). The former route operates at a high temperature under atmospheric or reduced pressure using dry feedstock (thus, it requires costly dewatering/drying processes), while the latter operates at a mild temperature (200–400°C) but elevated pressure in water or water-containing solvent, where high-water-content feedstocks can be used directly. Thus, HTL has attracted an increasing interest due to its mild operating conditions and its suitability for conversion of high-water-content feedstocks such as wastewater sludge and microalgae.

3.1.1 Wastewater sludge management

Sewage sludge is generally the solid residue in suspension from the wastewater stream and can be produced in two steps during the treatment of the effluent. Primary sludge is generated during the removal of insoluble matters such as grit, grease, and scum from wastewater by screening followed by coagulation and sedimentation [13]. The settled primary sludge contains mainly water (between 97% and 99%) and highly putrescible organic matters [10]. The secondary sludge or waste-activated sludge (WAS) is generated as a result of biological treatment of the effluent from primary treatment containing different types of dissolved organic matters, and it consists of a complex activated sludge floc structure. Generally, sludge is a mixture of organics (such as proteins, carbohydrates, and lipids) or volatile matters, inorganic matters, and associated water [11]. Typical characteristics of the two different sludge types are listed in Table 3.1. As shown in the table, primary sludge typically has higher percentage of total solids (TS) compared with the activated sludge. The two types of sludge contain resources such as nitrogen (1.5%–4% for primary and 2.4%–5% for WAS) and phosphorous (0.17–0.6 for primary and 0.6%–2.3% for WAS), which could be recovered and used as fertilisers or soil conditioners. Primary sludge, captured due to gravity, usually has a higher energy content, while a secondary sludge has a lower energy content as during the secondary treatment, the microorganisms consume most of the energy content of the sludge.

Table 3.1 Characteristics of primary and secondary sludge

Item	Primary sludge	WAS
Total dry solids (TS), %	2.0–8.0	0.83–1.16
Volatile solids (% of TS)	60–80	59–88
Grease at fats (% of TS)	13–65	5–12
Protein (% of TS)	20–30	32–41
Nitrogen (N, % of TS)	1.5–4	2.4–5
Phosphorous (P, % of TS)	0.17–0.6	0.6–2.3
Potash (K, % of TS)	0–0.41	0.2–0.29
pH	5.0–8.0	6.5–8.0
Alkalinity (mg dm^{-3} as $CaCO_3$)	500–1500	580–1100
Organic acids (mg dm^{-3} as acetate)	200–2000	1100–1700
Energy content (MJ/kg)	23.2–29	18.6–23.2

Modified from Weemaes MPJ, Verstraete WH. Evaluation of current wet sludge disintegration techniques. J Chem Technol Biotechnol 1998;73(2):83–92.

The biological processes such as activated sludge treatment depend on the dissolved organic strength or organic load for different wastewater treatment plants, which are determined by three types of analytical measurements. The total organic carbon (TOC), measured by combustion at very high temperatures and calculated by the resultant CO_2, includes both dissolved biodegradable and recalcitrant organic compounds that cannot be broken down biologically [14]. Biological oxygen demand (BOD) is the total amount of biodegradable organics, measured by the oxygen uptake over a 5-day period by a small 'seed' of bacteria confined in a dark bottle containing the wastewater. During this time, the biodegradable organic carbon is taken up with a corresponding decrease in the dissolved oxygen [14]. Chemical oxygen demand (COD) is the total amount of oxidisable organics (biodegradable and nonbiodegradable and both dissolved and particulate), measured by the amount of oxygen in the form of oxidising agent required for the oxidation of organic matters by heating the sample in strong sulphuric acid containing potassium dichromate [14].

Sludge management is one of the most difficult and challenging tasks of wastewater treatment plants due to its high water content and poor dewaterability and strict regulation for sludge reuse or disposal. Sludge management can cost as high as 60% of the total cost for a wastewater treatment plant [15]. The United States is currently producing 6.9 million dry tons of sewage sludge annually, and Canada produces >0.66 million dry tons (2.5 million wet tons) of biosolids and sludges per year [16,17]. In Canada, the use or disposal of the sludge is regulated by the provincial government, and thus, there are major differences between the methods used to manage biosolids from one province to another [18]. The main objectives of sludge management are to reduce the volume and remove its pathogens. Conventionally, sludge is disposed by three main approaches: land applications as soil conditioner, landfill disposal, and incineration [11,15,19]. Sludge should be first dewatered to 20–40 wt% solids to meet the requirements of landfill or incineration, which is costly. There is also a large energy requirement in evaporating the water content of the sludge during combustion or incineration. Landfill sludge has undesirable secondary emissions to water, air, and soil. For instance, heavy metals in the sludge can contaminate soil if sludge is used as soil conditioner [20]. Previously, Ontario regulations allowed untreated septage to be land applied under a certificate of approval. However, as of 1 January 2011, only treated septage may be applied to agricultural land under the general nutrient management regulation as a nutrient [21].

At present, wastewater treatment plants are required to achieve a lower carbon footprint by developing environmentally friendly processes to reduce the volume of sludge for disposal and/or convert sludge into bioenergy. This has shifted the current view on sewage sludge from a waste to be treated and disposed of to a renewable resource for energy recovery. The potential for energy recovery from a sludge is a function of its composition and energy content, depending mainly on volatile solid content that can be subdivided into readily degradable organics (approx. 50% in primary sludge and 25% in WAS) and not readily degradable organics (approx. 30% in primary sludge and 55% in WAS) [11].

Currently, there are several options for energy recovery from sludge, of which the most important ones are sludge-to-biogas and sludge-to-bio-oil processes. This review chapter provides an overview of current research on production of bio-oils and biogas from waste materials, while focusing on the current technologies for energy recovery from sludge based on hydrothermal treatments.

3.2 Bio-oil production from wastewater sludge through HTL

3.2.1 Fundamentals of HTL

Hydrothermal liquefaction (HTL) is a promising technique for conversion of biomass, in particular wet biomass, into biofuels. This process operates at high pressure (5–20 MPa) and relatively low temperature (<400°C) and uses water as the main solvent, mostly in subcritical or near-critical conditions of water (T_c = 374°C and P_c = 22.1 MPa) [4,12,19]. The characteristics of subcritical water, as listed in Table 3.2, play important roles in hydrothermal liquefaction of biomass.

Subcritical water behaves differently from water at ambient conditions. As temperature and pressure of water increase in the reactor, viscosity and dielectric constant of water decrease significantly, which enhances its solubility for organic compounds [22]. Table 3.2 shows the characteristics of subcritical water in comparison with those at ambient conditions. The ionic product of subcritical water ($KW = [H^+][OH^-]$) is much higher than that at ambient temperature and pressure [23]. When ionic product increases, more H^+ and OH^- are released in water, which promote the acid- or base-catalysed reactions such as hydrolysis of cellulose, hemicellulose, and lignin [22].

During hydrothermal liquefaction, biomass is subjected to a series of complicated reactions. The basic reaction mechanism proposed by researchers can be summarised as follows (Fig. 3.2): [22,24,25]

Table 3.2 Water properties at ambient and subcritical conditions

	Ambient water	Subcritical water	
Temperature (°C)	25	250	350
Pressure (MPa)	0.1	5	25
Density (g/cm^3)	1	0.8	0.6
Dielectric constant ($F\,m^{-1}$)	78.5	27.1	14.07
Ionic product (K_w)	10^{-14}	$10^{-11.2}$	10^{-12}
Heat capacity ($kJ\,kg^{-1}\,K^{-1}$)	4.22	4.86	10.1
Dynamic viscosity (mPas)	0.89	0.11	0.064

Modified from Toor SS, Rosendahl L, Rudolf A. Hydrothermal liquefaction of biomass: a review of subcritical water technologies. Energy 2011;36(5):2328–2342.

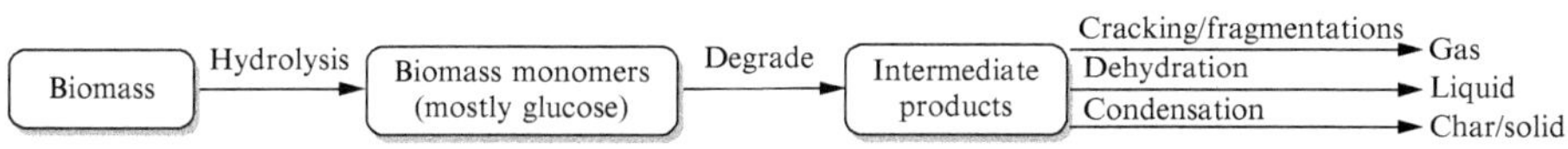

Fig. 3.2 HTL reaction pathway.

1) Biomass hydrolysis and depolymerisation to monomers and unit structure
2) Decomposition/degradation of the produced monomers (thermally and chemically) to form intermediates through reactions of dehydration, dehydrogenation, deoxygenation, cleavage, deamination, decarboxylation, cracking/fragmentation, etc.
3) Removal of oxygen containing functional groups in the presence of hydrogen (if hydrogen is present)
4) Rearrangement of the reactive fragments/intermediates through repolymerisation, condensation, and cyclisation to form products

Biomass consists of different components such as carbohydrates, lignin, proteins, and lipids. Each of these components when subject to hydrothermal conditions can undergo different conversion pathways, as briefly described below.

Cellulose, hemicellulose, and starch are the most common carbohydrates in biomass. Carbohydrates are mostly found in plant biomass. In HTL, biomass is exposed to water at elevated temperature and pressure, when cellulose and hemicellulose will be hydrolysed into glucose and other saccharides by hot-compressed or subcritical water [26].

Cellulose is normally the largest fraction of plant biomass (40–50 wt%) followed by hemicellulose and lignin [27,28]. It is a linear polysaccharide with an average molecular weight of around 100,000 g/mol made of D-glucose units and crystalline structure, making it water insoluble. However, subcritical water with high ionic product can break its crystallinity, and decomposition of cellulose increases with increasing temperature. Starch is also a polysaccharide consisting of glucose monomers. It is hydrolysed much faster than cellulose and can be completely hydrolysed at temperatures above 180°C [15].

Hemicellulose makes up 20%–40% of plant biomass [26–28]. It is a branched polymer with an average molecular weight of <30,000 g/mol with noncrystalline and random structure, making it easily hydrolysable [22]. In hydrothermal processing, hemicellulose is hydrolysed easily at temperatures above 180°C. At 220°C, hemicelluloses could completely dissolve in water. Hydrolysis of hemicelluloses produces a mixture of polysaccharides composed mainly of five-carbon (xylose and arabinose) and six-carbon sugars (glucose, galactose, and mannose), as well as formic/acetic acid, pyran derivatives, and aldehydes [15].

Lignin is a high-molecular-weight randomly cross-linked polymer of phenylpropane units with chemical formulas of $C_9H_{10}O_2$, $C_{10}H_{12}O_3$, and $C_{11}H_{14}O_4$, linked by C–C or C-O-C (mainly β-O-4) ether bonds. Lignin is commonly associated with hemicellulose both physically and chemically through covalent bonds [26]. It is relatively resistant to chemical or enzymatic degradation; however, in hydrothermal processing, many phenol and methoxyphenol compounds are formed through hydrolysis of the ether bonds [22,23,26].

Lipids are divided into different categories mostly fatty acids and glycerols. Lipids are not water-soluble at normal temperatures; however, at subcritical or supercritical conditions, the dielectric constant of water decreases and makes these molecules more soluble. Lipids can hydrolyse in hot-compressed water without catalyst. Glycerol is the product of triglyceride hydrolysis and is commonly converted to aqueous fraction when subjected to hydrothermal treatment. Fatty acids are relatively thermally stable, but in hydrothermal processing, they could partially degrade and produce long-chain hydrocarbons. They degrade faster in alkaline conditions mainly via a mechanism involving fatty acid decarboxylation [22].

Proteins are polymers of amino acids mostly found in animal or microbial type of biomass. Proteins contain a large fraction of nitrogen that can affect the bio-oil properties such as smelling and combustion properties, so the degradation of proteins is also of great importance to HTL of biomass. Proteins can be slightly hydrolysed at temperatures below 230°C. At temperatures above 250°C, proteins are hydrolysed to amino acids, which will then be degraded fast to hydrocarbons, amines, aldehydes, and acids through decarboxylation and deamination reactions [22].

Biomass in HTL is mostly converted to bio-crude oil, water-soluble product (WSP), biochar, and noncondensable gases. The characteristics and yields of the products highly depend on HTL operating conditions such as type of biomass, biomass-to-water ratio, temperature, pressure, residence time, process gas, and the presence or absence of catalyst, which are discussed as follows.

Operating parameters such as reaction temperature, retention time, the presence of a catalyst, and water-to-biomass ratio can affect the yields and characteristics of HTL products. In this section, previous studies on the effects of operating conditions on HTL of biomass are reviewed.

3.2.2 Effects operating conditions on HTL of biomass

3.2.2.1 Effects of type and composition of feedstock

Feedstock type and composition can largely affect the produced bio-oil characteristics. As a general trend, the conversion rates for different biomass constituents under HTL conditions are in the order of lipids > proteins > hemicellulose > cellulose > lignin. The more lignin the biomass contains, the higher operating temperature is needed [19,29].

Lignin and cellulose hydrolyse to oxygen-containing compounds such as phenolic compounds, carbohydrates (sugars), pyran derivatives, ketones, aldehydes, and formic/acetic acids, contributing to the high oxygen content of the bio-oil. Carbohydrates, lignin, and protein would form phenolic compounds, while lipids produce fatty acids/esters and cholesterols [15,19,29].

Karagoz et al. tested the effects of biomass type on liquefaction yield, using two types of waste biomass, sawdust and rice husk, in a 200-mL autoclave reactor at 280°C for 15 min without catalyst at the biomass-to-solvent ratio of 5:30 (g/mL). Their obtained oil yield with sawdust was 8.6 wt%; although low, it was still higher than that obtained from rice husk [30].

In another work published by Demirbas, HTL of different types of woody biomass with different compositions was studied with biomass-to-solvent ratio of 10:100 (g/mL) without using any catalyst at 277–377°C. The maximum bio-oil yield (28 wt%) was obtained from beech wood powder (with the lowest lignin and highest cellulose and hemicellulose contents) with 34.9 MJ/kg higher heating value (HHV). They concluded that the oil yield is inversely proportional to the lignin content of the feedstock: with increasing lignin content, the oil yield decreases, while the water-soluble product yield increases [31]. However, as per a previous study [32], a different conclusion was drawn; Tymchyshyn and Xu investigated HTL of lignocellulosic wastes (sawdust and cornstalks) and two model biomass compounds (pure lignin and pure cellulose as references) in hot-compressed water at temperatures from 250°C to 350°C in the presence of 2 MPa H_2. The liquefaction operations at 250°C for 60 min produced bio-oil with a yield of about 53, 32, 32, and 17 wt% for lignin, sawdust, cornstalk, and cellulose, respectively, suggesting the bio-oil yield is proportional to the lignin content of the feedstock [32]. Notably, the conditions used (type of feedstock, HTL temperature, time, and atmosphere) in these two literature studies [31,32] are not the same; hence, more research in this regard is needed in order to draw more definite conclusions.

In another work by Vardon et al., the effects of different feedstock and their composition on HTL were studied in a 2 L batch reactor at 300°C, 10–12 MPa pressure using nitrogen, 30 min residence time with three kinds of waste feedstock (*Spirulina* algae, swine manure, and anaerobically digested sludge) at a substrate concentration of 20% (*w*/w). These feedstocks have different compositions of protein, lipid, and carbohydrate. The organic matter content of the feedstocks was in the order: microalgae > swine manure > digested sludge. Microalgae contained the highest amount of protein and lipid and lower amount of carbohydrate; manure contained high amounts of carbohydrates and moderate lipid and protein contents; and the digested sludge had high carbohydrates, moderate protein, and very low lipid content. The highest bio-oil yield (32.6 wt%) was obtained with microalgae as the feedstock, while the digested sludge resulted in only 9.4 wt% oil yield, suggesting that the bio-oil yield is higher with high lipid-containing feedstocks but lower for feedstocks with a high carbohydrate content, although the HHV of all bio-oils obtained were almost the same [19].

It can be concluded that the type and composition (lignin, carbohydrate, and lipid contents) of feedstock can greatly influence the product yields from the HTL process.

3.2.2.2 Effects of catalyst

It has been widely demonstrated in many studies that the use of a catalyst in HTL of biomass could effectively promote liquefaction efficiency, leading to higher biomass conversion and bio-oil yields and improved oil quality (heating values) [33–38]. Different kinds of homogeneous and heterogeneous catalysts have been employed in hydrothermal liquefaction process, but the most common catalysts used were alkali metal compounds, for example, Na_2CO_3, NaOH, K_2CO_3, KOH, LiOH, RbOH, and CsOH [3,15].

The effects of using an alkaline homogeneous catalyst (K_2CO_3) in HTL of pine wood sawdust were investigated by Karagoz et al. in a 200-mL autoclave reactor at 280°C and 15 min residence time [33]. The bio-oil yield without using catalyst was reported to be as low as 8.6 wt% and increased to 33.7 wt% when using 0.94 M of K_2CO_3 solution as a catalyst. In another work performed by Zou et al. [34], effects of alkali catalyst (Na_2CO_3) on HTL of microalgae were investigated. Increasing catalyst dosage was found to raise the biomass conversion and increase bio-oil production. Ross et al. [35] studied the effects of different alkali and organic acid catalysts on HTL of microalgae in a 75-mL batch reactor and concluded that the catalytic activities for bio-oil yields increased in the order of $HCOOH < KOH < CH_3COOH < Na_2CO_3$. With Na_2CO_3, the bio-oil yield attained 27.3 wt% at 350°C and 1 h residence time, and the oil has an HHV of 39.9 MJ/kg. Similar results were reported by many other researchers, demonstrating good catalytic activities of Na_2CO_3 for HTL of various biomass feedstock (such as microalgae, MSW, and paulownia) [25,36,37,39].

Zhang et al. [40] studied the effects of both homogeneous and heterogeneous catalysts (i.e. formic acid (HCO_2H), KOH, and FeS) on HTL of mixtures of secondary pulp/paper mill sludge and waste newspaper (a common municipal solid waste that contains a high fraction of volatile material and lignin). Effects of using hydrogen as a reducing gas were also investigated. The study concluded that the presence of H_2 had no significant effects on bio-oil yield. In terms of oil yield, the performance of catalysts was $FeS > KOH > HCO_2H$, and in terms of conversion of biomass, the order of the catalysts was $KOH > FeS > HCO_2H$. The highest oil yield obtained was 29.9 wt% with FeS catalyst at 300°C and 2 MPa initial pressure of nitrogen and 20 min residence time [40].

A wide variety of heterogeneous catalysts including noble metals supported on carbon, Pd/C, Pt/C, Ru/C (5 wt%), sulphided $CoMo/\gamma\text{-}Al_2O_3$ (4.4–11.9 wt%), and $Ni/SiO_2\text{-}Al_2O_3$ (65 wt%) were tested for HTL of microalgae in a 31-mL batch reactor by Duan and Savage [41]. In this comprehensive study, they also investigated the effects of He and H_2 as inert and reducing environments, respectively, under 70 kPa initial pressure. The presence of a reducing atmosphere (H_2) did not increase the yield of bio-oils, but using hydrogen, produced bio-oils were with a higher H/C ratio and a higher heating value as well as a lower nitrogen content attributed to hydrodeoxygenation of the bio-oils. For instance, the HHV of bio-oil was as high as 40.1 MJ/kg with Pt/C in the presence of hydrogen. Pt/C, Pd/C, and Ru/C catalysts were found to be very effective for producing bio-oils with increased H/C ratio, and Pt/C, $Ni/SiO_2\text{-}Al_2O_3$, and sulphided $CoMo/\gamma\text{-}Al_2O_3$ were effective for decreasing O/C ratio in the obtained bio-oils. The highest bio-oil yield was 57 wt% with Pd/C in the absence of hydrogen and 52 wt% in the presence of hydrogen at the optimum conditions of 350°C and 60 min [41].

3.2.2.3 Effects of reaction temperature

Temperature is the most influencing parameter for HTL process. Generally, increasing temperature in an appropriate range promotes biomass fragmentations and produces mainly liquid products [28]. As previously mentioned, biomass conversion pathway includes biomass hydrolysis, bio-oil formation, and bio-oil decomposition. When the reaction temperature is sufficiently above the activation energy of bonds cessation, biomass depolymerises to its unit structures such as glucose and phenolic compounds and gradually transforms into bio-oil. At temperatures higher than 350°C, bio-oil yield decreases due to secondary decomposition or condensation/repolymerisation reactions, producing more gaseous products and char. On the other hand, at relatively low temperatures (<280°C), incomplete conversion of biomass and hence a lower bio-oil yield occur [23,28]. Thus, selection of an optimum temperature is very important for HTL bio-oil production. As commonly reported in literature, the temperature range of 300–330°C is optimal for production of HTL bio-oil at a higher yield [28], though the optimal temperature also depends on biomass type.

Zhong and Wei studied HTL of woody biomass at different temperatures of 280°C, 300°C, 320°C, 340°C, and 360°C. They reported that the bio-oil yield increased with increasing reaction temperature up to 320–340°C. Further increasing the temperature resulted in a decline in bio-oil yield, as described previously. The maximum bio-oil yield of 32.2% and lowest solid residue yield of 10.34% were obtained at 340°C and in the presence of K_2CO_3 catalyst [42]. Similarly, Ogi et al. studied the effects of temperature on HTL of woody biomass in the presence of K_2CO_3 catalyst. The highest oil yield of 28.4% was obtained at 300°C and a higher temperature above 300°C resulted in lower bio-oil production [43]. Xu and Lad reported an initial increase in heavy oil production up to 300°C followed by a continuous drop with increasing temperature during HTL experiments in water with sawdust without catalyst. The drop in bio-oil yield was attributed to the formation of gases, water, and char as a result of enhanced cracking and dehydration reactions and condensation reactions at higher temperatures [44].

Shuping et al. suggested that the increase in bio-oil yield with increasing temperature is mainly due to the enhanced ionic characterisation of subcritical water, which turns water into an acid/base precursor that can catalyse hydrolytic degradation of the biomass [34].

3.2.2.4 Effects of residence time

Duration of reaction also defines overall conversion of biomass and the distribution of products. Thus, optimisation of residence time is necessary for effective conversion of organic compounds in biomass. It has been commonly observed that bio-oil yield is higher with a shorter residence time due to the restriction of secondary or tertiary reactions that would convert the produced heavy oils into water, gaseous products, or solid residue [23,28].

Yin et al. observed a decrease in bio-oil yield from HTL of cattle manure at 310°C and in the presence of NaOH as catalyst while increasing reaction time. Increasing the reaction time from 15 to 40 min resulted in a decrease in bio-oil yield from 39.57% to 12.95%, which was attributed to the decomposition and condensation of bio-oil products to form residual solids [23].

A different trend was however observed by Xu and Lancaster [15]. The authors observed an increase in bio-oil yield at longer reaction times from 15 to 120 min, which might be due to conversion of water-soluble products (WSP) to bio-oils. They proposed that the increased dehydration of carbohydrates and other WSP compounds in the HTL products or intermediates at longer retention time could account for the conversion of the WSP to bio-oil [15].

3.2.2.5 Effects of water to biomass ratio

In HTL, mass ratio of biomass to water is considered as one of key parameters (along with others, i.e. temperature, time, and catalysts as discussed previously). From economic perspective, more concentrated solutions are desirable for HTL bio-oil production; however, a low water content of the reactor feed was found to restrict the hydrolysis of biomass leading to a smaller yield of the bio-oils and a higher yield of the solid residue or char [15,28]. For instance, Yin et al. reported that when mass ratio of cattle manure to water was increased from 0.5 to 2, the corresponding bio-oil yield decreased sharply from 48.7 to 1.46 wt%. The decrease was believed due to the conversion of bio-oil/intermediates into solid residue through self-condensation reactions [23]. Boocock and Sherman also observed greatly reduced oil yields at high wood-to-water ratios during HTL of polar wood at 350°C, suggesting the critical solvent/reactant roles of water in biomass HTL process, which contribute to hydrolytic degradation of lignocellulosic biomass and stabilising the product oils [45].

However, decreasing biomass-to-water ratio does not always produce higher bio-oil yields. Xu and Lancaster reported a greater yield of bio-oil at a higher initial biomass concentration [15]. The bio-oil yield increased from 15% to 22% as the initial biomass concentration increased from 4.8% to 16.7%. They suggested that this result was due to the promotion of the dehydration reactions of the water-soluble intermediates or products at lower water concentrations, resulting in a greater yield of bio-oil [15].

3.2.3 HTL of wastewater sludge

HTL of different types of biomass has been widely reported in literature [2,4,33,37,40,41,46–49]. An early study was first reported by Berl in 1940s who suggested that cornstalks, corn cobs, sugar cane, bagasse, seaweed, algae, sawdust, Irish moss, molasses, sorghum, and grasses could be converted into petroleum-like products [25]. In 1990, the Shell Research Laboratory in Amsterdam published a work on hydrothermal upgrading process (HTU) of waste biomass, especially the high-water-content materials. In HTU, the biomass is treated at temperatures ranging from 300°C to 350°C and pressure of 12–18 MPa in water for 5–15 min without using any reducing gas environment or catalyst. The product consists of around 45% bio-oil, 25% gas,

20% H_2O, and 10% dissolved organic materials (i.e. water-soluble products such as acetic acid and ethanol). The heating value of the produced bio-oil from HTU can be in the range of 30–35 MJ/kg with an oxygen content of 10%–18% [24,50].

Many researchers demonstrated that HTL can be an alternative to traditional technologies for waste management such as disposal and valorisation of wastewater sludge. Hydrothermal liquefaction could be an effective technical route for treatment of wastewater sludge for recovery of energy in the form of bio-oil. The process is even more promising if it is made energy-self-sufficient by using a part of the bio-oil and biochar to provide heat for the HTL process. HTL of wastewater sludge was first reported by Kranich and Eralp in 1984 [26]. Sewage sludge was converted to oil at different reaction temperatures in the presence of hydrogen as reducing gas and catalysts such as Na_2CO_3, $NiCO_3$, and Na_2MnO_4. The oil yields were <20 wt% with water as the reaction medium [15,51].

In 1986, a pilot plant process to produce bio-oil from sludge was patented as sludge-to-oil reactor system (STORS) by Battelle Memorial Institute in the United States [11,18,52]. The process was carried out at 300°C, 10 MPa, and 90 min residence time with initial substrate concentration of 20 wt% of the sludge feed and Na_2CO_3 as a catalyst. The technology was patented as sludge-to-oil reaction system (STORS) with oil yields ranging from 10 to 20 wt% and char from 5 to 30 wt% [11,52].

Itoh et al. reported a demonstration plant operating at 300°C with the capacity of 5 t/day of the dewatered sludge as the feedstock. The total heavy oil yield was reported to be 47.9% with about half of the organic materials in the sludge converted. The oils were separated from the reaction mixture by high-pressure distillation. The heating values of the heavy oils were found to be 37–39 MJ/kg [53]. In a more recent study, HTL of algal biomass, swine manure, and digested anaerobic sludge (total solids = 20%) was investigated by Vardon et al. at 300°C for 30 min. The organic matter content of the feedstocks was reported to be in the order of algae > manure > sludge. The sludge feedstock, which was mainly composed of cellulose and lignin, resulted in a bio-oil containing a high percentage of ester, phenolic, and nitrogenous compounds. The bio-oil yield from sludge was 9.4% and had the highest amount of high-boiling-point compound compared with the bio-oil from other feedstocks, as well as had the highest molecular weight of 1870 g/mol [19].

Huang et al. studied HTL of sewage sludge at 350°C for 20 min. The sludge was composed of 33.6% protein, 20.3% carbohydrates, 6.9% lipids, and 39.2% ash. The bio-oil yield was reported to be 49.6%, however, accompanied by a large char yield (45.73%), which could be attributed to the proteins and carbohydrates content of the sludge [29].

3.2.4 HTL of biomass in continuous operation processes

Most of the HTL studies were performed in batch reactors. However, continuous flow reactors are desirable for large-scale applications of the HTL technologies in order to promote the energy utilisation efficiency and make the process economically feasible [54]. With batch reactors, it is almost impossible to have very short residence time for better oil yields (as discussed in the previous section). Thus, the advantage of using continuous flow reactor with controllable residence time is that maximum bio-oil yield could be achieved in a shorter reaction time [54].

Jazrawi et al. [54] studied HTL of *Chlorella* and *Spirulina* microalgae in a continuous flow reactor of 2 L volume at 250–350°C and 150–200 bar for a residence time of 3–5 min. The optimal operating conditions were found at 350°C and 3 min residence time, giving 41.7 wt% bio-oil yield. Hammerschmidt et al. [39] also performed HTL of three types of sludge (two food industry sludge with 12 and 6.7 wt% solid concentration, respectively, and one wastewater sludge with 7.7 wt% solid concentration) in a 127-mL continuous flow reactor. The reactor was operated at 300–350°C and 25 MPa with a feed rate of 5.2 and 10.5 g/min, respectively. The feed was first mixed with K_2CO_3 as a homogeneous catalyst, and then, the mixture was pumped to the reactor filled with ZrO_2 as a heterogeneous catalyst. Although the residence time in the catalytic bed was controlled at 10 min in most experiments, a longer residence time and higher catalyst concentration of the feed resulted in greater bio-oil yields.

From the literature, it is clear that many parameters during HTL would influence the yield and composition of liquefaction products, especially bio-oils. Table 3.3 presents the results of selected previous studies using various reactor types at their optimum operating conditions for bio-oil yield.

3.2.5 HTL of biomass in pilot or demonstration scale

HTL processes in pilot and demonstration scales have been practiced at several places of the world. In order to make HTL economically viable, dry matter content in the feedstock should be sufficiently high to obtain a higher oil production rate. This however renders challenges in feeding the feedstock to the HTL reactor and increased potential of plugging in the reactor system. In addition, there should also be an appropriate solution to handle the aqueous waste stream generated from the process [22]. Some previous and current large-scale applications of the HTL technologies are summarised below.

The pioneered work in this area was performed in 1970s by Appell et al. at the Pittsburgh Energy Research Centre, where a pilot plan was demonstrated in Albany, Oregon. The process used woodchips as feedstock and converted them into bio-oil at temperatures of 330–370°C and pressure of 200 bar for a residence time of 10–30 min. The oil yield obtained was 45%–55% based on the dry matter of the feed. The operation of the unit was stopped after 1981 due to some serious technical problems [22,24].

The Lawrence Berkeley National Laboratory (LBNL) introduced hydrolysis of biomass with sulphuric acid followed by neutralisation by sodium carbonate and subsequent liquefaction at 330–360°C and 100–240 bar. The research was halted in the early 1980s as the price of petroleum dropped [22,24].

Shell research laboratory in Amsterdam (the Netherlands) started to develop hydrothermal upgrading (HTU) process in 1982. In their process, a number of different types of biomass with a moisture content of preferably between 55 and 85% such as forestry, agricultural waste, wood, paper/pulp waste, and algae can be liquefied at temperature of 300–350°C and 120–180 bar for 5–20 min, producing bio-oil at 45% yield with HHV of 30–35 MJ/kg [22,24].

Table 3.3 Summary of the best operation conditions and oil yield for HTL of biomass with various types of reactor

Feedstock	T (°C)	Time (min)	Catalyst	Reactor type	Reactor volume (mL)	Oil yield (wt%)	Ref.
Sawdust	280	15	–	Batch	200	8.6	[30]
Beech wood powder	377	25	–	Batch	250	28	[31]
Swine manure	300	30	–	Batch	2000	30.2	[19]
Lignin	250	60	–	Batch	14	53	[32]
Swine manure	295	120	–	Batch	1800	70.2[a]	[55]
Pine wood sawdust	280	15	K_2CO_3	Batch	200	33.7	[33]
Microalgae	360	50	Na_2CO_3	Batch	100	25.8	[34]
Microalgae	350	60	Na_2CO_3	Batch	75	27.3	[35]
Garbage	340	30	Na_2CO_3	Batch	300	27.6	[36]
Paulownia	340	10	Iron powder	Batch	1000	36.3	[37]
Secondary pulp/paper mill sludge	300	20	FeS	Batch	75	29.9	[38]
Microalgae	350	60	Pd/C	Batch	31	57	[41]
Cattle manure	310	15	NaOH	Batch	3000	48.9[a]	[23]
Dairy manure	350	15	Na_2CO_3	Batch	300	24[b]	[25]
Microalgae	350	3	–	Continuous	2000	41.7	[54]
Sludge	330	10	K_2CO_3 and ZrO_2	Continuous	127	NP	[39]
Swine manure	300	60	–	Continuous	NP[c]	70[a]	[56]

[a] Based on organic matter content of the biomass feedstock.
[b] 4.8 g bio-oil was produced from 20 g dry manure feedstock.
[c] NP, not provided in the paper.

The high-pressure hydrogenation, known as direct liquefaction of organic substances (DOS) process, developed by Hochschule für Angewandte Wissenschaften (HAW; Hamburg, Germany), is a direct one-step process for liquefaction of lignocellulosic biomass (e.g. wood and straw). The process operates under 80 bar and between 300°C and 500°C, and the thermal efficiency of the whole system can be around 70% based on the heating value of the reactor feed [22,24].

A Danish company SCF Technologies has developed the CatLiq technology that operates on a 20 L/h capacity pilot plant in Copenhagen, Denmark. Unlike other combustion and gasification technologies, the CatLiq technology is able to generate energy from wet waste and biomass containing >50% water such as sewage sludge; manure from livestock production; and residues from food processing, ethanol production, and biodiesel production. This process converts organic waste to oil in the presence of K_2CO_3 and zirconia as homogenous and heterogeneous catalysts, respectively. The operating conditions of the process are 280–350°C and 22.5–25 MPa, and the process achieved 30%–35% oil yields with 70%–75% energy recovery from biomass to bio-oil [22]. Another Danish company, Steeper Energy, is commercialising Hydrofaction, a proprietary 'biomass-to-liquid' technology using supercritical water chemistry. After the successful execution of the continuous bench-scale unit based at Aalborg University in Denmark (over 3000 operating hours at approx. 20 L/h feeding rate or ½ barrel bio-oil per day (BPD)), Steeper Energy is now actively developing a continuous pilot-scale (CPS) 100 bar/day plant in Alberta.

The sludge-to-oil reactor systems (STORS) developed in Cincinnati, Ohio, the United States, is capable of processing undigested municipal sewage sludge with 20% solids at the rate of 30 L/h. The process operates at 300°C with 73% energy recovery in the forms of oil and char. In the early 1990s, a demonstration plant capable of handling 5 t/day of dewatered sludge, operated at 300°C and 10 MPa, was constructed in Japan. The oil yield was 38% (ash included) with HHV of 31–39 MJ/kg. In 2001, a STORS demonstration project sponsored by the US Environmental Protection Agency was successfully finished. The plant was located in Colton, California, the United States, for bio-oil production from raw sewage sludge. The technology was further advanced by a company called ThermoEnergy [22].

Thermal depolymerisation technology has been designed by Changing World Technologies Inc. (CWT). In 1998, CWT started a subsidiary, thermal depolymerisation process, LLC (TDP). A test plant was built in Philadelphia, Pennsylvania, to evaluate the TDP technology. The first large-scale plant was inaugurated in Carthage, Missouri, the United States, in 2004 and converted approximately 250 t/day of turkey offal and fats into approximately 500 bar bio-oil [22].

Table 3.4 presents a summary of the HTL processes in pilot or demonstration scale.

3.3 Biogas production from wastewater sludge through anaerobic digestion

Anaerobic digestion (AD) is the most common method for sludge stabilisation resulting in the reduction of solids and production of biogas that can be combusted to generate electricity and heat or can be processed into renewable gas and transportation

Table 3.4 **Overview of HTL processes in pilot or demonstration scale**

Process name	Developer/supplier of the process	Raw material	Temperature (°C)	Pressure (MPa)	Plant scale	Oil yield (%)
PERC	Pittsburgh Energy Research Centre (the United States)	Woodchips	330–370	20	N/A	53
LBL	Lawrence Berkeley Laboratory (the United States)	Woodchips	330–360	10–24	N/A	33
HTU	Shell Research Institute (NL)	All types of biomass; domestic, agricultural, and industrial residues; wood	300–350	12–18	100 kg/h (wet) pilot plant	45
Direct liquefaction of organic substances (DoS)	HAW (GER)	Lignocellulosic biomass (e.g. wood and straw)	350–500	8	5 kg/h semicontinuous test plant	Not known
STORS, the United States	EPA's Water Engineering Research Laboratory, Cincinnati, Ohio, the United States	Sewage sludge	300	10	30 kg of sludge/h	10–20
STORS, Japan	Organo Corp.	Sewage sludge	300	10	5 t of sludge per day	38
CatLiq	SCF Technologies A/S (DK)	Dried distiller grain with solubles (DDGS)	280–350	22.5–25	20 L/h capacity pilot plant	34
Thermal depolymerisation process (TDP)	Changing World Technologies Inc. (the United States)	Turkey offal and fats	200–300	4	250 t/day	Not known

Reprinted with permission from Toor SS, Rosendahl L, Rudolf A. Hydrothermal liquefaction of biomass: a review of subcritical water technologies. Energy 2011;36(5):2328–2342. Copyright (2011) Elsevier.

fuels. It is usually carried out at mesophilic (30–40°C) or thermophilic (50–55°C) conditions [57] and involves stages of hydrolysis, acidogenesis, acetogenesis, and methanogenesis [10,58]. During hydrolysis (Eq. 3.1), which is normally the rate-limiting step of AD, both insoluble organic material and high-molecular-weight compounds such as lipids, proteins, polysaccharides, and nucleic acids are degraded into soluble organic substances (e.g. amino acids and fatty acids) [10]. In the second stage, the components formed during hydrolysis are converted into volatile fatty acids (VFAs) along with NH_3, CO_2, H_2, and other by-products by acidogenic (fermentative) bacteria. During the third stage (Eq. 3.2), acetogens further digest the higher organic acids and alcohols from the acidogenesis stage to produce mainly acetic acid, CO_2 and H_2. These three steps are usually called acid fermentation [59]. In the last stage, methane is produced by two groups of methanogenic consortia: The first group produces methane and carbon dioxide from acetate cleavage (Eq. 3.3), and the second group (Eq. 3.4) uses hydrogen as electron donor and carbon dioxide as acceptor to produce methane [59]. The stages of AD are shown in Fig. 3.3 [59].

$$nC_6H_{10}O_5 + nH_2 \xrightarrow{\text{Hydrolysis}} nC_6H_{12}O_6 \tag{3.1}$$

$$nC_6H_{12}O_6 \xrightarrow{\text{Acid from bacteria}} 3nCH_3COOH \tag{3.2}$$

$$CH_3COOH \xrightarrow{\text{Methane forming bacteria}} CH_4 + CO_2 \tag{3.3}$$

$$CO_2 + 4H_2 \xrightarrow{\text{Reduction}} CH_4 + H_2O \tag{3.4}$$

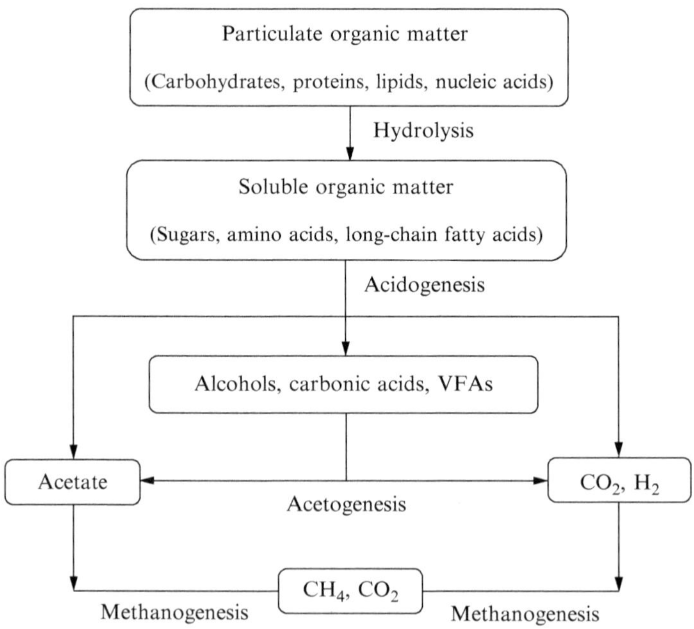

Fig. 3.3 Anaerobic digestion stages.

Various parameters such as temperature, pH, alkalinity, and types of substrates can affect the rate of the different stages of the digestion process. Some of the optimal operating conditions for anaerobic digestion include the following:

- No dissolved oxygen
- No inhibitory substances (such as heavy metals and sulphides)
- pH range of 6.6–7.6 [60]
- Sufficient nitrogen and phosphorous
- Carbon-to-nitrogen ratio of 25–30:1 [59]
- VFA concentration in the substrate between 2000 and 3000 mg/L [59]

Anaerobic digestion of wastewater sludge is often practiced at long retention times (20–30 days) in order to ensure sludge stabilisation (i.e. pathogen reduction and reduction in vector attraction) and maximise volatile solids reduction and methane production. This long duration is required often as a result of low overall degradation efficiency of the organic dry solids (30%–50%) that is due to the rate-limiting hydrolysis stage [10]. Biosolids are composed of diverse microorganisms and organic and inorganic compounds agglomerated together in a polymeric network termed extracellular polymeric substances (EPS), composed of proteins, carbohydrates, lipids, and VFAs. During hydrolysis, cell walls are ruptured and EPS is degraded, resulting in the release of organic materials for the acidogenic microorganisms [10]. It is known that hydrolysis of EPS and/or microbial biomass limits the rate and extent of degradation. Thus, applying a disintegration process as a pretreatment step for breaking the EPS network can potentially enhance anaerobic biodegradability and dewaterability of the digested sludge. Thermal, chemical, biological, and mechanical processes, as well as combinations of these, have been studied as possible pretreatment approaches to accelerate sludge hydrolysis, as reviewed below.

3.3.1 Effects of thermal pre-treatment

Thermal pretreatments were initially applied to sludge to improve their dewaterability [59,61]. They can destroy cell walls and make organic compounds accessible for biological degradation [62–64] as well as remove pathogens and decrease digestate viscosity [65]. A wide range of temperature from 60°C to 270°C, which can be further divided into low-temperature (< 100°C) and high-temperature treatments (>130°C, mostly referred to as thermal hydrolysis) has been studied. The most common pretreatment temperatures are between 60°C and 180°C with the reported optimum reported condition of 160–180°C and 30–60 min for sewage sludge [59,66]. However, the optimum temperature and duration of the pretreatment depend on the nature of the sludge [64]. Temperature above 200°C had been found to lead to production of refractory and toxic compounds and thus reduction in the biodegradability [62,64,67,68]. The formation of these refractory compounds is reported to be as a result of Maillard reaction in which sugars and amino acid react to form melanoids that are known to be inhibitory to the methanogens [69]. The formation of poorly degradable materials such as dioxins, at temperature of ≥200°C, has also been reported [69]. Due to lower energy requirements and applicability, low pretreatment temperatures are more advantageous over high temperatures. At low temperature, treatment time plays a more dominant role than temperature, and the low-temperature pretreatment required a longer contact time than the high-temperature treatment. Hiraoka et al. investigated low-temperature thermal pretreatment and revealed an increase of >30% in gas production by pretreatments at 60°C and 80°C [70]. The main effect of thermal pretreatment is the disintegration of cell membranes and thus solubilisation of organic compounds. A direct correlation between COD solubilisation and temperature has been reported [65].

However, there have been different observations reported for the effects of pretreatments on biogas production. In one study, Ferrer et al. reported that the proportion of soluble to total organic matter increased by almost 10 times after 70°C pretreatment of the mixture of thickened primary and secondary sludge. After sludge pretreatment, almost 50% increase in methane volume and higher methane content in biogas was observed with conducting treatments at 9, 24, and 48 h [67]. Appels et al. investigated the impact of thermal treatment of WAS at 70–90°C on anaerobic digestion. They reported that organic and inorganic compounds were efficiently solubilised as a result of the treatment. A negligible increase in biogas production was observed from sludge pretreated at 70°C for 60 min, whereas the biogas production was improved by 20 times when applying a 60 min pretreatment at 90°C [64]. On the other hand, Dhar et al. investigated thermal pretreatments of municipal WAS at 50°C, 70°C, and 90°C for 30 min to examine the effects of pretreatment on COD solubilisation, solid reduction, and methane production potential. [71] They reported that the pretreatments caused significant increase in the ratios of soluble to total COD (SCOD/TCOD), compared with the control, but the increase in methane production was only in the range of 13%–19% for all thermal pretreatments.

The major disadvantage of high-temperature treatment is the high energy requirement. However, increased biogas production may compensate, to some extent, for the high energy consumption. In a study by Bougrier et al., thickened sludge was thermally treated at 135°C and 190°C for 30 min, leading to COD solubilisation of 34% and 46%, respectively. A 25% increase in methane production was observed after the pretreatments at 190°C, compared with only 12% increase in methane production at 130°C. Formation of refractory soluble COD was reported in the treatments at 190°C [72]. Nielsen et al. reported only 13% and 9% increase in methane production for the pretreatments of WAS at 130°C and 170°C, respectively, and negligible effect on methane yield was observed for pretreatment at 80°C [62]. Rafique et al. compared the effects of low-temperature and high-temperature pretreatments on AD by thermally treating pig manure at temperature ranges of 25–150°C [73]. Enhancement in biogas and methane production was observed for the samples treated at 25–100°C with the highest amount of cumulative biogas obtained for the pretreatment at 100°C, where the biogas produced was 30%, 29%, and 30% higher than that of the untreated sludge after AD of the thermally treated sample for 7, 19, and 29 days AD, respectively. The sludge thermally treated at 130°C and 150°C led to the least production of biogas that might be due to the formation of complex organic compounds or refractory soluble COD at high temperatures [73].

Optimal treatment conditions for methane production from wastewater sludge depend on the sludge composition that can vary over time and between different wastewater treatment plants. Bougrier et al. hypothesised that the initial biodegradability of the sludge that is linked to its organic matter content defines the thermal treatment effectiveness. For a sludge with good initial biodegradability, the gain obtained by thermal treatment will be lower [72]. A summary of the previous studies on thermal treatment and their effects are listed in Table 3.5.

Currently, the best known commercial thermal sludge pretreatment technology is Cambi, which has developed a system based on thermal hydrolysis, involving batch treatment of both undigested primary and secondary sludges. It consists of three closed reactors: a pulping vessel, a hydrolysis reactor, and a flash tank. Dewatered sludge (with approximately 16% solid) is first preheated with steam to 80°C in the pulping vessel and then sent to the reactor, operating at 160–180°C, 6 bar, and 30 min residence time. Then, the sludge is transferred to the flash tank operating at atmospheric pressure. The change in the pressure from the reactor to flash tank causes cell lysis. The pretreated sludge is then cooled and sent for AD. Approximately 30% solid solubilisation was reported with an associated increase in biogas production by 150%. Tests of dewaterability also showed 60%–80% increase. The main challenges with Cambi installations are the moderately complex technology with three reactors, one of which operates at elevated temperature and pressure, necessity of sludge dewatering prior to the process, and a medium-pressure steam supply. Odour problems have also been reported for this process [10,18,63].

The BioThelys process, another commercial thermal sludge pretreatment technology, uses a single reactor to treat sludge with solid concentration higher than 10% dry solids. The sludge is treated in the reactor at 150–180°C and 8–10 bar for 30–60 min retention time. Reduction in sludge volume by up to 80% has been reported by the vendor. Two full-scale facilities have been in operation in France since 1998. Similar to the Cambi process, the BioThelys technology also involves a high-pressure and high-temperature reactor for sludge treatment, and it may also be associated with odour concerns [18].

The continuous Turbotec process is another thermal pretreatment process for the sludge patented by Sustec company in the Netherlands. In this process, sludge is pretreated at 140–180°C using steam and 4–6 bar of pressure. Increase in biogas production up to 35% and improved final dewatering >30% TS are reported by the vendor [86].

3.3.2 Effects of chemical pre-treatment

Chemical pretreatments are used to hydrolyse the cell wall and membranes by means of strong acids, alkalis, or oxidants to increase the solubility of the organic matters contained within the cells [10,65]. The effects of chemical pretreatment depend on the type of method applied and the sludge characteristics. It has been demonstrated that this pretreatment method is not suitable for easily biodegradable substrates that contain a high amount of carbohydrates, since the degradation of carbohydrates may result in subsequent accumulation of VFA and lead to failure of the methanogenesis step in AD [65]. Various chemical methods have been developed based on different operating principles, and the major chemical methods are (1) acid pretreatment, (2) alkaline pretreatment, (3) ozonation, and (4) advanced oxidation methods.

Table 3.5 Summary of the literature studies on impacts of thermal pretreatment on AD of sludge or other types of biomass

Waste type	Temperature (°C)	Results	Ref.
WAS and primary sludge	100–250	– 60%–70% increase in methane production compared with untreated sludge at 175°C – Higher temperatures resulted in decreased gas production	[74–76]
Dewatered sewage sludge	50–70	– Increase in the ratio of SCOD/TCOD by 2%–21% – A great improvement in SCOD concentration (by up to 27%) was translated in only 8% increase in the methane yield	[69]
WAS and primary sludge	170	– 40%–60% and 20%–35% solubilisation of organic matters from thermal treatment of the mixture of WAS and primary sludge and WAS, respectively – Increased dewaterability	[77]
Sewage sludge	>150	– Dewaterability was improved at temperatures above 150°C – The effect of temperature was more pronounced at temperatures higher than 180°C; however, refractory COD compounds were formed	[78]
WAS	175	– 42% increase of convertibility of COD to CH_4	[79]
WAS	62–175	– Organic particulates in WAS were solubilised by thermal pretreatment – 100% increase of convertibility of COD to CH_4 at 170°C	[80]
WAS	115–180	– Maximum of 30% VSS solubilisation at 180°C – 90% increase in methane production after treatment at 180°C	[66]
WAS and primary sludge	80	– The methane production rate was mostly influenced by the pretreatment of secondary sludge followed by mesophilic and thermophilic digestion – Increase of CH_4 production from 8.30 to 10.45 mmol/g VS_{in}	[81]
Sewage sludge	121	– Increase of biogas production from 350 to 420 mL/g VSS_{in}	[82]

WAS	90–210	– COD solubilisation was found to increase linearly with treatment temperature for temperatures lower than 200°C – Sludge dewaterability was improved by pretreatment at above 150°C of temperature but deteriorated for lower temperatures – Thermal treatments at up to 190°C allowed the biogas production to increase and biogas volume enhancement during batch anaerobic digestion of sludge, attributed to sludge COD solubilisation	[61]
WAS	70–134	– A 50% increment in biogas production was observed for 70°C thermally treated sludge	[83]
Microalgae biomass	70	– At the tested temperature, solubilisation of organic matter occurred – No significant effect on anaerobic biodegradability was observed	[84]
Household waste	70	– Did not result in enhancement of methane potential	[85]

However, alkaline pretreatment is usually the preferred chemical method since it is more compatible with subsequent anaerobic digestion [65,87].

3.3.2.1 Acid pre-treatment

Acid pretreatment can break down both cellulose and lignin, so it is more suitable for lignocellulosic substrates. The hydrolytic microbes are also capable of acclimating to acidic conditions [65]. The main reaction that occurs during acid treatment is the hydrolysis of polymers into monomers and oligomers; thus, it could result in increase in the rate of digestion time as the hydrolysis step has been partially carried out [65,88]. The acid could interfere with the charge of sludge floc and reduce the overall negative charge of the sludge particles and thus prevents large colloidal sludge flocs formation [65]. The main disadvantages of acid pretreatment are the possibility of formation of inhibitory by-products such as hydroxymethylfurfural (HMF) at strong acidic conditions, loss of fermentable sugar due to the increased degradation of complex substrates, high costs of acids for the pretreatment and alkali for neutralising the acidic substrates before AD [65].

Devlin et al. investigated the effects of acid pretreatment (pH 1–6) on AD of WAS by adding 37% HCl and keeping the samples at the desired pH for 24 h prior to neutralisation for AD. The solubilisation of carbohydrates, proteins, and COD as a result of cell lysis or EPS solubilisation was found to increase by decreasing pH; pH of 3–6 did not affect the biogas production, while pH of 2 and 1 showed an increase of biogas production by 17% and 32%, respectively. They also reported that the acid pretreated WAS required 40% less cationic polymer addition to achieve the same cake solid content in the dewatering process [88].

3.3.2.2 Alkali pre-treatment

Alkali treatment is one of the best methods for complex organic matter solubilisation. The main reactions that occur during alkali treatment are solvation and saponification that result in the swelling of the particulate organics, making the cellular substances more susceptible to enzymatic attack and thus improving the biodegradability of the solid and liquid phases [65,87,89]. COD solubilisation increases through saponification of uranic acids, acetyl esters, and lipids, which leads to membrane solubilisation and the release of intercellular material, as well as neutralisation of various acids formed from the degradation of the particulate organics [65]. Electrostatic repulsion increases as a result of increasingly negatively charged bacterial surfaces due to the increase of pH, causing desorption of some part of extracellular polymers [90]. The efficiency of the alkali added to the sludge has been found to be in the order of NaOH > KOH > Mg $(OH)_2$ and $Ca(OH)_2$ [91]. However, too high concentrations of Na^+ or K^+ may result in inhibition of AD [59]. It should be noted that during alkali treatment some of the alkali is consumed by the biomass itself, increasing the consumption of the alkali reagents in the process [65].

Rajan et al. studied alkali pretreatment of WAS at 20°C and 38°C using NaOH and $Ca(OH)_2$ and observed over 45% solubilisation of particulate COD. The experiments showed better solubilisation with NaOH compared with $Ca(OH)_2$ [87]. Chen et al.

investigated the effects of pH from 4 to 11 on the hydrolysis and acidification of WAS [60]. The SCOD, soluble proteins, and carbohydrates were found to be significantly higher at alkaline pH (9, 10, and 11) than those at near-neutral pH (6, 7, and 8) or acidic pH (4 and 5). This was attributed to the dissociation of acidic groups in EPS and repulsions between the negatively charged EPS, which resulted in increased solubility of proteins and carbohydrates at alkaline conditions. Methane production was found to increase with the increase of pH from 4 to 6, while it decreased when the pH increased from 7 to 10. No methane production was observed at pH 10 and 11, and it was concluded that the activity of methanogens decreased or was lost in extremely strong alkaline environment [60]. Table 3.6 shows a summary of the previous studies on alkaline pretreatments and their effects on AD.

Table 3.6 **Summary of the literature work on impacts of various alkaline pretreatments on AD**

Waste type	Chemical	Results	Ref.
Pig manure	$Ca(OH)_2$ (5% of the dry weight of manure)	– Cumulative biogas production was 28%, 14%, and 12% higher than the untreated sample after 7, 19, and 29 days, respectively	[73]
Kraft mill sludge	NaOH and KOH (20–200 meq/L)	– Up to a 32% increase in the soluble COD/TCOD ratio with possible improved methane yield and the efficiency of the AD process	[92]
Microbial biomass from an industrial plant	0–26.1 g NaOH/L	– The COD solubilisation reached 63%, and total solid elimination reached 33% when 4.6 g of NaOH/L were added – COD solubilisation levelled off by further increase in alkali from 4.6 to 26.1 g NaOH/L – The COD solubilisation was the consequence of protein solubilisation	[93]
WAS	1.68–3.65 KOH/L	– Addition of an alkaline agent led to an increase of COD solubilisation (after 1 h, soluble COD reached 9.3% at pH 10 and 30.7% at pH 12) – Using alkali did not improve the biodegradability. This might be due to the inhibition of methanisation by refractory molecules solubilised by the base of such high pH values	[94]

3.3.2.3 Ozonation

Ozonation is an oxidative treatment process to enhance sludge hydrolysis and improve solids biodegradability [18]. During ozonation pretreatment, sludge flocs break down into fine and dispersed particles. Ozonation is one of the effective techniques to disintegrate sludge; however, the production of ozone requires a high energy input. Weemaes and Verstraete reported an increase from 110 to 220 mL/g COD_{in} in methane production as a result of ozonation pretreatment for mixed sludge at 33°C [95]. Also, increased methane production from 82 to 183 mL/g COD_{in} was observed after ozonation pretreatment of a sewage sludge [96].

3.3.2.4 Oxidation processes

Oxidation processes such as wet air oxidation (WAO) and Fenton's peroxidation operate at 150–300°C and 20–200 bar, where aqueous materials are oxidised by an oxidant such as O_2 or H_2O_2. Many hazardous organic compounds can be destroyed during under these conditions [97]. The main application of WAO is in sludge conditioning for improving its dewaterability and for reducing the sludge volume [11]. Similar to WAO, supercritical water oxidation (SCWO) operates at 600°C and above the critical pressure of water (i.e. 22.1 MPa). Up to 99.9% of the organic materials in the sludge can decompose in the SCWO process, but the process is not yet available in large scale due to many technical challenges associated with the process such as the high-pressure reactor and piping requirements and reactor corrosion problems associated with SCW [11].

Currently, the established and emerging technologies for chemical pretreatment include the MicroSludge technology, using caustic soda to treat thickened WAS (5%–10% dry solids) for about an hour to weaken the cell membranes. The sludge is then transferred to a homogeniser operating at high pressure (80 bar), followed by sudden depressurisation to atmospheric pressure to lyses the sludge bacterial cells. The pretreated sludge is then mixed with the primary sludge prior to AD. The reported retention times in anaerobic digester were found to be <9 days after the pretreatment. The first full installation of this technology was at Chilliwack MWTP near Vancouver, Canada, in 2004, with a second full-scale demonstration installation at Los Angeles County, Sanitation District, in October 2005 [18].

3.3.3 Combined thermal and chemical pre-treatment

Combined low thermal and chemical pretreatment has the advantage of avoiding the necessity of high temperature and has been reported to give the best result in the solubilisation compared with either chemical or thermal pretreatment [63,98]. It is an effective technique for breaking down the microbial cells or compounds that are difficult to hydrolyse to easy biodegradable compounds [99]. Acids and alkalis act as catalysts in thermal hydrolysis of organic macromolecules. Acidic or alkaline pH has different effect on protein or polysaccharide components of EPS. They result in loosing of the natural shape of EPS proteins. Polysaccharides are unstable in strong acids due to the breakdown of the glycosidic linkages by acid hydrolysis reactions. However, they are stable in alkali especially at high temperatures [100].

Almost all of the previous studies on thermochemical pretreatments showed an increased solubilisation of organic matters. However, the results on biodegradability are contradictory. Haug et al. studied the combined thermal and chemical pretreatment of WAS and primary sludge at 175°C and different pH values of 12 and 1.2 and reported a decrease in biodegradability by 60% due to the formation of inhibitory materials that adversely affected the digester performance [75]. Penaud et al. reported COD solubilisation up to 70% of a microbial biomass treated at 140°C for 30 min at pH 12, but the biodegradability batch tests revealed that the thermochemical pretreatment did not vary the biodegradability. This was considered to be due to the formation of some inhibitory molecules or the induction of some intramolecular reactions during the thermochemical pretreatment, leading to the formation of refractory compounds [93]. On the other hand, Tanaka et al. showed that pretreatment of WAS at 130°C for 5 min after addition of 0.3 g NaOH/g VSS led to a significant increase in biodegradability, which could reach 230% [66,101]. Effects of different pretreatments on VSS solubilisation and methane production were found to be in the order of chemical < thermal < thermochemical and increased with increasing the alkali dosage (pH) and temperature [66].

Thermochemical treatments were also found to affect the dewaterability of the sludge. Neyens et al. studied the alkaline thermal hydrolysis of WAS at temperatures ranging from 20°C to 120°C for 30, 60, and 90 min by addition of NaOH, KOH, $Ca(OH)_2$, and $Mg(OH)_2$. They reported improved sludge dewaterability measured by capillary suction time (CST) when WAS was pretreated with $Ca(OH)_2$ and $Mg(OH)_2$ at 120°C and pH 10. $Ca(OH)_2$ gave better results, reducing the sludge volume by 40%. Sludge volume and CST values decreased (dewaterability was improved) by increasing temperature that could be explained by hydrolysis of extracellular and intracellular materials leading to destruction of colloidal particles of these macromolecules. Increasing the treatment time from 30 to 60 min decreased the CST values, but further increasing treatment time from 60 to 90 min had detrimental effects on sludge dewaterability [90]. In another study by Guan et al., WAS was treated at mild temperatures (50–90°C) in $CaCl_2$ solution. CST values were found to increase with increasing temperature to 80°C, indicating a significant deterioration in the dewaterability that was considered to result from the breakdown of sludge and the rise of smaller flocs that are hard to be dewatered. However, when $CaCl_2$ was added to the process, the CST of 60°C and 80°C treated samples decreased, and further reduction was observed by increasing the chemical dosage [98].

Table 3.7 presents a summary of results from the previous studies on thermochemical treatment of sludge or other wastes.

Since the mid-1980s, a growing number of thermochemical pretreatment technologies are emerging, but none of the developed processes has been successfully commercialised because of high costs involved and poor quality of the biogas products [63]. Among them, three representative processes are introduced here.

The Protox process was developed to improve WAS dewaterability by low-temperature acid hydrolysis. It was trailed by Wassex Water in the mid-1990s. However, it was not a true hydrolysis operation, and the dewaterability improvement was probably due to the changing of the isoelectric point of the sludge [63].

Table 3.7 Summary of results from the previous studies on thermochemical pretreatments of sludge or other wastes and their effects on AD

Waste type	Conditions	Results	Ref.
Pig manure	$T = 70°C$ $Ca(OH)_2$ (5% of the dry weight of manure)	– Cumulative biogas production was 127, 81, and 72% higher than the untreated sample after 7, 19, and 29 days of AD, respectively	[73]
WAS	$T = 130°C$ and 150°C KOH to obtain pH 10 and 12	– 63.1% COD solubilisation at 130°C and pH 12 – 59.5% COD solubilisation at 170°C and pH 12 – 7% increase in biogas production was observed when KOH was added at 130°C (pH 12) compared with the thermal treatment at 130°C – No significant changes in biogas production for treatments at 170°C – 130°C and pH 10 led to 71% of COD degradation and 59% of TS degradation and biogas production improvement by 54%	[94]
WAS	$T = 170°C$, pH 10	– 2% improvement in final methane yield	[62]
WAS	$T = 60°C$, 0.6 mg H_2O_2 and 1.5 mg $FeCl_2$	– Increase in SCOD, soluble proteins, and carbohydrates – 9% and 17% decrease in TSS and VSS, respectively – 20% increase in methane production rate – No significant effects on methane yield	[102]
WAS from dairy plant	$T = 50–80°C$ pH 10, 11, 12 (NaOH)	– At 60°C and pH 12 after 50 days, biogas production was 51% higher than the untreated raw sludge – At 60°C with pH 12, COD solubilisation and suspended solid reduction was 23% and 22% higher than that of the control – An improvement by 103% in biogas production was observed	[99]

WAS	$T = 20–40°C$ NaOH and $Ca(OH)_2$	– 45% COD solubilised, gas production increased by 112% over the control	[87]
WAS	$T = 130°C, 150°C, 170°C$ pH 10	– Methane yields increased from 145 L/kg VS to 238 and 256 L/kg VS in for the treatment at 150°C and 170°C, respectively – The KOH added led to an increase in TS concentration before and after AD – Thermal treatment improved AD performance. Increase in AD seems to be linked to sludge solubilisation – The treatment at 170°C led to the best results: improvement in biogas yield by 80%	[103]

The Synox process was piloted at Jacksonville, Florida. The process was very compact and produced a stable product that resembled brown boot polish. However, it suffered from high chemical costs and formation of complex products [63].

The Krepro process was developed in Sweden and employed heat, pressure, and sulphuric acid to dissolve particulate phosphates, metal oxides, and a large fraction of organic compounds from thickened sludge. The main steps involved in this technology were initial acidification, heating with steam, hydrolysis in a pressurised reactor, organic sludge separation, and precipitation of iron phosphate from the concentrate. The process aimed to separate sludge into three fractions: (1) an organic and dewatered fraction to be used as a fuel, (2) a liquid fraction to be used as a carbon source for denitrification, and (3) a ferric phosphate to be used as fertiliser. However, the process suffered from lack of markets for its complex products [18,63].

3.3.4 Mechanical pre-treatment

Mechanical pretreatment methods are based on disruption of microbial cells by shear stress generated by pressure, translational, or rotational energy [63,67]. The shear forces decrease the particle size and increase the surface area available for enzymatic degradation [68]. Mechanical pretreatments are effective in solubilising microbial cells, but they are expensive due to the high energy consumption. The most used mechanical pretreatment technique is sludge sonication [67,104–106]. Ultrasonication has shown to improve the solubilisation of particulate COD and subsequent increased biogas production for primary sludge, waste-activated sludge, and hog manure, and the efficiency of the process increased with increased initial total suspended solid (TSS) concentration. Hog manure was found to be more amenable to ultrasonication than waste-activated sludge, as it took only 3000 kJ/kg total solids to cause 15% more solubilisation as compared with 25,000 kJ/kg total solids for waste-activated sludge. An economic evaluation indicated that only a specific energy of 500 kJ/kg TS was economical, with a net energy output valued at $4.1/t of dry solids, due to a 28% increase in methane production.

3.4 Concluding remarks

Converting sludge into bioenergy has shifted the view to sewage sludge from a waste to be treated and disposed of to a renewable resource for energy recovery. Bio-oil and biogas are two main types of energy that could be obtained from sludge through hydrothermal treatments and anaerobic digestion, respectively. During hydrothermal treatment, sludge is first hydrolysed and depolymerised to monomers and unit structures. Intermediates are formed by decomposition/degradation of the produced monomers. The reactive fragments/intermediates then rearrange through repolymerisation, condensation, and cyclisation to form bio-oil products. Operating parameters, such as temperature, residence time, the presence of a catalyst, biomass-to-water ratio, and type of biomass can affect the product distribution and characteristics. Shell HTU process, the high-pressure hydrogenation (DOS) process developed by HAW in Germany,

CatLiq technology developed by SCF Technologies, and Hydrofaction commercialised by Steeper Energy are examples of successful pilot- and demonstration-scale plants of hydrothermal liquefaction of biomass that have been practiced in the world. Anaerobic digestion is the most common method for sludge stabilisation, resulting in the reduction of volatile solids (VS) and production of biogas. Hydrolysis is commonly believed to be the rate-limiting step of AD; however, this step can be enhanced by thermal, chemical, thermochemical, and mechanical pretreatment of the feedstock to improve its anaerobic biodegradability and dewaterability. The commercialised and emerging technologies for sludge pretreatments include Cambi and BioThelys for thermal treatments; MicroSludge technology for chemical treatments; and Protox, Synox, and Krepro processes for thermochemical treatments.

Acknowledgements

The authors would like to acknowledge the funding from BioFuelNet Canada, a Network of Centres of Excellence, and from NSERC through the Discovery Grants.

References

[1] Azadi P, Afif E, Foroughi H, Dai T, Azadi F, Farnood R. Catalytic reforming of activated sludge model compounds in supercritical water using nickel and ruthenium catalysts. Appl Catal B Environ 2013;134–135:265–73.
[2] Yang C, Jia L, Chen C, Liu G, Fang W. Bio-oil from hydro-liquefaction of Dunaliella salina over Ni/REHY catalyst. Bioresour Technol 2011;102(6):4580–4.
[3] Xu C, Etcheverry T. Hydro-liquefaction of woody biomass in sub- and super-critical ethanol with iron-based catalysts. Fuel 2008;87(3):335–45.
[4] Yang Y, Gilbert A, Xu C(C). Production of bio-crude from forestry waste by hydro-liquefaction in sub–/super-critical methanol. AIChE J 2009;55(3):807–19.
[5] Demirbas A. Competitive liquid biofuels from biomass. Appl Energy 2011;88(1):17–28.
[6] Ma L, Wang T, Liu Q, Zhang X, Ma W, Zhang Q. A review of thermal-chemical conversion of lignocellulosic biomass in China. Biotechnol Adv 2012;30(4):859–73.
[7] Ho DP, Ngo HH, Guo W. A mini review on renewable sources for biofuel. Bioresour Technol 2014;742–9.
[8] Wen D, Jiang H, Zhang K. Supercritical fluids technology for clean biofuel production. Prog Nat Sci 2009;19(3):273–84.
[9] Devi L. Catalytic removal of biomass tars; olivine as prospective in-bed catalyst for fluidized-bed biomass gasifiers (Thesis). Technische Universiteit Eindhoven; 2005.
[10] Appels L, Baeyens J, Degrève J, Dewil R. Principles and potential of the anaerobic digestion of waste-activated sludge. Prog Energy Combust Sci 2008;34(6):755–81.
[11] Tyagi VK, Lo S-L. Sludge: a waste or renewable source for energy and resources recovery. Renew Sust Energ Rev 2013;25(71):708–28.
[12] Xiu S, Shahbazi A. Bio-oil production and upgrading research: a review. Renew Sust Energ Rev 2012;16(7):4406–14.
[13] Rushdi AI, Al-Mutlaq KF, Sasmal SK, Simoneit BRT. Alteration of sewage sludge biomass into oil-like products by hydrous pyrolysis methods. Fuel 2013;103:970–9.

[14] Davies PS. The biological basis of wastewater treatment. 2005.
[15] Xu C, Lancaster J. Conversion of secondary pulp/paper sludge powder to liquid oil products for energy recovery by direct liquefaction in hot-compressed water. Water Res 2008;42(6–7):1571–82.
[16] Catallo WJ, Comeaux JL. Reductive hydrothermal treatment of sewage sludge. Waste Manag 2008;28(11):2213–9.
[17] Canadian Council of Ministers of the Environment. Canada-wide approach for the management of wastewater biosolids; 2012.
[18] Kalogo, Y.; Monteith, H. State of science report: energy and resource recovery from sludge; 2008.
[19] Vardon DR, Sharma BK, Scott J, Yu G, Wang Z, Schideman L, et al. Chemical properties of biocrude oil from the hydrothermal liquefaction of Spirulina algae, swine manure, and digested anaerobic sludge. Bioresour Technol 2011;102(17):8295–303.
[20] Cao Y, Pawłowski A. Sewage sludge-to-energy approaches based on anaerobic digestion and pyrolysis: brief overview and energy efficiency assessment. Renew Sust Energ Rev 2012;16(3):1657–65.
[21] Canadian Council of Ministers of the Environment. A review of the current Canadian legislative framework for wastewater biosolids; 2010.
[22] Toor SS, Rosendahl L, Rudolf A. Hydrothermal liquefaction of biomass: a review of subcritical water technologies. Energy 2011;36(5):2328–42.
[23] Yin S, Dolan R, Harris M, Tan Z. Subcritical hydrothermal liquefaction of cattle manure to bio-oil: effects of conversion parameters on bio-oil yield and characterization of bio-oil. Bioresour Technol 2010;101(10):3657–64.
[24] Behrendt F, Neubauer Y, Oevermann M, Wilmes B, Zobel N. Direct liquefaction of biomass. Chem Eng Technol 2008;31(5):667–77.
[25] Theegala CS, Midgett JS. Hydrothermal liquefaction of separated dairy manure for production of bio-oils with simultaneous waste treatment. Bioresour Technol 2012; 107:456–63.
[26] Ruiz Ha, Rodríguez-Jasso RM, Fernandes BD, Vicente Aa, Teixeira Ja. Hydrothermal processing, as an alternative for upgrading agriculture residues and marine biomass according to the biorefinery concept: a review. Renew Sust Energ Rev 2013;21:35–51.
[27] Goyal HB, Seal D, Saxena RC. Bio-fuels from thermochemical conversion of renewable resources: a review. Renew Sust Energ Rev 2008;12(2):504–17.
[28] Akhtar J, Amin NAS. A review on process conditions for optimum bio-oil yield in hydrothermal liquefaction of biomass. Renew Sust Energ Rev 2011;15(3):1615–24.
[29] Huang H, Yuan X, Zhu H, Li H, Liu Y, Wang X, et al. Comparative studies of thermochemical liquefaction characteristics of microalgae, lignocellulosic biomass and sewage sludge. Energy 2013;56:52–60.
[30] Karagoz S, Bhaskar T, Muto A, Sakata Y. Comparative studies of oil compositions produced from sawdust, rice husk, lignin and cellulose by hydrothermal treatment. Fuel 2005;84(7–8):875–84.
[31] Demirbaş A. Thermochemical conversion of biomass to liquid products in the aqueous medium. Energy Sources 2005;27(13):1235–43.
[32] Tymchyshyn M, Xu CC. Liquefaction of bio-mass in hot-compressed water for the production of phenolic compounds. Bioresour Technol 2010;101(7):2483–90.
[33] Karagöz S, Bhaskar T, Muto A, Sakata Y. Hydrothermal upgrading of biomass: effect of K2CO3 concentration and biomass/water ratio on products distribution. Bioresour Technol 2006;97(1):90–8.

[34] Shuping Z, Yulong W, Mingde Y, Kaleem I, Chun L, Tong J. Production and characterization of bio-oil from hydrothermal liquefaction of microalgae Dunaliella tertiolecta cake. Energy 2010;35(12):5406–11.
[35] Ross AB, Biller P, Kubacki ML, Li H, Lea-Langton A, Jones JM. Hydrothermal processing of microalgae using alkali and organic acids. Fuel 2010;89(9):2234–43.
[36] Minowa T, Masanori M, Yutaka D, Tomoko O, Shin-Ya Y. Oil production from garbage by thermochemical liquefaction. Biomass Bioenergy 1995;8(2):117–20.
[37] Sun P, Heng M, Sun S, Chen J. Direct liquefaction of paulownia in hot compressed water: influence of catalysts. Energy 2010;35(12):5421–9.
[38] Zhang L, Xu CC, Champagne P. Energy recovery from secondary pulp/paper-mill sludge and sewage sludge with supercritical water treatment. Bioresour Technol 2010; 101(8):2713–21.
[39] Hammerschmidt A, Boukis N, Hauer E, Galla U, Dinjus E, Hitzmann B, et al. Catalytic conversion of waste biomass by hydrothermal treatment. Fuel 2011;90(2):555–62.
[40] Zhang L, Champagne P, Xu C(C). Bio-crude production from secondary pulp/paper-mill sludge and waste newspaper via co-liquefaction in hot-compressed water. Energy 2011;36(4):2142–50.
[41] Duan P, Savage PE. Hydrothermal liquefaction of a microalga with heterogeneous catalysts. Ind Eng Chem Res 2011;50:52–61.
[42] Zhong C, Wei X. A comparative experimental study on the liquefaction of wood. Energy 2004;29(11):1731–41.
[43] Ogi T, Yokoyama S-Y, Koguchi K. Direct liquefaction of wood by catalyst (part 1) effects of pressure, temperature, holding time and wood/catalyst/water ratio on oil yield. Sekiyu Gakkaishi (Jl Jpn Pet Inst) 1985;28(3):239–45.
[44] Xu C, Lad N. Production of heavy oils with high caloric values by direct liquefaction of woody biomass in sub/near-critical water. Energy Fuel 2008;22(1):635–42.
[45] Boocock DGB, Sherman KM. Further aspects of powdered poplar wood liquefaction by aqueous pyrolysis. Can J Chem Eng 1985;63(4):627–33.
[46] Anastasakis K, Ross AB. Hydrothermal liquefaction of the brown macro-alga Laminaria saccharina: effect of reaction conditions on product distribution and composition. Bioresour Technol 2011;102(7):4876–83.
[47] Cheng S, D'Cruz I, Wang M, Leitch M, Xu C(C). Highly efficient liquefaction of woody biomass in hot-compressed alcohol–water co-solvents. Energy Fuel 2010; 24(9):4659–67.
[48] López Barreiro D, Prins W, Ronsse F, Brilman W. Hydrothermal liquefaction (HTL) of microalgae for biofuel production: state of the art review and future prospects. Biomass Bioenergy 2013;53:113–27.
[49] Yin S, Tan Z. Hydrothermal liquefaction of cellulose to bio-oil under acidic, neutral and alkaline conditions. Appl Energy 2012;92:234–9.
[50] Goudriaan F, Peferoen DGR. Liquid fuels from biomass via a hydrothermal process. Chem Eng Sci 1990;45(8):2729–34.
[51] Kranich, W. L.; Eralp, A. E. Conversion of sewage sludge to oil by hydroliquefaction; Cincinnati, OH, 1984.
[52] Molton, P. M.; Fassbender, A. G.; Brown, M. D. STORS: the sludge-to-oil reactor system; Cincinnati, OH, 1986.
[53] Itoh, S.; Suzuki, A.; Nakamura, T.; Yokoyama, S. Production of heavy oil from sewage sludge by direct thermochemical liquefaction. Desalination 1994, 98 (1–3), 127–133.
[54] Jazrawi C, Biller P, Ross AB, Montoya A, Maschmeyer T, Haynes BS. Pilot plant testing of continuous hydrothermal liquefaction of microalgae. Algal Res 2013;2(3):268–77.

[55] He BJ, Zhang Y, Funk TL, Riskowski GL, Yin Y. Thermochemical conversion of swine manure: an alternative process for waste treatment and renewable energy production. Am Soc Agric Eng 2000;43(6):1827–33.
[56] Ocfemia KS, Zhang Y, Funk T. Hydrothermal processing of swine manure to oil using a continuous reactor system: effects of operating parameters on oil yield and quality. Am Soc Agric Biol Eng 2006;49(6):1897–904.
[57] Athanasoulia E, Melidis P, Aivasidis A. Optimization of biogas production from waste activated sludge through serial digestion. Renew Energy 2012;47:147–51.
[58] Kim D, Lee K, Park KY. Enhancement of biogas production from anaerobic digestion of waste activated sludge by hydrothermal pre-treatment. Int Biodeterior Biodegrad 2015;101:42–6.
[59] Kondusamy D, Kalamdhad AS. Pre-treatment and anaerobic digestion of food waste for high rate methane production—a review. J Environ Chem Eng 2014;2(3):1821–30.
[60] Chen Y, Jiang S, Yuan H, Zhou Q, Gu G. Hydrolysis and acidification of waste activated sludge at different pHs. Water Res 2007;41(3):683–9.
[61] Bougrier C, Delgenès JP, Carrère H. Effects of thermal treatments on five different waste activated sludge samples solubilisation, physical properties and anaerobic digestion. Chem Eng J 2008;139(2):236–44.
[62] Nielsen HB, Thygesen A, Thomsen AB, Schmidt JE. Anaerobic digestion of waste activated sludge-comparison of thermal pretreatments with thermal inter-stage treatments. J Chem Technol Biotechnol 2011;86(2):238–45.
[63] Neyens E, Baeyens J. A review of thermal sludge pre-treatment processes to improve dewaterability. J Hazard Mater 2003;98(1–3):51–67.
[64] Appels L, Degrève J, Van der Bruggen B, Van Impe J, Dewil R. Influence of low temperature thermal pre-treatment on sludge solubilisation, heavy metal release and anaerobic digestion. Bioresour Technol 2010;101(15):5743–8.
[65] Ariunbaatar J, Panico A, Esposito G, Pirozzi F, Lens PNL. Pretreatment methods to enhance anaerobic digestion of organic solid waste. Appl Energy 2014;123:143–56.
[66] Tanaka S, Kobayashi T, Kamiyama K, Bildan MLNS. Effects of thermochemical pretreatment on the anaerobic digestion of waste activated sludge. Water Sci Technol 1997;35(8):209–15.
[67] Ferrer I, Ponsá S, Vázquez F, Font X. Increasing biogas production by thermal (70°C) sludge pre-treatment prior to thermophilic anaerobic digestion. Biochem Eng J 2008;42(2):186–92.
[68] Strong PJ, Gapes DJ. Thermal and thermo-chemical pre-treatment of four waste residues and the effect on acetic acid production and methane synthesis. Waste Manag 2012;32(9):1669–77.
[69] Nges IA, Liu J. Effects of anaerobic pre-treatment on the degradation of dewatered-sewage sludge. Renew Energy 2009;34(7):1795–800.
[70] Hiraoka M, Takeda N, Sakai S, Yasuda A. Highly efficient anaerobic digestion with thermal pre-treatment. Water Sci Technol 1985;17:529–39.
[71] Dhar BR, Nakhla G, Ray MB. Techno-economic evaluation of ultrasound and thermal pretreatments for enhanced anaerobic digestion of municipal waste activated sludge. Waste Manag 2012;32(3):542–9.
[72] Bougrier C, Delgenès JP, Carrère H. Impacts of thermal pre-treatments on the semi-continuous anaerobic digestion of waste activated sludge. Biochem Eng J 2007;34(1):20–7.
[73] Rafique R, Poulsen TG, Nizami AS, Asam ZUZ, Murphy JD, Kiely G. Effect of thermal, chemical and thermo-chemical pre-treatments to enhance methane production. Energy 2010;35(12):4556–61.

[74] Haug RT. Sludge processing to optimize digestibility and energy production. J Water Pollut Control Fed 1977;49(7):1713–21.
[75] Haug RT, Stuckey DC, Gossett JM, Mccarty PL. Effect of thermal pretreatment on digestibility and dewaterabilty of organic sludges. J Water Pollut Control Fed 1978;50(1):73–85.
[76] Haug RT, LeBrun TJ, Tortorici LD. Thermal pretreatment of sludges: a field demonstration. J Water Pollut Control Fed 1983;55(1):23–34.
[77] Brooks RB. Heat treatment of sewage sludge. Water Pollut Control 1970;69(2):221–31.
[78] Fisher WJ, Swanwick JD. High temperature treatment of sewage sludges. Water Pollut Control 1971;71(3):225–370.
[79] Stuckey DC, McCarty PL. Thermochemical pretreatment of nitrogenous materials to increase methane yield. Biotechnol Bioeng Symp (US) 1978;8:219–33.
[80] Li Y-Y, Noike T. Upgrading of anaerobic digestion of waste activated sludge by ultrasonic pretreatment. Water Sci Technol 1992;26(3–4):857–66.
[81] Gavala HN, Yenal U, Skiadas IV, Westermann P, Ahring BK. Mesophilic and thermophilic anaerobic digestion of primary and secondary sludge. Effect of pre-treatment at elevated temperature. Water Res 2003;37(19):4561–72.
[82] Barjenbruch M, Kopplow O. Enzymatic, mechanical and thermal pre-treatment of surplus sludge. Adv Environ Res 2003;7(3):715–20.
[83] Climent M, Ferrer I, Baeza MdM, Artola A, Vazquez F, Font X. Effects of thermal and mechanical pretreatments of secondary sludge on biogas production under thermophilic conditions. Chem Eng J 2007;133(1–3):335–42.
[84] Gonzalez-Fernandez C, Sialve B, Bernet N, Steyer JP. Thermal pretreatment to improve methane production of scenedesmus biomass. Biomass Bioenergy 2012;40:105–11.
[85] Chamchoi N, Garcia H, Angelidaki I. Methane potential of household waste. J Environ Resour 2011;33:13–26.
[86] Sustec. http://sustec.nl/.
[87] Rajan RV, Lin J, Ray BT. Low-level chemical pretreatment for enhanced sludge solubilization. Res J Water Pollut Control Fed 1989;61(11/12):1678–83.
[88] Devlin DC, Esteves SRR, Dinsdale RM, Guwy aJ. The effect of acid pretreatment on the anaerobic digestion and dewatering of waste activated sludge. Bioresour Technol 2011;102(5):4076–82.
[89] Vlyssides aG, Karlis PK. Thermal-alkaline solubilization of waste activated sludge as a pre-treatment stage for anaerobic digestion. Bioresour Technol 2004;91(2):201–6.
[90] Neyens E, Baeyens J, Creemers C. Alkaline thermal sludge hydrolysis. J Hazard Mater 2003;97(1–3):295–314.
[91] Kim J, Park C, Kim T-H, Lee M, Kim S, Kim S-W, Lee J. Effects of various pretreatments for enhanced anaerobic digestion with waste activated sludge. J Biosci Bioeng 2003;95(3):271–5.
[92] Navia R, Soto M, Vidal G, Bornhardt C, Diez MC. Alkaline pretreatment of kraft mill sludge to improve its anaerobic digestion. Bull Environ Contam Toxicol 2002; 69(6):869–76.
[93] Penaud V, Delgenes JP, Moletta R. Influence of thermochemical pretreatment conditions on solubilization and anaerobic biodegradability of a microbial biomass. Environ Technol 2000;21(1):87–96.
[94] Valo A, Carrère H, Delgenès JP. Thermal, chemical and thermo-chemical pre-treatment of waste activated sludge for anaerobic digestion. J Chem Technol Biotechnol 2004; 79(11):1197–203.
[95] Weemaes MPJ, Verstraete WH. Evaluation of current wet sludge disintegration techniques. J Chem Technol Biotechnol 1998;73(2):83–92.

[96] Yeom IT, Lee KR, Ahn KH, Lee SH. Effects of ozone treatment on the biodegradability of sludge from municipal wastewater treatment plants. Water Sci Technol 2002;46:421–5.

[97] Abelleira J, Pérez-Elvira SI, Sánchez-Oneto J, Portela JR, Nebot E. Advanced thermal hydrolysis of secondary sewage sludge: a novel process combining thermal hydrolysis and hydrogen peroxide addition. Resour Conserv Recycl 2012;59:52–7.

[98] Guan B, Yu J, Fu H, Guo M, Xu X. Improvement of activated sludge dewaterability by mild thermal treatment in CaCl2 solution. Water Res 2012;46(2):425–32.

[99] Uma Rani R, Adish Kumar S, Kaliappan S, Yeom IT, Rajesh Banu J. Low temperature thermo-chemical pretreatment of dairy waste activated sludge for anaerobic digestion process. Bioresour Technol 2012;103(1):415–24.

[100] Neyens E, Baeyens J, Dewil R, De Heyder B. Advanced sludge treatment affects extracellular polymeric substances to improve activated sludge dewatering. J Hazard Mater 2004;106(2–3):83–92.

[101] Tanaka S, Kamiyama K. Thermochemical pretreatment in the anaerobic digestion of waste activated sludge. Water Sci Technol 2002;46(10):173–9.

[102] Dhar BR, Elbeshbishy E, Hafez H, Nakhla G, Ray MB. Thermo-oxidative pretreatment of municipal waste activated sludge for volatile sulfur compounds removal and enhanced anaerobic digestion. Chem Eng J 2011;174(1):166–74.

[103] Bougrier C, Delgenès J-P, Carrère H. Combination of thermal treatments and anaerobic digestion to reduce sewage sludge quantity and improve biogas yield. Process Saf Environ Prot 2006;84(4):280–4.

[104] Elbeshbishy E, Nakevski A, Hafez H, Ray M, Nakhla G. Simulation of the impact of SRT on anaerobic digestability of ultrasonicated hog manure. Energies 2010;3(5):974–88.

[105] Aldin S, Elbeshbishy E, Nakhla G, Ray MB. Modeling the effect of sonication on the anaerobic digestion of biosolids. Energy Fuel 2010;24(9):4703–11.

[106] Elbeshbishy E, Aldin S, Hafez H, Nakhla G, Ray M. Impact of ultrasonication of hog manure on anaerobic digestability. Ultrason Sonochem 2011;18(1):164–71.

Hydrothermal liquefaction of aquatic Feedstocks

4

P. Biller
Aarhus University, Aarhus, Denmark

4.1 Aquatic biomass in a hydrothermal liquefaction context

The development of third-generation biofuels from microalgae has seen increasing research efforts over the last two decades. Aquatic biomass has been identified as a promising renewable biomass feedstock for the production of biofuels due to its high photosynthetic efficiency and area specific yields [1]. Historically, microalgae have been under investigation due to their potentially high lipid content for biodiesel production. One of the big advantages of microalgae compared with terrestrial biomass is their higher photosynthetic efficiency, which results in higher growth rates and potentially improved CO_2 mitigation [2]. There are two functional groups of algae, phototrophic and heterotrophic. Phototrophic algae perform photosynthesis using CO_2 and light, whilst heterotrophic algae require an organic carbon source and oxygen for their growth via aerobic respiration. Both of these types of algae have been under investigation for hydrothermal liquefaction (HTL) processing. Similar to microalgae, there has also been research into cyanobacteria and diatoms, which are similar to microalgae regarding thermochemical applications but differ in biology.

Macroalgae, also known as seaweed, is a group of eukaryotic photosynthetic marine organisms. They differ significantly from microalgae in physical appearance and biochemical composition. They are composed of diverse species and are abundant in the world's oceans and coastal waters. Macroalgae are typically composed of a blade or lamina, a stipe, and a holdfast (haptera) for anchoring and support in marine environments [3]. Macroalgae are of larger size than microalgae (up to several metres in length), and only certain species such as those of the genus *Ulva* do not require pretreatment; the majority of macroalgae require grinding before pumping due to their large size.

Other aquatic biomass such as duckweed and water hyacinth are not classed as algae but aquatic plants. These are often invasive species, which can take over entire water bodies, thereby blocking out any sunlight to the lower water, and are of particular concern in China [4]. They have been regarded as a major disturbance in many waterways and have to be removed manually or with machinery in many parts of the world to maintain an intact biological system. This can be very costly, and the disposal of the wet biomass poses an issue due to its high moisture content, as it cannot simply be burnt. Therefore, HTL has been proposed to process such invasive aquatic species. Such aquatic plants can however also been grown specifically for biofuel production

Direct Thermochemical Liquefaction for Energy Applications. https://doi.org/10.1016/B978-0-08-101029-7.00003-5

and have been reported to be able to double its mass in 24 h, yield up to 20 t (dry) per hectare per year, and can utilise waste water nutrients [5].

Despite research into macroalgae and aquatic plants, the research community has largely concentrated on the utilisation of microalgae in the development of HTL research and commercialisation. Traditionally, the idea behind microalgae biofuels was to target the lipid fraction of algal cells. Similar to traditional biodiesel manufacture from terrestrial oil crops, the lipid fraction is extracted and transesterified to fatty acid methyl esters (FAME). The motivation lies in the fact that algae can contain up to ~70% lipids by weight and have theoretical oil yields per acre surpassing those from terrestrial oil crops. For example, a review states that potential oil yields of 8000–32,000 L/ha/year could be possible, with current technology achieving a maximum productivity of around 8000 L/year [6]. This surpasses the highest yielding terrestrial oil-producing plants [7]; palm and rape seed oils can be produced at 6000 and 1500 L/ha/year, respectively [6]. Additionally, microalgae can be harvested all year around, whilst terrestrial plants are usually seasonal and often only allow one harvest per year.

Terrestrial biomass consists mainly of cellulose, hemicellulose, and lignin, whilst algae consist of varying amounts of protein, starchlike carbohydrates, and lipids. If the algae is from a marine origin, it typically contains large amounts of ash as salts and other mineral matters; certain strains can exhibit ash contents as high as 60 wt% on a dry basis. Macroalgae are typically higher in carbohydrate and much lower in lipid content compared with microalgae and if from a marine environment can be very high in ash content [8]. This can affect hydrothermal reactions and presents a significant difference to terrestrial biomass, which is typically low in ash content.

All microalgae produce lipids, proteins, and carbohydrates in varying proportions. The composition is heavily influenced by culturing conditions and nutrient supply. The lipid content is the primary component for traditional biodiesel production, and this fraction can make up anywhere between approximately 5 and 70 wt% [9]. There are also valuable by-products that can be harvested from microalgae, which can significantly increase the economics of biofuel production. Many algal strains can produce polyunsaturated fatty acids (omega-3) in the form of EPA and DHA, which are used as food supplements [7]. The cyanobacteria *Spirulina* has been a successful food supplement for years due to its high protein content and essential oils. There are also a number of valuable recombinant proteins found in microalgae and valuable pigments such as β-carotene, astaxanthin, and phycocyanin [10–12]. *Dunaliella salina* is commercially cultivated for its high levels of the pigment β-carotene [11]. There have been numerous suggestions of extraction of valuable coproducts from micro- and macroalgae in a biorefinery context to make fuel production from residues more economically favourable. Macroalgae have been grown in the Far East as a foodstuff for human consumption for centuries, particularly in Japan, China, and Korea. They are also cultivated industrially for extraction of valuable hydrocolloids such as agar, alginate, and carrageenan. These water-soluble polysaccharides are commonly used as thickening agents. Macroalgae generally have low lipid content but are high in carbohydrates. Adams et al. studied the seasonal variation in the chemical composition of the macroalgae, *Laminaria digitata*, for thermochemical conversion and reported that summer harvests contained the highest proportion of carbohydrate and lowest ash content making them most suitable for conversion to biofuels [3,8].

Extraction of above-mentioned valuable phytochemicals from algae is often suggested in literature to be a potential stepping stone to make algal biofuels cost-competitive. There has been some research of extracting, for example, lipids [13] or β-carotene [14] before hydrothermal treatment of the residual biomass. Such approaches may give an economic advantage over processing the entire feedstock for fuel production, but to date, research into such pathways is limited.

For the dedicated cultivation of algae, it is advantageous to use natural sunlight due to cost at the expense of diurnal cycles and seasons resulting in a limiting availability of light. A location with high solar irradiation is therefore beneficial for algal growth. Inorganic nutrients have to be present in the growth media and include primarily nitrogen and phosphorus. Grobbelaar presented a formula for nutrient requirements based on the molecular composition of typical algae, $CO_{0.48}H_{1.83}N_{0.11}P_{0.01}$, but this varies for different algae strains [15]. One way of providing nutrients to microalgae being researched is the use of waste water or nutrient-rich agricultural run-off. Pittman et al. conclude in their review that algae cultivation based on current technologies is unlikely to provide a positive energy return or to be economically viable unless it is combined with waste water treatment, as this reduces the energy cost, GHG emissions, and the nutrient and freshwater resource costs [16].

Algae cultivation can be achieved in either open-pond or photobioreactor systems. The most common method for outdoor cultivation is the use of raceway ponds. These structures are arranged in a closed loop of various designs with a paddle wheel for circulation and mixing. The floor is lined with a plastic liner or with nonpermeable clay at a depth of approximately 20–30 cm. The depth is limited by the penetration of sunlight to algae cells deeper in the pond. The culture is fed continuously in front of the paddlewheel, and the algae broth is harvested just behind the paddlewheel.

Open raceway ponds are advantageous due to cheap construction, low energy input during operation, easy maintenance, and the fact that they can be built on nonarable land. One of the biggest problems with raceway ponds is the contamination of other algae and protozoa. To achieve monoculture cultivation, it is therefore necessary to have extreme growing environments. For example, *Chlorella* can grow in extremely nutrient-rich media, *Dunaliella salina* is adaptable to very high salinity, and *Spirulina* can withstand high alkalinity [17]. Another problem of open raceway ponds is the evaporation of water particularly in hot environments. Particularly, when salt water is used in the growth medium, this will lead to build-up of salt in the ponds.

The more sophisticated culturing systems used are known as photobioreactors. These incorporate the use of transparent tubes or plates, which act as the culturing environment. The most common design is the tubular design, but flat plate and column photobioreactors are also used, each with different advantages and limitations. The main advantage of closed systems is that contamination of a desired monoculture without contamination is possible for a substantial period without cleaning or other maintenance. CO_2 and light utilisation efficiency are higher than with open systems that drastically increase the biomass production per volume. This has the effect that harvesting also becomes much cheaper as the biomass concentration is up to 30 times higher than in open systems [9]. The major drawback is that photobioreactors have a higher construction cost and therefore increase the cost of biomass. Estimations of growth yields for microalgae in both open ponds and photobioreactors vary widely. Lab-scale results are often not transferable to large-scale outdoor cultivation making the assessment of potentials and costs difficult. In open-pond systems, an annual average productivity of around 20–25 g/m^2/day is realistic if there is sufficient sunlight. The algae cell concentration in such a system is typically around 0.1 g/L. Photobioreactors can achieve much higher concentrations of around 1 g/L [7]. For both cases, this is not high enough for subsequent HTL that requires solid loadings of around 200 g/L to make the process energy-efficient. It is therefore evident that water removal is still required after microalgae cultivation. This is still a major cost factor on top of the cultivation itself, which entails costs for nutrients, capital, operating costs, etc. Realistically, a dry ton of microalgae can currently be bought for around $1000/t based on the current state of technology [18]. It is however possible that this cost will drop significantly in coming years with advances in genetics, cultivation, and overall optimisation. Nevertheless, currently, the largest bottleneck for microalgae biofuels is quite simply the excessive cost of cultivation to deliver a feedstock with 20% solids for HTL processing. It is apparent that there are large uncertainties in estimations in literature and this is an area where additional research efforts should focus on.

4.2 Conversion characteristics, effect of process parameters on major product properties

4.2.1 Aquatic HTL conceptual layout

One of the challenges for producing microalgae-derived biofuels in the traditional biodiesel pathway is performing economic extraction of the lipids from the wet biomass before conversion. The lipids can be extracted by solvent extraction or by physical extraction following rupture of the cell wall. These lipids are then transesterified to biodiesel in the same manner as terrestrial seed oils [19,20]. One of the problems of this approach is that the wet aquatic biomass requires drying before it can be processed. The main economic and energetic drawbacks in the processing of microalgae are the dewatering stage as microalgae typically only grow to a solid concentration of 0.5–5 g/L [2]. This makes concentration and drying challenging and energy-intensive. It can account for as much as 25% of the energy contained in the algae [21]. Hydrothermal processing avoids this step as the algae is processed as a slurry in hot-compressed water and does not require drying. Concentration of the microalgae to a slurry of around 20 wt% is still required, but this can be performed by mechanical means with relatively little energy input compared with thermal drying required if a slurry with <10% water is required as it is for dry lipid extraction and pyrolysis.

The general steps anticipated in a stand-alone HTL system for aquatic biomass facility are summarised in Fig. 4.1. Microalgae require no pretreatment due to their small particle size and are pumped more easily as a slurry compared with macroalgae or other aquatic plants. The processing of other aquatic biomass such as seaweed may require grinding or maceration. Dewatering is required for any aquatic biomass

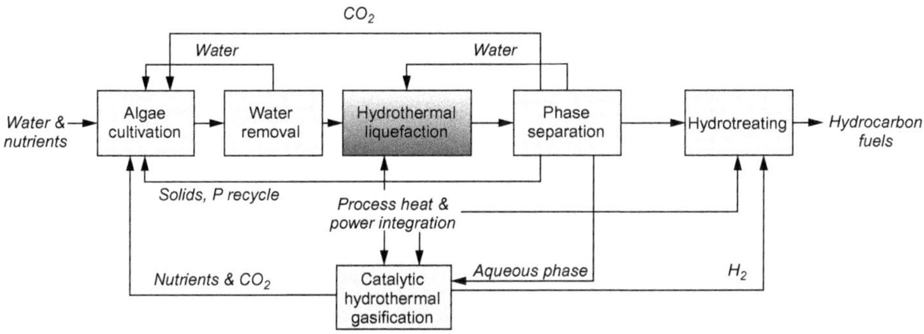

Fig. 4.1 Schematic layout of the HTL process for algae as feedstock.

feedstock to obtain a pumpable slurry at high solid loadings. The water obtained during concentration of the algae slurry to the desired solid concentration can be recycled back to the cultivation medium or open waters in the case of water plants and macroalgae.

The feedstock slurry is processed via HTL at temperatures of around 350°C and pressures of ~200 bar for approximately 15–30 min. Phase separation is achieved gravimetrically after the reaction resulting in a gaseous phase rich in CO_2 and solid residue, the desired bio-crude fraction, and an aqueous phase. The process water can be recycled back to the HTL reactor. Water phase recycling is achieved after separation of bio-crude and solids. The recycling of the water phase leads to the concentration of water-soluble organics that increases bio-crude yields in subsequent runs. However, there is always a surplus of water produced that cannot be recycled due to the high initial water content of the feedstock [22]. This surplus water after recycling needs to be treated before it can be returned into the environment. The options for water treatment investigated to date include anaerobic digestion and catalytic or noncatalytic hydrothermal gasification. Anaerobic digestion is used to produce a biogas rich in methane, whilst hydrothermal gasification (~500°C) and catalytic hydrothermal gasification (~350°C) produce a hydrogen-rich and methane-rich product gas, respectively [23,24]. There is potential for the recovery of nitrogen and phosphorous compounds following treatment of the water by either AD or HTG.

The other option for water treatment, which is unique for microalgae liquefaction, is to feed it back to the microalgae cultivation step. The water phase contains a large proportion of nitrogen, which is a required nutrient for algae cultivation. Recycling this nutrient can result in significant cost reductions, as nutrients are typically a major cost factor in algae cultivation. Finally, after separation, the bio-crude requires upgrading via hydrotreatment to produce final fuel products or refinery-ready feedstock. Each of these processing steps outlined in Fig. 4.1 is described in more detail in the coming sections.

4.2.2 Process parameter optimisation

Processing parameters during HTL of aquatic feedstocks are the same as they are for any other feedstock and include temperature, residence time, heating rate, cooling rate, solid loading, and addition of catalysis. There have been numerous publications in the literature investigating these effects on various micro- and macroalgae species. One problem in the assessment of processing parameters is that most studies used unique batch reactors, which are not necessarily transferable to others due to the difference in heating rates, size, etc. Another source of uncertainty is the sample work-up procedure. Many studies employ solvents at lab scale to recover bio-crude. This extracts some of the polar organics in the water phase and can change the yields and composition of bio-crude and aqueous phase [25]. Nevertheless, there have been some very significant and repeatable studies published in the literature on the optimisation of processing conditions.

One of the first and widely cited studies on HTL of algae was performed by the Savage group in the United States using *Nannochloropsis* as a feedstock. The authors investigated the effects of temperature on the bio-crude and gas formation of at a constant residence time of 60 min. Temperatures of 200–500°C with 50°C steps were investigated [26]. The highest bio-crude yield was observed at 350°C, whilst higher temperatures favoured gasification reactions. At 350°C, the bio-crude yield was found to be 43 wt% with a heating value of 39 MJ/kg, leading to a maximum energy recovery of 78%.

Jena et al. also observed that deoxygenation increases at higher temperatures but the levels of nitrogen in the bio-crude increase due to the onset of protein breakdown; there is therefore a trade-off between high O and high N in the bio-crude [27]. Many studies have shown that higher temperatures not only favour deoxygenation of the bio-crude but also result in increased nitrogen concentration of the bio-crude.

Garcia Alba et al. published a detailed study of the hydrothermal treatment of microalgae (*Desmodesmus* sp.); processing conditions with varying temperature (175–450°C) and reaction time (0–60 min) [28] were investigated [29]. It was concluded that a maximum bio-crude yield is achieved at 375°C and 5 min reaction time, achieving a bio-crude yield of 49 wt%. The bio-crude exhibited an HHV of 35 MJ/kg, corresponding to an energy recovery of 75%. The largest step change in bio-crude yield was achieved from 225°C to 250°C that was shown to coincide with a major visual breakage of cells after 5 min. Microalgae cells at 175°C appeared virtually intact; above 250°C, no individual cells were recognisable. It was concluded that the thermal breakage of cells at 250°C allows additional bio-crude to be formed and extracted leading to a large decrease in solid residue. In terms of elemental composition, the largest step change in nitrogen content in the bio-crude for shorter reaction time (5 min) occurred at 200–225°C, whilst at the longer reaction residence time (60 min), it was at the lower temperature of 175–200°C. This indicates the conditions where the onset of protein breakdown begins. The nitrogen content continues to increase with increasing operating temperature, whilst the oxygen content decreases in agreement with previous work [26,27]. Garcia Alba et al. conclude that operating conditions are largely dependent on the desired product; if a lipid-rich oil is preferred, operating temperatures should not exceed 250°C, but if crude-like oil is desired, temperatures as high as 375°C could be used.

In general, most parametric studies on operating conditions tend to agree that the optimal processing conditions for microalgae lie at around 350°C and 15 min reaction time. However, more recently, the Savage group has shown that heating rates during HTL also play an important role. Using a small batch reactor set-up, very high heating rates approaching 300°C/min could be achieved [30]. Very short reaction times of only 1 min were investigated. It was shown that the highest bio-crude yields were obtained at these very high heating rates and low reaction times. A yield of 66% was achieved compared with around 40% using more conventional heating rates, temperature, and holding times of 350°C and 20 min. In subsequent publications, the positive effect of high heating rates on bio-crude yields was confirmed [31,32]. It was shown that up to 90% of the calorific value of the feedstock could be recovered in the bio-crude. A quantitative kinetic model for both fast and isothermal HTL was developed and showed that similar bio-crude yields can be obtained using fast heating and low residence time compared with using slow heating and longer residence time. Therefore, the results that were shown by other researchers of identifying optimum conditions of 350°C and 15 min are also correct, but these can equally be achieved at much lower residence times when using high heating rates.

These findings by the Savage group could potentially have influences on industrial applications where continuous systems are used. Continuous processing of microalgae has been performed in several laboratories and is technically easily feasible, and no pretreatment of the microalgae is required due to its small particle size facilitating the formation of pumpable slurries. Jazrawi et al. and Biller et al. investigated short residence times and high heating rates in continuous systems, similar to the batch work by the Savage group [33,34].

Both these continuous studies showed that lower residence times and higher heating rates resulted in higher bio-crude yields but less deoxygenation, in agreement with batch results presented by Savage et al. [28]. In the longer-residence-time experiments, bio-crude has a lower oxygen and higher carbon content, leading to an increased HHV. The oxygen content of the bio-crudes was affected by the varying residence times, reducing from 16% to 11%. This is in agreement with HTL batch results by Faeth et al. (2013) who observed a reduction in O content from 12% to 8% with increasing residence time from 10 to 90 min [28]. This means that the quality of the bio-crude is superior at the longer residence times and more suitable for upgrading due to lower H_2 demand for hydrodeoxygenation at the cost of a lower yield.

As the residence time in a continuous reactor decreases by increasing the flow rate, the reactor output increases, and energy requirements decrease. This reduces the capital cost of a continuous system and could improve the energy balance by improving the net energy return. As the chemical energy recovery is reduced however at the lower residence time, the optimum conditions for continuous HTL need to be carefully adjusted. The effect of this trade-off should become more apparent with the transition from lab-scale to pilot- and full-scale HTL systems. Careful consideration should be given to flow rates, heating rates, and residence times in terms of bio-crude yield, chemical energy recovery, capital reactor cost, and energy requirements.

The PNNL group has extensively run microalgae and macroalgae slurries continuously using various strains. Their operating conditions were kept similar at 350°C and an LHSV of around 1.5–2.2, corresponding to a residence time of around 25–40 min. These operating conditions have been very successful for the group, and they were able to demonstrate the best carbon and energy conversions on continuous systems to date, with many continuous hours on stream and good repeatability and bio-crude quality.

4.2.3 Effect of species

Microalgae are a very diverse set of organisms and can vary widely in their composition. This in turn affects their liquefaction behaviour. Microalgae consist of proteins, lipids, and carbohydrates in varying proportions. This depends on both culturing condition and strain. There has been a lot of work performed on studying model compounds of algae during HTL and trying to relate their behaviour to actual microalgae. Each biochemical fraction can be converted to bio-crude during liquefaction but each component to a different degree with different product properties. Lipids, for example, are converted to bio-crude almost entirely; therefore, it is beneficial to have an algae strain with high lipid content. However, algae with high lipid contents generally have lower growth rates and are more challenging to cultivate. This is one of the great benefits of HTL over lipid extraction technologies as the other biochemical components can also be converted to bio-crude. Proteins are converted to bio-crude relatively well with yields ranging from 20% to 40%. Carbohydrates on the other hand are less favourable to convert to bio-crude; their yields are low and resulting oxygen contents relatively high. Addition of alkali catalysts helps in processing carbohydrates to ensure higher yields and less excessive char formation, similar to what is observed in liquefaction of lignocellulosic feedstocks [35]. Researchers have tried to model the bio-crude yields and properties of bio-crude via the biochemical composition of microalgae. Biller and Ross were the first to investigate this and derived a simple equation after processing separate model compounds and four microalgae strains. The equation is based on additive model compound behaviour for predicting bio-crude yields that form real algae [36]; the format of the equation is given in Eq. (4.1):

$$\text{Bio}-\text{crude yield}\left(\%dw\right) = x \times L + y \times P + z \times C \tag{4.1}$$

where L, P, and C are the lipid, protein, and carbohydrate content of algae, respectively, and x, y, and z are the relative bio-crude yields as outlined in Table 4.1. The values show that lipids are converted to bio-crude most efficiently with a contribution of 0.8–0.97, meaning a pure lipid feedstock would yield 80%–97% bio-crude yield. Carbohydrates are least efficiently converted with yields ranging from 6% to 17%, whilst proteins are converted well with yields ranging from 18% to 42%.

The prediction equation developed by Biller and Ross worked well for some strains but not for others, indicating that the liquefaction behaviour is not that straightforward. Such predictive work has since been improved, and more complex models have been derived that are able to predict product distributions and some bio-crude quality

Table 4.1 Fractional conversion constants of microalgae biochemical components to bio-crude yield

Lipids	Carbohydrate	Protein	Reference
0.97	0.17	0.42	Leow et al. [37]
0.95	0.06	0.33	Teri et al. [38]
0.80	0.06	0.18	Biller and Ross [36]

parameters accurately [37,38]. The Savage group investigated the liquefaction behaviour of six model compounds by processing them alone and in mixtures to derive a more accurate predictive equation [38]. These results were not compared with actual microalgae samples in their study, but a latter study by Leow et al. compared the literature results and found an $r^2 = 0.15$ for the results [37]. Leow et al. study derived a similar, improved correlation that resulted in a correlation of predicting bio-crude from their 10 *Nannochloropsis* microalgae samples with varying biochemical composition of $r^2 = 0.97$. Applying the equation to literature data still resulted in a relatively good correlation of $r^2 = 0.46$. Within the boundaries of their own sample set, Leow et al. were also able to predict parameters such as CHNS content, energy balance, and bio-crude boiling point highly accurately. The study went as far as suggesting optimal harvesting periods of microalgae cultivation for best economic performance. The biochemical composition changes during algal growth cycle so that it can be beneficial to harvest earlier by reducing cultivation costs at the expense of increased lipid yields.

HTL has been carried out on vast number of different microalgae species. The species is, however, less of an important factor compared to the biochemical composition discussed above. Microalgae species that have been utilised frequently include *Chlorella*, *Nannochloropsis*, and the cyanobacteria *Spirulina*. For different algae species with the same biochemical composition, the results in terms of bio-crude yields and properties are insignificant. For example, Barreiro et al. processed eight different microalgae species and found that the differences amongst species at 375°C are not pronounced. At 250°C, however, high lipid contents were more favourable, and the effect of protein content was also more pronounced as this fraction contributes to bio-crude formation only at higher temperatures [39].Other factors such as marine or freshwater origin have a larger effect; marine species have a much higher ash content, which can influence liquefaction behaviour due to the high amounts of alkali salts present. Ash content can vary from <5% for freshwater species to >50% for marine species. In terms of total mass into the HTL reaction, this has a large effect on bio-crude yields; however, in terms of bio-crude yields on an organic basis, the effects are less pronounced. PNNL has processed several different algae species from different growth conditions, also including lipid-extracted algae [23,40]. Lipid-extracted algae (LEA) is the residue obtained from solvent-extracting algae for biodiesel production. It has been suggested that this could be an economic advantage, but generally, researchers are concentrating their efforts on applying HTL on the entire algae feedstock. Most microalgae result in a bio-crude yield of around 40%–60% on an organic basis. If the algae have a high lipid content, levels of 60% are routinely achieved. High-lipid strains also result in higher energy and carbon recoveries due to the higher affinity of lipids to form bio-crude as can be seen in Table 4.1. Albrecht et al. were able to demonstrate bio-crude yields of 71% with a carbon recovery of 87% for a heterotrophic algae strain, high in lipids. Comparatively, using the same reactor but a low-lipid strain resulted in a bio-crude yield of 38% and a carbon recovery of 50% [40]. Algae with low lipid contents generally have a higher protein content, which means that bio-crude is also produced from protein. This results in higher nitrogen contents in the bio-crude as seen in Table 4.2. For example, 4% versus the 0.5% is obtained from low- and high-lipid algae by PNNL, respectively. Table 4.2 shows examples of different

Table 4.2 **Bio-crude yields and properties from different aquatic feedstocks under HTL conditions**

Feedstock/strain	Description	Bio-crude yield (%)	Oxygen content (%)	Nitrogen content (%)	Carbon recovery (%)	Reference
Nannochloropsis	Low lipid, microalgae	38	5.3	4.2	50	Elliott et al. [23]
Chlorella	High lipid, heterotroph, microalgae	71	10.1	0.5	87	Albrecht et al. [40]
Saccharina spp.	Macroalgae	28	6–8	4	58	Elliott et al. [41]
Duckweed	Aquatic plant	30	13.3	5.4	56	Xiu et al. [42]

types of aquatic plants processed under HTL conditions with their yields, resulting bio-crude oxygen and nitrogen content, and the carbon recovery.

Whilst the main focus has been directed towards microalgae, some notable investigations have been performed on macroalgae. PNNL processed the macroalgae *Saccharina*, harvested at six different times throughout a year and were able to show widely varying yields from 9% to 28%. This also affected the carbon recovery strongly that varied from 17% to 58%, whilst energy yields varied from 20% to 53% [41]. Anastasakis and Ross investigated the influence of reaction conditions on the HTL of the brown macroalgae *Laminaria saccharina* [43]. Macroalgae generally exhibit a low lipid content and higher carbohydrate content, and they typically have a lower HHV due to the high ash content [8]. Optimum reaction conditions were found to be 350°C and 15 min (19.3 wt% bio-crude yield).

4.2.4 Catalysis

The idea of incorporating catalysts in HTL rests on the potential of increasing yields and reducing oxygen and nitrogen content and viscosity. The majority of research into catalytic HTL was performed using homogenous catalysts, that is, catalysts soluble in water. The most widely investigated homogeneous catalysts are the alkali Na_2CO_3, KOH, and K_2CO_3. It is common practice in the liquefaction of lignocellulosic biomass to incorporate alkali catalysts. It is generally accepted that HTL requires alkaline conditions if the amount of solid residue is to remain small since alkali catalysts reduce char formation, particularly from carbohydrates [22,35]. Most microalgae contain a significant amount of protein, usually at least 30%; this means that during HTL, the protein is broken down to ammonium that results in an alkaline pH during reaction and has therefore the same effect as adding an alkali catalysts. Therefore, for the liquefaction of microalgae with protein content, the addition of homogenous alkali is not necessary. Biller and Ross were able to show that the addition of alkali is only beneficial in the liquefaction of algae feedstocks with high carbohydrate content and not for the ones high in protein and lipid [36]. The lipid fraction can undergo saponification reactions in alkali conditions, which results in lower bio-crude yields. Therefore, generally, HTL of algae does not incorporate catalysts.

Recovery of the homogenous catalysts poses a problem, and its cost is a major factor to consider at large scale. Therefore, some groups have looked at heterogeneous catalysts. These could potentially have advantages over homogeneous catalysts as they are easily recoverable and can be used in a fixed-bed fashion in continuous systems, although stability and poisoning of catalysts under hydrothermal conditions pose a challenge [44]. The Savage group investigated six different catalysts supported on carbon, Al_2O_3 support, and a zeolite [45]. Palladium, platinum, and ruthenium on carbon and nickel and cobalt-molybdenum on alumina and a zeolite were investigated. It was shown that the bio-crude yields increased for all catalysts. The maximum yield was achieved for the Pd/C, 57% compared with 37 wt% with only water. The bio-crude yields of all the other catalysts ranged from 45 to 50 wt%. The effect on heating value and heteroatom removal was negligible. The potential of heterogeneous catalysis in HTL of algae appears promising when the carbon recoveries reported are considered;

the elemental carbon recovery in the bio-crude using a heterogeneous catalyst exceeded 90% compared with 62% without catalysts. However, there are still significant challenges that need to be addressed: recovery, stability, and reusability of heterogeneous catalysts have not been reported and are areas where further work is required [44]. Moreover, none of the catalysts were able to significantly reduce the amount of N in the bio-crude.

4.3 Utilisation and composition of product streams

4.3.1 Bio-crude

Bio-crude is the primary product produced from HTL. It is a black, highly viscous material at room temperature often semisolid. It is often compared with petroleum crude but has a higher density, viscosity, and particularly heteroatom content. The higher heating value is generally found to be around 35 MJ/kg, whilst petroleum is in the range of 40 MJ/kg. Table 4.3 compares a typical microalgae bio-crude with a North Sea petroleum crude. It is evident that the oxygen content is undesirably high and sulphur in this specific case is also high but largely depends on the microalgae feedstock. In addition, the nitrogen content microalgae bio-crudes can be very high, depending on the protein content of the microalgae feedstock. Similar to the discussion about bio-crude yields, the composition of bio-crude very much depends on the biochemical composition of the starting feedstock. Higher viscosity and lower HHV are additional reasons why bio-crude should be upgraded via hydrotreatment in order to resemble petroleum crude more closely. This not only could potentially allow them to be cofed to conventional refineries but also increases the energy density, which makes transportation of bio-crude more cost-efficient.

As mentioned above, the quality and composition of bio-crude largely depends on the biochemical composition of the feedstock. Table 4.4 shows a comparison of bio-crude derived from a strain with very high lipid content and one with normal lipid content. It is evident that the high-lipid strain results in a much lower nitrogen content due to the lower original protein content. This is highly beneficial for the bio-crude as much less hydrogen is required to hydrodenitrogenate the bio-crude. On the other

Table 4.3 Comparison North Sea petroleum crude with HTL-derived bio-crude

	HTL bio-crude	Petroleum crude
Sulphur, total (wt%)	0.5	< 0.1
Oxygen, total (wt%)	5–15	0.3
Density at 15°C (kg/m^3)	0.94–0.96	0.846
Viscosity at 40°C (cSt)	100–350	5–100
Net heat of comb. (MJ/kg)	33–38	44.4
H/C ratio molar	1.50–1.63	1.8
References	[23]	[46]

Table 4.4 Physico-chemical properties of bio-crude derived from microalgae with normal ('standard algae') and high ('high-lipid algae') lipid content

Parameter	Standard algae	High-lipid algae
C (%)	81	78
H (%)	9.9	11.2
O (%)	2.6	10.1
N (%)	5.7	0.5
S (ppm)	0.4	0.1
Moisture (%)	4.8	1.4
Density (g/L)	0.97	0.91
Viscosity (cSt)	178	40
TAN (gKOH/L)	60	180

Data from Albrecht, K.O., et al., Impact of heterotrophically stressed algae for biofuel production via hydrothermal liquefaction and catalytic hydrotreating in continuous-flow reactors. Algal Res 2016;14: 17–27.

hand, it is easier to grow microalgae with high protein content compared with high lipid contents. This is because lipid production is at its maximum when algae cells are stressed that results in lipid production but reduced growth rate. The high-lipid strain is also shown to have a lower viscosity and sulphur content. Sulphur is primarily, just like nitrogen, derived from the protein fraction. Oxygen content in this instance is however higher for the high-lipid strain; this is due to the lipids primarily being present as fatty acids after HTL and not being fully deoxygenated to alkanes.

Bio-crudes are highly complex mixtures of many different organic compounds of varying volatility. This makes them relatively difficult to analyse via conventional GC–MS techniques. The high molecular weights and boiling points of many compounds found in bio-crude mean they are not volatile enough to be detected via GC–MS. Often, as little as 30%–50% of the bio-crude is actually volatile enough to be analysed via GC–MS [47]. Even in this relatively small fraction of crude, there are still a large number of different compounds found. The composition depends largely on the original composition of the feedstock. For example, lipids mainly form free fatty acids under HTL conditions, glycerol, and some alkanes. Protein results in the formation of nitrogen-containing compounds such as pyrroles, indole, and pyrazines. Carbohydrate on the other hand forms oxygenates such as phenols, furans, and cyclopentanones. Biller and Ross showed the formation of such bio-crude components from model compounds and microalgae samples [36]. Cross-reactions amongst biochemical components also occur during HTL, complicating the formation pathways and products found in the bio-crude [29]. Fig. 4.2 shows a typical GC–MS chromatogram of the microalgae strain *Chlorella*; it shows a large number of peaks and a high abundance of oxygen- and nitrogen-containing compounds.

In order to analyse the non-volatile fraction of bio-crude, techniques such as pyrolysis-GC–MS have been performed that employ stepwise micropyrolysis of the bio-crude to break down non-volatile components to smaller volatile fractions, which are amenable to GC–MS analysis. Torri et al. have used this technique extensively to

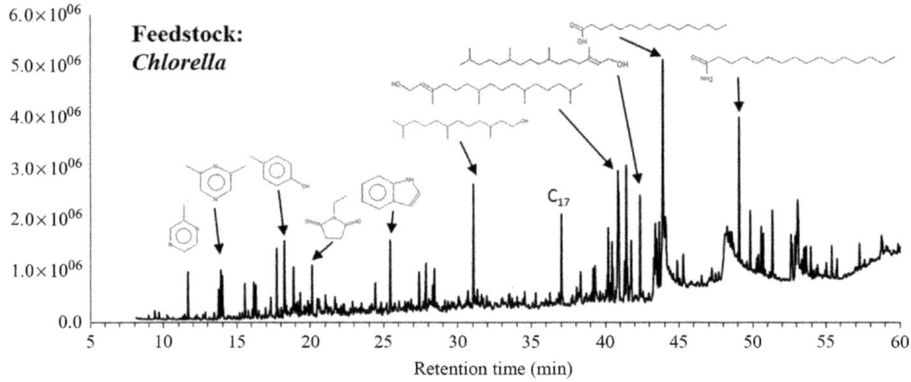

Fig. 4.2 GC–MS chromatogram of *Chlorella* bio-crude.

characterise microalgae bio-crudes. They found that there is a large fraction of bio-crude that is asphaltene-like and of high molecular weight. It was shown that the difference between different algae strains becomes less pronounced at high processing temperatures (375°C) and the composition of bio-crudes is fairly similar. The main compound classes identified include phytols, free fatty acids, fatty amides, alkanes, pyrrolidones, amines, phenols, peptides, methyl indoles, and a large fraction that is still unidentified (~30%–40%) [48]. This exemplifies the difficulties in analysing the entire bio-crude from HTL and further research efforts should be addressed at analysing the full composition of bio-crude via advanced analysis techniques.

4.3.2 Upgraded bio-crude

The overall aim of HTL of microalgae is the production of liquid transportation fuels. Bio-crude itself is not suited to replace conventional gasoline, diesel, or kerosene. The viscosity is too high, and it exhibits undesirably high heteroatom content. Therefore, researchers have concentrated on upgrading the fuel to a fuel with higher quality and reduced heteroatom content. A range of technologies have been employed to upgrade bio-crude and include hydrotreatment, catalytic hydrotreatment, hydrothermal treatment, and distillation. Distillation is the simplest method of upgrading and has been shown to yield fuels with significantly improved quality. Eboibi et al. (2014) produced bio-crude from the two microalgae *Tetraselmis* and *Spirulina*. The bio-crude contained between 8%–17% oxygen and 4%–5% nitrogen. The fuels were vacuum distilled into a residue and a distillate, yielding a distillate that contained significantly less oxygen (0.4%–1.5%) and slightly reduced nitrogen. This was achieved with distilled fuel yields ranging from 63% to 73%, carbon recoveries of ~80%, and energy receives of ~70%–80%. The distillation process further resulted in almost complete removal of metallic constituents in the fuel, which could be a significant advantage if the bio-crude is mixed with petro crude. Despite this promising research of applying the relatively simple technique of distillation with very good results, the majority of research into bio-crude upgrading has been carried out via catalytic hydrotreatment.

This process is largely based on the process of petroleum hydrotreatment and has therefore been performed with typical hydrotreating catalysts such as NiMo and CoMo. This approach has been shown by several researchers to reasonably well on batch reactors; oxygen removal from the bio-crude is typically 60%–80%, but nitrogen is removed less successfully. A recent review summarises recent advances in heterogeneous catalytic upgrading of microalgae bio-crudes [49]. The main issue that becomes apparent when consulting recent literature on upgrading of bio-crude is that there have not been sufficient studies using continuous upgrading in a trickle-bed-type catalytic reactor. Researchers at PNNL are the only ones to have used such a continuous system and were able to demonstrate superior results to what has been achieved on batch reactors. This mostly has to do with the increased catalytic mass transfer on such as system. The PNNL laboratories processed at least six different microalgae strains on their continuous HTL reactor and subsequently performed catalytic hydrotreatment using a Co-promoted, sulphided catalyst to upgrade the bio-crude in a continuous reactor. For each upgraded bio-crude, the nitrogen content was below 0.25%; this is something that is rarely achieved using batch reactors where the nitrogen content after upgrading is commonly around 1.5%. Good deoxygenation was also demonstrated with low residual water and acid numbers in the upgraded bio-oil. Importantly, the yields of upgraded fuel were found to be >80 wt%, higher than most batch studies. Overall carbon recoveries from algae to upgrade fuel ranged from 54% to 85%. The PNNL work highlights the necessity of moving to continuous hydrotreating with improved heteroatom removal and increased oil yields.

The composition of upgraded fuels is primarily alkanes and some low-molecular-weight aromatic hydrocarbons as exemplified in Fig. 4.3, which shows a chromatogram of hydrotreated bio-crude from *Chlorella*. Carbon chain lengths range from C8 to C20 with the largest abundance being present in the diesel and kerosene range (C12–C20) [47]. Upgraded fuels after hydrotreating are very promising as a future renewable fuel substitute; heteroatom content is sufficiently low for coprocessing with petroleum crude or distillation into fuel cuts as renewable drop-in gasoline and diesel blendstocks. Future research should focus on catalyst stability on continuous reactor systems to fully validate the suitability of such fuels in existing infrastructure.

4.3.3 Water phase

The water phase is a very important aspect to consider in a HTL system. During HTL, a large fraction of the organic matter is fractionated to the water phase. It contains approximately 30%–60% of the starting feedstocks carbon and over 50% of the nitrogen. In order to operate more efficiently in terms of carbon and cost, the use of this water phase is important. There have been several suggestions on how to best utilise the HTL process water such as the following:

- Recycling water phase nutrients for algae growth
- Catalytic hydrothermal gasification
- Anaerobic digestion
- Heterotrophic microbial growth
- Conversion to valuable chemicals

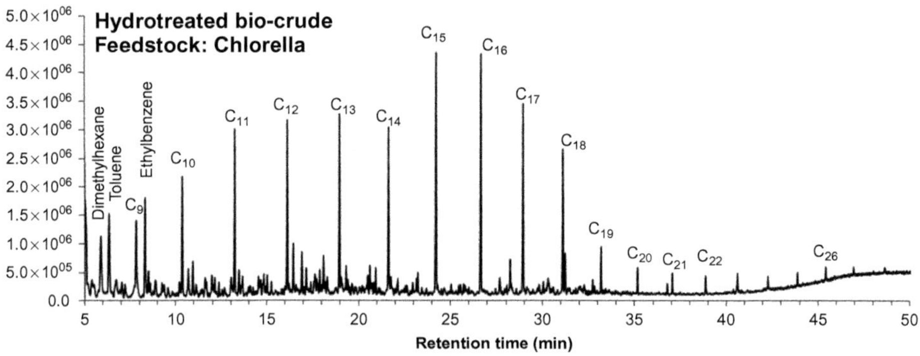

Fig. 4.3 Chromatogram of a microalgae (*Chlorella*) catalytically hydrotreated bio-crude.

In terms of overall elemental distribution from the feedstock to the water phase, it contains significant amounts of the nitrogen, all the potassium, and calcium and varying amounts of phosphorus and sulphur [50]. Nitrogen is typically found with a selectivity of around 50% in the aqueous phase, carbon 15%–45%, and phosphorus 40%–75% [23,50]. However, these numbers vary from system to system and are affected by operating conditions.

Researchers at PNNL used a combination of different analysis techniques for the quantitative characterisation of the aqueous fraction from four salt water and four freshwater microalgae strains. They found the weight percent of carbon in the aqueous phase to lie around 1.5%–3.5%, representing 10%–44% of the feedstock carbon. An additional 0.5 wt% inorganic carbon was also measured. Total nitrogen content ranged from 0.8 to 1.6 wt%. With their analysis techniques, they were able to identify and quantify 45%–72% of the organic carbon and 41%–85% of the organic nitrogen. Ammonia was the single compound measured in the highest concentration of up to 0.9 wt%. Acetic acid was the most prevalent organic compound with up to 0.7 wt% with propanoic and butanoic acid also present in high concentrations. They also showed the presence of alcohols (e.g. ethanol and butanone), ketones (e.g. acetone), pyridine, pyrazine, acetamides, and many more [51]. The PNNL analysis of the water phase is currently seen as the state of the art in the field but still leaves a large fraction of the nitrogen and carbon species unidentified. In a different study, the advanced analysis technique FT-ICR-MS, a mass spectral technique with extremely high mass accuracy, was applied to the organic residue of HTL algal process water. Over 6000 unique peaks were observed, highlighting the complexity of HTL process waters [52]. A deeper understanding of the HTL process water could be beneficial in order to best assess its processing options.

The recycling of nutrients from the process water from hydrothermal processing has been investigated in several studies and proposed as an advantage of hydrothermal routes over other alternative routes such as biodiesel production. Macroalgae is usually cultured in marine environments where nutrient recycling would be more challenging. Significant amounts of the feed nitrogen, phosphorous, and potassium have been shown to concentrate in the process water after hydrothermal processing.

Minowa and Sawayama were the first to recognise this potential and attempted to cultivate microalgae in the process water from the HTG of *Chlorella vulgaris* [53].

Jena et al. performed growth trials using the process water from the HTL of *Spirulina* to grow a strain of *Chlorella* [54]. It was shown that growth was possible in dilutions of the process water. When using a dilution factor of 10, no growth occurred, and this was attributed to the presence of growth inhibition, possibly by nickel, phenols, or fatty acids. These have all previously been shown to adversely affect algae growth [55–58] and are common constituents of the water phase. Highest growth was observed in a 500-fold dilution, reaching a maximum of around 80% compared with a standard growth medium. A mass balance on an integrated HTL system with nutrient recycling was performed, and the results suggest that from 1 t of dry biomass, 0.4 t of bio-crude could be produced. The process water would contain 3.4 kg of P and 70 kg of N and significant amounts of miner nutrients [54].

One issue of recycling nutrients in an algae HTL context could be the progressive build-up of compounds toxic to algae growth. When using a continuous closed-loop system, integrating nutrient recycling, the process water becomes enriched in contaminants such as nickel. It has been suggested that if nutrient recycling is incorporated into a hydrothermal system, nickel concentrations in the effluent need to be monitored closely, and if required, the stream should be diluted or contaminants removed [56]. Biller et al. performed nutrient recycling research on four different microalgae strains. Each strain is able to use the recycled water as a nutrient source for growth. Different dilutions of the process water were necessary due to growth-inhibiting compounds such as phenols and nickel. Additionally, this research showed that some strains were able to use some of the organic carbon in the water phase to grow mixotrophically. During sunlight hours, the algae used inorganic CO_2 for photosynthesis and grew heterotrophically during dark hours using organic carbon sources [59].

The potential of recycling of nutrients to cultivate algae is significant due the cost associated with supplying large amounts of nutrients. The fossil energy input for the production of growth nutrients is significant and reduces the life-cycle energy balance. This is particularly true for phosphorous that is a finite non-renewable resource with a high carbon footprint. The limited studies on recycling of nutrients for algae cultivation are promising, but additional research needs to be performed to identify problems occurring with progressive build-up of growth inhibitors. The studies investigating nutrient recycling to date have also only really been performed at lab scale, and full-scale validation would be required to make a bigger impact. Nutrient recycling really is a necessity when trying to make HTL of microalgae commercially viable. However, there is the remaining issue of the high carbon content in the water phase. If this is not utilised efficiently, the overall carbon efficiency of the system will be lower. Heterotrophic growth can be one solution. Another solution, investigated primarily by PNNL, is the use of catalytic hydrothermal gasification of the water phase to produce methane. This would still result in most of the nitrogen being available for algae growth in the form of ammonium and can utilise the organic carbon in the water phase efficiently. Their studies have shown a COD reduction of 98%–99% and a methane content of around 60 vol% in the produced gas [23]. The majority of phosphorus was found in the solid residue for their work and it was suggested that this could be recycled to algal growth.

An alternative is the production of hydrogen from the water-soluble organics via supercritical water gasification. Cherad et al. demonstrated the supercritical water gasification of *Chlorella* HTL process water and achieved near-complete gasification of organics (98%). 0.3 g of H_2 was produced per gram of bio-crude produced, which is more than required for complete hydrogenation of the bio-crude [60]. Therefore, this presents an alternative to CHG where methane is produced, for on-site power generation, with the aim of producing hydrogen for hydrotreating of the bio-crude.

The issue of phototrophic algae not being able to utilise the organic carbon in the water phase post HTL has further been addressed by an alternative to microalgae cultivation. Limited studies have looked at the use of purely heterotrophic organisms such as yeast, bacteria, and fungi. The idea is that these can utilise the organic carbon more efficiently whilst also recycling other nutrients such as nitrogen. The produced biomass could then be reused for bio-crude production via HTL, thereby improving the overall carbon efficiency of the system. It is however unlikely that microalgae cultivation would then be economically feasible if all the nutrients have been used for heterotrophic bacterial growth. It is however viable to have a purely bacterial system not using algae. There are some distinct advantages of this route such as much higher cell growth densities and light not being a limiting factor so that cultivation can be performed in large tanks rather than shallow ponds or thin tubes. Similar to microalgae such bacteria are also able to produce lipids and can have significantly higher growth rates [61]. Nelson et al. showed that *E. coli* and *P. putida* were able to grow in less diluted HTL water phases compared with microalgae and achieved a maximum aqueous carbon reduction of 40% after 48 h [62]. He et al. were able to grow the heterotrophic bacteria *Rhodococcus* in algal HTL process waters with a maximum COD reduction of 93% after 11 days. Simultaneously, the grown biomass had a high lipid content of around 40% that resulted in a lipid production of around 0.4 g/L. This could then potentially be fed back to the HTL reaction for increased bio-crude production whilst having produced water with significantly less COD [63].

An alternative to biological conversion of the water-soluble carbon via heterotrophic bacteria for biomass or lipid production and thermochemical processing (CHG and SCWG) is the aerobic microbial utilisation for methane production, that is, anaerobic digestion. This is an established technology for sludge and manure treatment and biogas production. Its application for HTL waste waters is very limited, but one study demonstrated some promising results. This technology could potentially convert some of the carbon in the water to methane for on-site power generation whilst still yielding a water phase with residual nitrogen, potentially as a nutrient recycle source. Tommaso et al. used HTL process water from mixed algal culture and performed biomethane potential tests [64]. It was found that some compounds in the water such as cyclic hydrocarbons and phenols induced a lag phase to the anaerobic digestion process. However, after adaption issues, an anaerobic biodegradability of up to 84% was achieved at the highest HTL temperature of 320°C. Relative methane production was found to be 89% using this HTL water, and the overall COD removal rate was 61%. Despite the initial lag phase, these results are promising for recovery of some of the organic carbon in the water phase especially as AD is an established technology.

Summarising the water phase characterisation and utilisation research, it is apparent that the HTL water phase is highly complex and compositionally not fully understood. It contains significant amounts of feedstock carbon, nitrogen, and phosphorus. Nutrient recycling for algal growth is most likely a necessity to make algal HTL biofuels commercially viable. However, the build-up of inhibitory compounds could pose a problem during prolonged application. A variety of technologies for energy recovery from organic carbon in the water phase have been investigated and include anaerobic digestion, CHG, and SCWG. The best option for dealing with the HTL water phase from aquatic biomass is currently not apparent, and future research should focus on demonstrating technologies at large scale.

4.4 Outlook and economic and environmental assessment

Table 4.5 shows a summary of some recent techno-economic assessment (TEA) studies on the HTL technology using aquatic feedstocks. Unfortunately, not all studies use the same basis, units, conversion yields, feedstock, and capital costs, but nevertheless, it provides an estimate of the minimum fuel selling price (MFSP) of upgraded HTL fuels. In most studies, this represents a combination of gasoline, diesel, and aviation fuel. As a comparison, the conventional market prices of these were quoted in Ref. [67] as 0.34, 0.36, and 0.38 €/L, respectively, as of Dec. 2014. Table 4.5 shows the cost summary from three studies using different feedstocks including two microalgae, wet waste from the milk industry, and yeast. The dairy effluent was used as a feedstock for growth of oleaginous microbial biomass with subsequent HTL rather than a feedstock directly. Therefore, the datapoint is included as it is strictly speaking an aquatic biomass feedstock and there is generally a lack of published TEA studies in the literature. PNNL published their techno-economic analysis for algal biomass based on experimental results from their continuous HTL and upgrading system [13] [68]. The primary cost driver was determined to be the algae feedstock cost. The studies concluded that for the projected target case, $4.49 per gallon of gasoline equivalent (GGE) was the minimum fuel selling price. This amounts to a conversion only cost (excluding all feedstock-related costs) of $1.18/GGE. A similar study by Ou et al. (2015) evaluated

Table 4.5 **Summary of HTL TEA studies**

Feedstock	Cost $/GGE	Cost €/L[a]	Study
Algae	1.2–6.0	0.28–1.43	Albrecht et al. [40]
Oleaginous yeast	5.1	1.33	Jena et al. [65]
Dairy effluent/microbes	4.8	1.14	Summers et al. [41]
Defatted algae	2.6	0.62	Ou et al. [66]

[a] Converted from publications using 3.8 L/gal and $1.1/€.

the techno-economic analysis of transportation fuels via HTL of defatted microalgae [66]. Similar conclusions were drawn in regard to the MFSP that was calculated as \$2.6/GGE. A sensitivity analysis revealed that the price is most sensitive to the HTL product yield, highlighting the importance of conversion performance. The feedstock cost was also a major uncertainty towards the MFSP that was also highlighted by the study at the PNNL.

All studies agree that the cost of feedstock is a major if not the largest contributor to the cost of the final fuel product. The conversion cost of HTL including upgrading has been quoted as \$0.7–1.2/GGE [40,67] and the feedstock cost contributing as much as half of the total MFSP. The overall MFSP values range from a low value of 0.32 €/L when feedstock is assumed to be free, to a high value of 1.33 €/L when a microbial yeast platform is considered. In another study, the HTL conversion and fuel upgrading process was quoted as 23% of the overall cost with the remaining being the biomass production cost [65]. The main cost driver for a microalgae-to-biofuel production scenario using aquatic feedstocks is clearly the microalgae cost. It has been estimated at anywhere from \$0 to 1200/t, but some current estimations report values as high as 3200 €/t [69]. If this type of technology is ever going to be economically viable, the feedstock production cost should be lowered significantly. An alternative would be the use of macroalgae; these have been estimated to cost around \$100–200/t [70].

The greenhouse gas emissions from microalgae biofuels have been assessed in various studies and are summarised in Fig. 4.4. These values all take into account the greenhouse gas (GHG) emissions associated with cultivating and harvesting of microalgae, which is a highly energy-intensive process. Therefore, these values may overestimate the GHG emissions from aquatic feedstocks such as macroalgae or duckweed, but unfortunately, no LCA studies have been published on such feedstocks. For example, Liu et al. study estimated the conversion GHG allocation to be below half of the total GHG contribution and Fortier et al. around half [71,72]. The summarising figure shows a wide spread of different GHG values and includes all three fuel types (gasoline, diesel, and kerosene) produced via HTL. The value of current GHG emissions of low-sulphur diesel is also shown as 93 g CO_{2e}/MJ. The values for

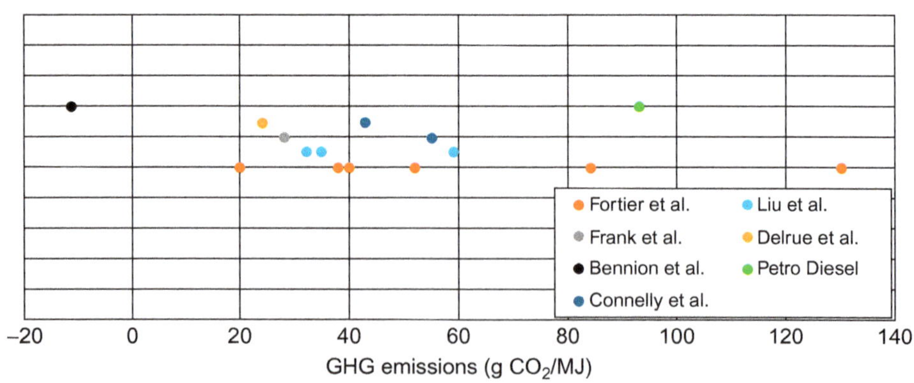

Fig. 4.4 GHG emission summary of microalgae-derived HTL fuels from different studies [71–76] (petroleum diesel for comparison at 93 g CO_{2e}/MJ).

microalgae HTL fuels range from 20 to 130 g CO_{2e}. The higher values, around 60 g, only represent a minor improvement to the petroleum-derived alternative, whilst the more optimistic values at around 20 g represent a reduction of over 80%. One study even finds a negative CO_2 value, but this is unrealistic with the current state of technology. There appears to be a clustering at around 40 g CO_{2e} that can therefore be seen as the most likely actual emission assessment. It has to be considered that these are only for microalgae and include the emissions associated with algae cultivation. It can therefore be estimated that these values can be reduced by around half to obtain GHG emissions for HTL conversion only and this could then be added to the GHG emissions associated with the transport and pretreatment of other aquatic feedstocks such as macroalgae. The GHG values in Fig. 4.4 are promising when compared with other technologies and give rise to optimism for future assessments using other feedstocks. A life-cycle analysis (LCA) was published based on HTL of microalgae to green diesel using data from a continuous pilot-scale plant. Three scenarios are analysed, namely, lab-scale, pilot-scale, and full-scale systems based on data and expertise from Sapphire Energy Inc. (the United States) [71]. The study concluded that pilot-scale facilities have life-cycle burdens on par with conventional biofuels. However, the results from extrapolating to full-scale facilities were more favourable; greenhouse gas emissions were lower compared with petroleum fuels and corn ethanol. The energy return on investment (EROI) was found to be between 1 and 3 with full-scale production facilities having an EROI of around 2.7. These EROI results are favourable compared with all conventional biofuels but not as high as petroleum-derived fuels.

Commercial applications of the algae HTL technology exist, but currently, limited information of capacities, reactor design, and operating conditions are available. Sapphire Energy Inc. (the United States) and Muradel Pty Ltd. (Australia) are examples of companies working on liquid fuel production using HTL of microalgae. Muradel currently has a continuous HTL facility in operation with a process capacity of 3 t of feedstock (30% solids) per day and is planning to increase the capacity to 6.5 t per day with a total output of 50 million litres per year by 2020. At the same time, Muradel is shifting their focus from microalgae to biosolids as a feedstock, due to the prohibitively high cost of microalgae cultivation and higher market value of algal lipids for the cosmetic market. Sapphire Energy was planning to construct a plant to produce ~200 million litres of bio-crude per annum from 2018 [77]. As of 2015, however, the company announced a shift in market emphasis to nutraceutical products and animal/aquaculture feed [78]. Algenol is a further company trying to commercialise algae biofuels. Their technology is based on ethanol-producing cyanobacteria. After periodic harvesting, the residual biomass is converted to bio-crude via HTL to add a second fuel revenue stream. The company claims that their process can yield ~75,000 L of fuel per year per hectare, 87% as ethanol and the remainder as upgraded HTL biofuels [78].

Despite the algae HTL technology being proven at lab and pilot scale, with good fuel yields and efficiencies, the technology is struggling to be commercialised. The major barrier for its breakthrough is the cultivation cost of microalgae biomass. It is therefore more likely that other wet feedstocks such as macroalgae, sewage sludge, or low-value lignocellulosics can pave the way towards the commercialisation of HTL biofuels.

References

[1] Pienkos PT, Darzins A. The promise and challenges of microalgal-derived biofuels. Biofuels Bioprod Biorefin 2009;3(4):431–40.

[2] Brennan L, Owende P. Biofuels from microalgae—a review of technologies for production, processing, and extractions of biofuels and co-products. Renew Sust Energ Rev 2009;14(2):557–77.

[3] Elliott, D.C., et al., Hydrothermal liquefaction of biomass: developments from batch to continuous process. Bioresour Technol, 2015. 178(0): p. 147–156.

[4] Duan P, et al. Hydrothermal processing of duckweed: effect of reaction conditions on product distribution and composition. Bioresour Technol 2013;135:710–9.

[5] Yan W-H, et al. Composition of the bio-oil from the hydrothermal liquefaction of duckweed and the influence of the extraction solvents. Fuel 2016;185:229–35.

[6] Wijffels RH, Barbosa MJ, Eppink MHM. Microalgae for the production of bulk chemicals and biofuels. Biofuels Bioprod Biorefin 2010;4(3):287–95.

[7] Demirbas, A. and M. Fatih Demirbas, Importance of algae oil as a source of biodiesel. Energy Convers Manag 52, 163–170 Corrected Proof.

[8] Adams JMM, et al. Seasonal variation in the chemical composition of the bioenergy feedstock Laminaria Digitata for thermochemical conversion. Bioresour Technol 2011;102(1):226–34.

[9] Chisti, Y., Biodiesel from microalgae. Biotechnol Adv. **25** 3: p. 294–306.

[10] Brennan L, et al. Phytochemicals from Algae. In: Biorefinery co-products. John Wiley & Sons, Ltd; 2012. p. 199–240.

[11] Spolaore P, et al. Commercial applications of microalgae. J Biosci Bioeng 2006;101(2):87–96.

[12] Biller P, Ross AB. Hydrothermal processing of algal biomass for the production of biofuels and chemicals. Biofuels 2012;3(5):603–23.

[13] Zhu Y, et al. Development of hydrothermal liquefaction and upgrading technologies for lipid-extracted algae conversion to liquid fuels. Algal Res 2013;2(4):455–64.

[14] Pirwitz K, Rihko-Struckmann L, Sundmacher K. Valorization of the aqueous phase obtained from hydrothermally treated Dunaliella Salina remnant biomass. Bioresour Technol 2016;219:64–71.

[15] Grobbelaar JU. Algal nutrition—mineral nutrition. In: Amos R, editor. Handbook of microalgal culture; 2007. p. 95–115.

[16] Pittman JK, Dean AP, Osundeko O. The potential of sustainable algal biofuel production using wastewater resources. Bioresour Technol 2011;102(1):17–25.

[17] Borowitzka MA. Commercial production of microalgae: ponds, tanks, tubes and fermenters. J Biotechnol 1999;70(1–3):313–21.

[18] Jones, S.B., et al., Whole algae hydrothermal liquefaction: 2014 state of technology 2014.

[19] Miao X, Wu Q. Biodiesel production from heterotrophic microalgal oil. Bioresour Technol 2006;97(6):841–6.

[20] Meng X, et al. Biodiesel production from oleaginous microorganisms. Renew Energy 2009;34(1):1–5.

[21] Xu L, et al. Assessment of a dry and a wet route for the production of biofuels from microalgae: energy balance analysis. Bioresour Technol 2011;102(8):5113–22.

[22] Biller P, et al. Effect of hydrothermal liquefaction aqueous phase recycling on bio-crude yields and composition. Bioresour Technol 2016;220:190–9.

[23] Elliott DC, et al. Process development for hydrothermal liquefaction of algae feedstocks in a continuous-flow reactor. Algal Res 2013;2(4):445–54.
[24] Cherad R, et al. Hydrogen production from the catalytic hydrothermal gasification of process water from microalgae liquefaction. Fuel 2015;166:24–8.
[25] Xu D, Savage PE. Characterization of biocrudes recovered with and without solvent after hydrothermal liquefaction of algae. Algal Res 2014;**6**(Part A(0)):1–7.
[26] Brown TM, Duan P, Savage PE. Hydrothermal liquefaction and gasification of Nannochloropsis sp. Energy Fuel 2010;24(6):3639–46.
[27] Jena U, Das KC, Kastner JR. Effect of operating conditions of thermochemical liquefaction on biocrude production from Spirulina Platensis. Bioresour Technol 2011;102(10):6221–9.
[28] Garcia Alba L, et al. Hydrothermal treatment (HTT) of microalgae: evaluation of the process as conversion method in an algae biorefinery concept. Energy Fuel 2012;26(1):642–657.
[29] Torri C, et al. Hydrothermal treatment (HTT) of microalgae: detailed molecular characterization of HTT oil in view of HTT mechanism elucidation. Energy Fuel 2012;26(1):658–671.
[30] Faeth JL, Valdez PJ, Savage PE. Fast hydrothermal liquefaction of Nannochloropsis sp *To Produce Biocrude*. Energy Fuel 2013;27(3):1391–8.
[31] Faeth JL, Savage PE. Effects of processing conditions on biocrude yields from fast hydrothermal liquefaction of microalgae. Bioresour Technol 2016;206:290–3.
[32] Hietala DC, Faeth JL, Savage PE. A quantitative kinetic model for the fast and isothermal hydrothermal liquefaction of Nannochloropsis sp. Bioresour Technol 2016;214:102–11.
[33] Jazrawi C, et al. Pilot plant testing of continuous hydrothermal liquefaction of microalgae. Algal Res 2013;2(3):268–77.
[34] Biller P, et al. Hydroprocessing of bio-crude from continuous hydrothermal liquefaction of microalgae. Fuel 2015;159.
[35] Zhu Y, et al. Techno-economic analysis of liquid fuel production from woody biomass via hydrothermal liquefaction (HTL) and upgrading. Appl Energy 2014;129:384–94.
[36] Biller P, Ross AB. Potential yields and properties of oil from the hydrothermal liquefaction of microalgae with different biochemical content. Bioresour Technol 2011;102(1):215–25.
[37] Leow S, et al. Prediction of microalgae hydrothermal liquefaction products from feedstock biochemical composition. Green Chem 2015;17(6):3584–99.
[38] Teri G, Luo L, Savage PE. Hydrothermal treatment of protein, polysaccharide, and lipids alone and in mixtures. Energy Fuel 2014;28(12):7501–9.
[39] López Barreiro D, et al. Influence of strain-specific parameters on hydrothermal liquefaction of microalgae. Bioresour Technol 2013;146:463–71.
[40] Albrecht KO, et al. Impact of heterotrophically stressed algae for biofuel production via hydrothermal liquefaction and catalytic hydrotreating in continuous-flow reactors. Algal Res 2016;14:17–27.
[41] Elliott DC, et al. Hydrothermal processing of macroalgal Feedstocks in continuous-flow reactors. ACS Sustain Chem Eng 2013;2(2):207–15.
[42] Xiu SN, et al. Oil production from duckweed by thermochemical liquefaction. Energy sourc A Recovery Util Environ Effects 2010;32(14):1293–300.
[43] Anastasakis K, Ross AB. Hydrothermal liquefaction of the brown macro-alga Laminaria Saccharina: Effect of reaction conditions on product distribution and composition. Bioresour Technol 2011;102(7):4876–83.
[44] Savage PE. A perspective on catalysis in sub- and supercritical water. J Supercrit Fluids 2009;47(3):407–14.
[45] Duan P, Savage PE. Hydrothermal liquefaction of a microalga with heterogeneous catalysts. Ind Eng Chem Res 2010;50:52–61.

[46] Jensen CU, Hoffmann J, Rosendahl LA. Co-processing potential of HTL bio-crude at petroleum refineries. Part 2: a parametric hydrotreating study. Fuel 2016;165:536–43.
[47] Biller P, et al. Hydroprocessing of bio-crude from continuous hydrothermal liquefaction of microalgae. Fuel 2015;159:197–205.
[48] Torri C, et al. Fast procedure for the analysis of hydrothermal liquefaction biocrude with stepwise Py-GC-MS and data interpretation assisted by means of non-negative matrix factorization. Energy Fuel 2016;30:1135–44.
[49] López Barreiro D, et al. Heterogeneous catalytic upgrading of biocrude oil produced by hydrothermal liquefaction of microalgae: State of the art and own experiments. Fuel Process Technol 2016;148:117–27.
[50] Christensen PS, et al. Hydrothermal liquefaction of the microalgae Phaeodactylum tricornutum: impact of reaction conditions on product and elemental distribution. Energy Fuel 2014;28(9):5792–803.
[51] Maddi B, et al. Quantitative characterization of the aqueous fraction from hydrothermal liquefaction of algae. Biomass Bioenergy 2016;93:122–30.
[52] Sudasinghe N, et al. High resolution FT-ICR mass spectral analysis of bio-oil and residual water soluble organics produced by hydrothermal liquefaction of the marine microalga Nannochloropsis salina. Fuel 2014;119:47–56.
[53] Minowa T, Sawayama S. A novel microalgal system for energy production with nitrogen cycling. Fuel 1999;78(10):1213–5.
[54] Jena U, et al. Evaluation of microalgae cultivation using recovered aqueous co-product from thermochemical liquefaction of algal biomass. Bioresour Technol 2011;102(3):3380–7.
[55] Scragg AH. The effect of phenol on the growth of Chlorella vulgaris and Chlorella VT-1. Enzym Microb Technol 2006;39(4):796–9.
[56] Haiduc A, et al. SunCHem: An integrated process for the hydrothermal production of methane from microalgae and CO_2 mitigation. J Appl Phycol 2009;21(5):529–41.
[57] Spencer DF, Nichols LH. Free nickel ion inhibits growth of two species of green algae. Environ Pollut Ser A, Ecol Biol 1983;31(2):97–104.
[58] Bosma R, et al. Growth inhibition of Monodus subterraneus by free fatty acids. Biotechnol Bioeng 2008;101(5):1108–14.
[59] Biller P, et al. Nutrient recycling of aqueous phase for microalgae cultivation from the hydrothermal liquefaction process. Algal Res 2012;1(1):70–6.
[60] Cherad R, et al. Hydrogen production from the catalytic supercritical water gasification of process water generated from hydrothermal liquefaction of microalgae. Fuel 2016;166:24–8.
[61] Kosa M, Ragauskas AJ. Lipids from heterotrophic microbes: advances in metabolism research. Trends Biotechnol 2011;29(2):53–61.
[62] Nelson M, et al. Microbial utilization of aqueous co-products from hydrothermal liquefaction of microalgae Nannochloropsis oculata. Bioresour Technol 2013;136:522–8.
[63] He, Y., et al., Biological conversion of the aqueous wastes from hydrothermal liquefaction of algae and pine wood by Rhodococci. Bioresour Technol.
[64] Tommaso G, et al. Chemical characterization and anaerobic biodegradability of hydrothermal liquefaction aqueous products from mixed-culture wastewater algae. Bioresour Technol 2015;178:139–46.
[65] Jena U, et al. Oleaginous yeast platform for producing biofuels via co-solvent hydrothermal liquefaction. Biotechnol Biofuels 2015;8(1):1–19.
[66] Ou L, et al. Techno-economic analysis of transportation fuels from defatted microalgae via hydrothermal liquefaction and hydroprocessing. Biomass Bioenergy 2015;72:45–54.

[67] de Jong S, et al. The feasibility of short-term production strategies for renewable jet fuels—a comprehensive techno-economic comparison. Biofuels Bioprod Biorefin 2015;9(6):778–800.
[68] Jones, S.B., et al., Process design and economics for the conversion of algal biomass to hydrocarbons: whole algae hydrothermal liquefaction and upgrading.2014 p. Medium: ED; Size: PDFN.
[69] Tredici MR, et al. Techno-economic analysis of microalgal biomass production in a 1-ha Green Wall panel (GWP®) plant. Algal Res 2016;19:253–63.
[70] Konda NVSNM, et al. An investigation on the economic feasibility of macroalgae as a potential feedstock for biorefineries. Bioenerg Res 2015;8(3):1046–56.
[71] Liu X, et al. Pilot-scale data provide enhanced estimates of the life cycle energy and emissions profile of algae biofuels produced via hydrothermal liquefaction. Bioresour Technol 2013;148:163–71.
[72] Fortier M-OP, et al. Life cycle assessment of bio-jet fuel from hydrothermal liquefaction of microalgae. Appl Energy 2014;122:73–82.
[73] Connelly EB, et al. Life cycle assessment of biofuels from algae hydrothermal liquefaction: the upstream and downstream factors affecting regulatory compliance. Energy Fuel 2015;29(3):1653–61.
[74] Frank E, et al. Life cycle comparison of hydrothermal liquefaction and lipid extraction pathways to renewable diesel from algae. Mitig Adapt Strateg Glob Chang 2012;1–22.
[75] Bennion EP, et al. Lifecycle assessment of microalgae to biofuel: comparison of thermochemical processing pathways. Appl Energy 2015;154:1062–71.
[76] Delrue F, et al. Comparison of various microalgae liquid biofuel production pathways based on energetic, economic and environmental criteria. Bioresour Technol 2013;136:205–12.
[77] U.S. Department of Energy, Sapphire energy integrated algal biorefinery (IABR). DOE/EE-0841, 2013.
[78] Elliott DC. Review of recent reports on process technology for thermochemical conversion of whole algae to liquid fuels. Algal Res 2016;13:255–63.

Hydrothermal liquefaction of protein-containing feedstocks

5

Y. Zhang, W.-T. Chen
University of Illinois at Urbana-Champaign, Urbana, IL, United States

Hydrothermal liquefaction (HTL), also refers to as hydrous pyrolysis, is a thermochemical depolymerisation process in an enclosed reactor to convert wet biomass into biocrude oil and chemicals at moderate temperature (typically 200–400°C) and high pressure (typically 10–25 MPa). Protein-containing feedstocks such as food processing waste, manure, and municipal sludge typically have high moisture content and thus make HTL an appropriate process. In HTL processes, water serves as an important reactant. As the reaction condition approaches to the critical point of water, several properties of water are drastically changed and able to bring about fast, homogeneous, and efficient reactions [1–3]. Because subcritical water has a leading role as a heat transfer and extracting medium, HTL is relatively independent of the size of biomass particles or heating rates. The product yield and physiochemical properties of an HTL are primarily affected by the types of feedstock, processing conditions (primarily reaction temperature and time), and existence of catalyst. HTL biocrude oil from protein-containing feedstocks typically contains high nitrogen content and many types of aromatics. Upgrading of the HTL biocrude is necessary for transportation grade fuel. Posthydrothermal wastewater (PHWW) typically has a very high chemical oxygen demand (COD) and various nutrients (primarily nitrogen) that need to be treated before discharge to waterways. Gaseous products of HTL contain malodours and thus must be treated before emitted to the atmosphere. In this chapter, important factors influencing HTL and HTL product characteristics are described.

5.1 Pretreatment of feedstocks

Protein-containing feedstocks, especially wet biowaste, often require pretreatment prior to entering an HTL reactor. There are three general pretreatment principles: physical [4–14], chemical [14–17], and biological [17,18] and a combination of the three. Pretreating HTL feedstocks has three general purposes: (1) extracting value-added products such as polysaccharides and crude lipids [4,5,8–15,17,18], (2) removing non-volatile components such as ash and metals [6,16], and (3) achieving reasonable properties such as solid contents suitable for an energy-efficient HTL conversion [7]. Table 5.1 summarises different pretreatment methods for HTL of protein-containing feedstocks.

Direct Thermochemical Liquefaction for Energy Applications. https://doi.org/10.1016/B978-0-08-101029-7.00004-7

Table 5.1 **Summary of pretreatment methods for HTL feedstocks that contain protein contents**

Category	Feedstock	Pretreatment mechanism	Value-added products during pretreatment	Benefit	Limitation
Physical pretreatment					
Maio et al. [4]	Yeast (*C. curvatus*)	Low-temperature liquefaction (160–300°C)	Polysaccharides and proteins derivatives	Produce value-added chemical and biocrude oil with lower N	Multistep processes, potential loss of organic matter
Eboibi et al. [5]	Microalgae (*Tetraselmis* sp.)	Low-temperature liquefaction (130–200°C)	Protein derivatives	Produce biocrude oil with improved yields and lower N	Multistep processes, organic carbon lost to the pretreatment process water, high energy output
Costanzo et al. [13]	Microalgae (*Spirulina*, *Nannochloropsis*, *Chlorella*, and *Scenedesmus*)	Low-temperature liquefaction (125–225°C)	Nitrogen-rich nutrient streams	Produce nutrient streams and biocrude oil with lower N	Multistep processes, organic matter lost to the pretreatment process
Chen et al. [6]	Mixed-culture algal biomass	Centrifugation and ultrasonication	N/A	Produce biocrude oil with improved yields and heating value	Multistep processes, require additional energy for pretreatment
Cheng et al. [10]	Microalgae (*Nannochloropsis*)	Microwave irradiation	Biodiesel	Produce biodiesel and biocrude oil simultaneously	Multistep processes, biocrude oil with lower HHV
Biller et al. [8]	Microalgae (*Nannochloropsis*, *Chlorogloeopsis*, *Pseudochoricystis*)	Microwave irradiation	Lipids and phytochemicals	Produce biocrude oil with lower N	Multistep processes, organic matter lost to the pretreatment process water
Miao et al., Chakraborty et al. [9,11,12]	Microalgae (*Chlorella*)	Low-temperature liquefaction (140–200°C)	Polysaccharides	Lower solid residue yield, produce biocrude oil with lower N	Multistep processes, loss of some organic matters

Minarick et al. [7]	Swine manure	Filtration and centrifugation	N/A	Achieve a higher solid content of feedstock, improve the energy recovery of HTL	Require additional energy for pretreatment
Chemical pretreatment					
Jazrawi et al. [14]	Microalgae (*Chlorella*)	Extracted with sulphuric or formic acids	Protein derivatives	Produce biocrude oil with lower N	Unselective removal of nitrogen-containing compounds
Barreiro et al. [17]	Microalgae (*Nannochloropsis* and *Scenedesmus*)	Soxhlet extraction with hexane	Crude lipids	Produce lipids and biocrude oil simultaneously	Produce biocrude oil with higher N, lower yield of biocrude oil
Vardon et al. [15]	Microalgae (*Scenedesmus*)	Soxhlet extraction with hexane	Crude lipids	Produce lipids and biocrude oil simultaneously	Produce biocrude oil with higher N, lower yield of biocrude oil
Leng et al. [16]	Sewage sludge	Extracted with acetic acid, hydroxylammonium chloride, and hydrogen peroxide	Solid residue or biochar-containing metals	Produce biocrude oil and biochar with lower metal concentrations	Required hazardous chemicals for demetalisation, lower biocrude oil yield
Biological pretreatment					
Kaushik et al. [18]	Food waste	Enzymatic hydrolysis (carbohydrase, protease, and lipase)	Hydrochars	Produce hydrochar with higher heating values	Enzymes are relatively expensive, difficult to recycle enzymes
Barreiro et al. [17]	Microalgae (*Nannochloropsis* and *Scenedesmus*)	Enzymatic hydrolysis (endoprotease alcalase and exoprotease flavourzyme)	Amino acid concentrates	Produce biocrude oil with lower N and higher heating values	Loss of algal biomass, unselective extraction of amino acids

Extraction of value-added coproducts from protein-containing feedstocks prior to HTL conversion has been extensively studied for augmenting the overall economics of the process [4,9,11,12]. For example, polysaccharides can be extracted at low liquefaction temperature (140–200°C) and produce biocrude oil and yeast at medium liquefaction temperature (220–300°C) from microalgae [4,9,11,12]. Extraction of crude lipids has also been investigated through microwave and Soxhlet extraction with hexane [8,10,15,17], whilst the protein derivatives (e.g. amino acids and nitrogen-rich nutrient streams) have been extracted with sulphuric or formic acids and enzymatic hydrolysis, at mild liquefaction (125–225°C) [5,13,14,18]. Pretreatment methods are often unselective for extracting polysaccharides and proteins. Ionic liquid extraction could be used for extracting proteins from protein-containing feedstocks with 75%–100% selectivity of proteins in a single-step extraction and reuse of the ionic liquid [19]. Pretreatment generally requires multistep processes to extract different types of value-added products, and the residual can be then subjected to HTL to produce biocrude oil. Such multistep processes often require multireactors, thus complicating the overall process. A continuous HTL reactor is critically needed for simultaneously obtaining value-added chemicals (at lower liquefaction temperature) and biocrude oil (at higher liquefaction temperature). In such a continuous reactor, the loss of organic matter during the pretreatment processes could be prevented. Further biochemical characterisations of value-added products are also significantly needed for different application potentials. For instance, the protein derivatives from algal biomass could be used as pigments [20], building blocks for polymer synthesis [21], and food additives [22].

Removing non-volatile components from protein-containing feedstocks, such as ash contents and metals, is another focus in the feedstock pretreatment. Excessive ash content in mixed-culture algal biomass would reduce the higher heating value (HHV) and hydrocarbon compositions in the HTL biocrude oil [6,23]. To resolve this issue, physical pretreatments were carried out to reduce the ash content (from 28.6% to 18.6%) by centrifugation, resulting in the biocrude oil yield increase from 30 to 55 wt% and improving the HTL biocrude oil quality by centrifugation plus ultrasonication [6]. However, how the ash contents interact with volatile components in the feedstock under HTL processes still remains unclear and requires further investigation to elucidate the role of the ash contents in HTL processes. Chemical pretreatments on protein-containing feedstocks to remove the ash content have also been investigated. Leng et al. [16] extracted metals from sewage sludge with acetic acid, hydroxylammonium chloride, and hydrogen peroxide and then converted the chemically pretreated sewage sludge via HTL. Although the biochar and biocrude oil with lower metal concentrations were produced, the chemical pretreatment led to a relatively lower biocrude oil yield, largely due to the loss of organic matter during multistep pretreatment processes. More selective method for extracting metallic compounds has to be investigated. Adsorbents, such as resins, zeolites, and activated carbon, were used to extract ash contents since these absorbents can be further modified to mitigate specific metallic and mineral elements and recycled through phase separation [24–26].

Achieving reasonable total solid contents for more efficient energy recovery in HTL conversion is another driving force to pretreat protein-containing feedstocks.

Oftentimes, the solid content in the feedstocks is <10 wt% in reality. HTL process requires elevating the total feedstock and water to desirable reaction temperature. Manure slurry from typical swine confinement facilities contains 1–2 wt% solid content [7], but the HTL process requires a solid content higher than 10 wt% in order to obtain a positive energy balance [27–31]. In order to produce biocrude oil with reasonable quality and quantity [7,29,32], the removal of moisture content is essential [15]. Minarick et al. [7] have dewatered swine manure using coarse filtration and centrifugation, thereby improving the energy recovery of HTL processes by 2–2.5 times. Additional costs and energy consumptions for the dewatering pretreatment processes should be incorporated for further upscaled technoeconomic analyses of the HTL process.

5.2 HTL process (yields, CHN & HHV, light oil fraction, functional groups)

Key parameters that affect HTL processes include feedstock types, process modes (such as batch and continuous), process conditions (such as temperature, retention time, and pressure), and catalysts. This section summarises the HTL process of different feedstocks in terms of conversion efficiency and product characteristics.

5.2.1 Feedstock type

Protein-containing feedstock includes biowaste (manure, municipal sludge, and food processing waste) and biomass such as algae and grains, and their biochemical compositions are summarised in Table 5.2 in terms of compounds (crude fat or lipid, protein, carbohydrate, and ash) and major elements (C, H, N, and O). These are the protein-containing feedstocks that have been hydrothermally liquefied in the past 20 years with some promising results in terms of biocrude yields and high heating values. Compositions of food processing wastes can vary widely among different plants such as a slaughterhouse and a cheese plant. Typically, the higher the amounts of crude fat in the feedstocks, the higher the biocrude oil yields [53,55,56], whilst the presence of lignin generally leads to char formation under HTL [57,58]. The highest HTL biocrude oil yield occurred with feedstocks containing high crude fat and nonfibrous carbohydrates at relatively lower temperatures (250–300°C) and shorter retention time (5–30 min) due to the fast decompositions of crude fat and nonfibrous carbohydrates [55,59]. Kabyemela et al. observed a 55% conversion of glucose after 2 s at 300°C [60]. Triacylglycerides (TAGs) were readily hydrolyzed in hot-compressed water without catalysts [59,61]. Feedstocks containing more proteins or fibrous carbohydrates require a higher reaction temperatures (300–350°C) and longer retention time (30–120 min). Sasaki et al. reported that the reaction rate of cellulose begins to accelerate at about 350°C at 25 MPa [59]. Rogalinski et al. [61,62] observed the decomposition of amino acids from bovine serum albumin in subcritical water hydrolysis at 250–330°C and 4–180 s retention time, with an almost complete degradation of all amino acids at 330°C and 200 s reaction time.

In addition to biocrude oil, other value-added chemicals can also be produced from HTL of food processing waste. Yoshida et al. converted fish meat at 240–350°C with a 5–30 min reaction time to produce biocrude oil and aqueous products containing value-added chemicals such as amino acids, lactic acids, and phosphoric acids [34]. Quitain et al. hydrothermally treated shrimp shells at 250–400°C for a 5–60 min reaction time to produce amino acids and glucosamine [33]. Changing World Technologies applied hydrothermal process to convert turkey offal into biocrude and fertilizer [63].

Compared with food processing waste, animal manure contains less crude fat and more crude proteins and carbohydrates. This results in a higher HTL reaction temperature to achieve an optimum biocrude oil yields. HTL of swine manure was reported at 240–350°C at reaction time of 5–180 min in batch reactor system [27–31,35,53]. The HTL biocrude from swine manure was found to have high heating values of 35–39 MJ/kg with 71%–78% carbon, 7%–14% oxygen, 8.9%–9.4% hydrogen, and 3.9%–4.6% nitrogen. Different process gases (e.g. CO, N_2, and CO_2) were also tested, with a CO as a reducing gas that improves the biocrude oil yield and quality. Study showed that the storage time changed the manure biochemical composition but did not affect the HTL biocrude oil yield [36]. Similar to swine manure, human faeces have been demonstrated as a promising HTL feedstock (Table 5.2). An additional benefit of HTL of human faeces is it completely destroys pathogens. However, HTL biocrude converted from swine manure and human faeces contains higher amounts of nitrogen and oxygen compared with petroleum; thus, further upgrading is needed if used for transportation fuel.

Swine manure and crude glycerol were coliquefied under hydrothermal conditions [38]. However, the heating value of the resulting coliquefied HTL biocrude oil was drastically decreased from 36 to 25 MJ/kg due to its dominant esterification and resulting high oxygen content in the biocrude.

HTL of cattle manure was investigated with the presence of NaOH as catalyst and initial process gas (air, N_2, CO, and H_2), for a 500 mL of cattle manure slurry (solid content was 23.6 wt%). The higher initial pressure, longer reaction time, and high solid content of the cattle manure led to the production of gases and char/tar instead of biocrude [39]. Dairy manure was studied as HTL feedstock at 250–350°C and 15 min reaction time with carbon monoxide as the process gas [40]. The highest energy recovery (68%), defined as the energy in the biocrude oil divided by that of raw manure, was achieved at 350°C [40].

Algae is considered as one of the promising HTL feedstocks for next-generation bioenergy because of their superior photosynthetic efficiency and lower cropland demand [64]. Algae can be classified as microalgae, macroalgae, and mixed-culture algae that may contain both micro- and macroalgae (Table 5.2). High-lipid microalgae has been favoured as HTL feedstocks [45]. Dote converted *Botryococcus braunii* into biocrude oil via HTL with biocrude oil yield higher than the original oil content in the algal biomass [65], indicating nonlipid fraction in algae contributed to the biocrude oil formation under HTL [2].

In contrast to high-lipid algae, low-lipid high-protein algae species, such as *Chlorella* and *Spirulina*, typically have higher biomass productivity under stressed environment such as wastewater [66,67]. Therefore, HTL of low-lipid algae has been

Table 5.2 **Compositions of different protein-containing feedstocks for HTL (wt%)**

Category	Crude fat	Crude protein	Carbohydrates[a]	Ash content	C	H	N	Optimum conditions for biocrude oil production	Optimum biocrude oil yields[b]	Optimum HHV of biocrude oil
Food processing waste										
Shrimp shells [33]	N/A	N/A	N/A	N/A	N/A	N/A	N/A	250°C/60 min	N/A	N/A
Fish meat [34]	N/A	N/A	N/A	N/A	58.7	11.1	8.66	300°C/5 min	16	N/A
Swine slaughtering house waste[a]	44.0	24.8	27.0	4.20	50.4	7.18	2.93	280°C/30 min	58	39
Cheese processing waste[a]	52.3	14.8	31.5	1.45	60.7	8.49	3.33	260°C/30 min	69	40
Salad dressing processing waste[a]	40.8	2.76	48.9	7.53	54.0	7.93	0.57	260°C/30 min	55	40
Animal manure										
Swine manure[c] [27–31,35–37]	20.3	24.5	38.9	16.3	41.1	5.42	3.36	295–305°C/ 15–30 min	40	39
Swine manure-14 day (G-F)[d] [36]	16.7	37.9	27.1	18.3	N/A	N/A	N/A	305°C/30 min	40	N/A
Swine manure-21 day (G-F)[d] [36]	17.9	43.3	20.5	18.3	N/A	N/A	N/A	305°C/30 min	38	N/A
Swine manure [38]	4.86	17.1	55.7	22.3	33.5	6.16	2.81	340°C/1 min	24	36
Cattle manure [39]	N/A	N/A	N/A	7.16	35.4	4.73	2.38	310°C/15 min	49	36
Separated dairy manure [40]	N/A	N/A	N/A	16.8	N/A	N/A	N/A	350°C/15 min	N/A	33
Human faeces[a]	24.4	34.5	25.0	16.0	45.5	6.5	5.7	340°C/10–30 min	34	41

Continued

Table 5.2 **Continued**

Category	Crude fat	Crude protein	Carbohydrates[a]	Ash content	C	H	N	Optimum conditions for biocrude oil production	Optimum biocrude oil yields[b]	Optimum HHV of biocrude oil
Algae Microalgae										
Tetraselmis [41]	14	58	22	6	42.0	6.8	8.0	350°C/5 min	65[g]	35–39
Nannochloropsis [42]	28	52	12	8	43.3	6.0	6.4	350°C/60 min	43	39
Dunaliella [43]	2.87	61.3	21.7	13.5	39.0	5.37	1.99		26	
Chlorella [44,45]	0.10	71.3	23.0	5.6	51.4	6.6	11.1	280–300°C/ 60–120 min	39–43	37–39
Spirulina [46]	5.10	64.4	21.0	9.5	49.3	6.4	11.0	280/120 min	45–50	33–37
Diatom [46,47]	5.57	40.0	10.0	32.6	33.7	4.90	5.51	320°C/45 min	33	33
Macroalgae										
Laminaria [48]	N/A	N/A	N/A	24.2	31.3	3.7	2.4	350°C/15 min	19.3	37
Alaria [49]	<0.7	15.0	<50.7	33.6	33.0	4.4	2.5	360°C/15 min	29.4	35
Enteromorpha [50]	N/A	N/A	N/A	30.1	28.8	5.2	3.7	300°C/30 min	20.4	29
Mixed-culture algae										
Mixed-culture algae from wastewater [23]	1.70	27.2	23.6	47.5	27.9	3.0	3.9	300°C/60 min	49.9[e]	33
Wastewater-fed mixed-culture algae [51]	14.0[e]	N/A	N/A	29.0[e]	48.9[e]	7.1[e]	8.4[e]	350°C/60 min	44.5[e]	39

High-ash algae from Dianchi Lake [52]	1.04	13.5	40.0	41.6	23.7	4.5	3.3	300°C/60 min	18.4[e]	37
Sewage sludge										
Sewage sludge from a sanitary plant [46]	0.10	41.6	32.0	26.3	38.0	7.8	7.3	N/A	33–35[f]	N/A
Anaerobic sludge [53]	<1	15.0	54.0	31.0	N/A	N/A	N/A	N/A	9.4	32[g]
Sludge from an urban wastewater treatment plant [54]	N/A	N/A	N/A	N/A	41.7	5.8	3.2	380–400°C/10 min	50–55	40

N/A, not available.
[a] Calculated by difference (i.e. carbohydrate (wt%) = 100 crude fat-protein-ash).
[b] Based on dry weight of feedstock.
[c] Fresh swine manure.
[d] Grower-finisher pig stored for different days.
[e] Based on dry ash-free of feedstock.
[f] HTL was carried out at 280°C for a 30 min reaction time.
[g] HTL was carried out at 300°C for a 30 min reaction time.

an increasing interest in the past 5 years [15,32,41,44,45,68]. HTL parameters including reaction temperatures, reaction time, initial pressure, biomass-to-water ratio, total volume of feedstocks, heating rate, reaction mode, catalysts, and reaction solvents were studied to enhance the algal biocrude oil yield and/or improve its heating value [23,32,42–45,48,68–74]. Among all the HTL parameters investigated, reaction temperature and reaction time are recognised as the two dominant factors affecting the biocrude oil yield and biocrude oil quality (heating value and elemental compositions).

Macroalgae were also investigated as HTL feedstocks due to their fast-growing rates and large quantity [48,50,75–77]. However, macroalgae generally has high contents of lignocellulose and ash; both adversely affect the HTL conversion efficiency [46] (Table 5.1).

Mixed-culture algae from wastewater environment have attracted increasing attention for HTL conversion. Coupling algal bioenergy production with wastewater treatment can alleviate the burden of nutrients in algae growing by uptaking nutrients from wastewaters instead of petroleum-based fertilizers [78,79]. In an environment-enhancing energy paradigm centred on HTL, nutrients in wastewater from post HTL can be reused from 3 (through laboratory study) to 10 times (through modelling) to amplify algae growth and biocrude oil production (Fig. 5.1) [23,31,79]. Wastewater algae was proved to be a promising HTL feedstock in a separate study [51]. In addition to mitigate eutrophication, high-ash low-lipid algal biomass from lakes was harvested for HTL [52]. Municipal sludge is another promising feedstock and presents comparable heating value of the HTL biocrude oil as those from animal manure and algae [46,53]. In the SlurryCarb process, one study reported that a scaled-up (a 20 t/day facility) HTL conversion of sewage sludge could produce a biofuel product up to 9500 Btu/lb. (dry basis) and can be utilised in conventional combustion infrastructure with <20% excess air [80]. The pretreatment processes for the municipal sludge in a wastewater treatment plant can substantially alter the biochemical compositions of the harvested

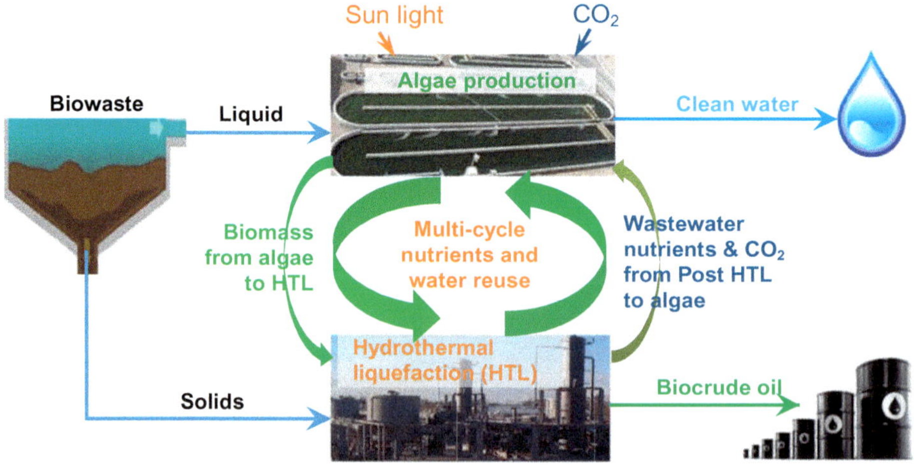

Fig. 5.1 A schematic environment-enhancing energy (E2-energy) system that can simultaneously produce biocrude oil and treat wastewater by reusing nutrients [23].

sludge, which in turn impact the resulting HTL biocrude oil yield and quality. Vardon et al. converted anaerobic sludge from a wastewater treatment plant into biocrude oil via HTL, but the biocrude oil yield was much lower than other types of sludge [46,53]. Systematic studies regarding HTL of different types of sewage sludge and process parameters such as reaction temperature and reaction time are identified as a research gap in the area of HTL of municipal waste [16,81,82].

5.2.2 Process mode

Hydrothermal liquefaction (HTL) of protein-containing biomass has been mostly investigated in a batch mode. There is only limited information yet available on continuous-flow HTL reactor design and experiment, even fewer studies on scale-up reactors [83]. Discussion about the reactor design and scale-up implementation is provided in the latter section (examples of plants/implementations).

Continuous HTL conversion has been investigated with algae and swine manure [70,72,77,84,85]. According to currently available literature, the algal biocrude oil yields with continuous plug-flow mode are usually higher than those with batch modes, whilst the carbon, hydrogen, nitrogen contents, and heating values of the algal biocrude oil are the same [70,72]. This reveals that the continuous plug-flow HTL systems can produce algal biocrude oil with the same quality as those from batch reactors. This may be that the swine manure was hydrothermally processed in a continuous-stirred tank reactor (CSTR), rather than a plug-flow reactor (PFR) [84]. Considering a reactor with the same volume, PFR typically can achieve a higher conversion yields than CSTR. Although CSTR systems could be more easily operated and maintained such as cleaning [86], it bears an inherent disadvantage of mixing mode compared with plug-flow system as part of the feedstock that will always be subjected to undesirable reaction time—some undercooked and some overcooked. Comparing continuous HTL systems with batch HTL systems is difficult as the products from the batch are often equilibrium limited [83].

Although the yields and heating values of the biocrude oil converted from algae and swine manure are similar under continuous modes, the optimum reaction conditions are significantly different for these two feedstocks in a continuous HTL reactor. For algal biomass, it is generally agreed that more severe reaction conditions can lead to higher yields and lower oxygen but increased nitrogen contents in the HTL biocrude [70,72,83]. In addition, very short residence time (e.g. 1–5 min) in a continuous reactor can produce algal biocrude oil yields similar to that of longer residence time under batch modes (e.g. 60–120 min) [70,72,83]. Both recent batch and continuous HTL studies also suggest that higher heating rate and lower residence time favour the production of biocrude. This feature would also help improve the energy efficiency and technoeconomic feasibility of biomass-to-fuel systems via a continuous HTL [71,83].

Continuous HTL of swine manure requires a longer retention time compared with algae feedstocks. Ocfemia et al. have conducted a continuous HTL of swine manure with a CSTR at 285–325°C for 40–80 min. The highest oil yield was achieved at about 285–305°C for a 60 min retention time, which is very similar to those obtained at a batch reactor [27,28]. In this continuous HTL study, it was reported that increasing the reaction temperature from 305°C to 325°C decreased the oil yield by 22–25 wt% at the same residence time whilst extending the residence time from 60 to 80 min only increased the oil yield by 1–2 wt% at 285–325°C [85]. According to the literature, swine manure is composed of 25 wt% proteins and 35 wt% of carbohydrate, whilst the algal feedstocks used in these continuous HTL studies contain much larger fraction of proteins (e.g. 60–68 wt% in *Chlorella* and *Spirulina*) [27,28,31,53,72,84]. Moreover, the continuous HTL of swine manure has been examined in a CSTR, but the continuous HTL of algal biomass has been investigated in a PFR or the combinations of CSTR and PFR.

5.2.3 Process conditions

Process conditions of HTL include reaction temperature, reaction (also refers to as retention) time, pressure, feedstock/water ratio (solid content), and catalyst applications. Fig. 5.2 summarises the effects of major process parameters (reaction temperature, reaction time, and total solid content of feedstocks) on the distribution of HTL product yields converted from algal biomass based on a number of studies.

Reaction temperature has been identified as the most significant factor affecting HTL product yields and the quality of biocrude oil converted from feedstocks such as animal manure, algae, and food processing waste [28,34,68]. In general, reaction temperatures from 250°C to 375°C were applied for biocrude oil production from protein-containing biomass. Reaction temperature varies largely depending on feedstock species. Yu et al. and Gai et al. converted microalgae (*Chlorella*) into biocrude oil via HTL without catalyst and suggested that 280–300°C is the optimal reaction temperature to achieve the highest biocrude oil yield, heating value, and energy recovery [44,68]. Brown et al. and Valdez et al. converted *Nannochloropsis* into biocrude oil via HTL and concluded that 300–350°C is the best reaction temperature to reach the highest biocrude oil yield [42,90]. Considering heating rates, mixing versus nonmixing, and reactor systems (e.g. batch vs. continuous), optimal reaction temperatures are quite different. Therefore, it is important to elucidate the reaction mechanisms of HTL in terms of feedstock composition, reaction temperature, reaction time, and catalyst.

Multiple studies have reported that more gas products were formed when the HTL reaction temperature is above 320°C [42,87,88,91]. Li et al. found that the gas product yields were drastically enhanced when the HTL reaction temperature increased from 320°C to 380°C [88]. Hydrolysis and depolymerisation of feedstock are the governing reactions at the initial stage, followed by repolymerisation between 220°C and 375°C and gasification beyond 375°C [92]. As the reaction temperature increased beyond the reaction regime of repolymerisation, more solid residues and char would be formed [23,28,87,90].

Retention time is another important factor on the formation of biocrude oil (Fig. 5.3). For protein-containing biomass such as swine manure, it takes at least 15 min reaction time to form asphalt-like biocrude products [37]. In the case of algal biomass, it typically takes 10 min to form self-separated biocrude products. Reaction time is an important factor to the reaction kinetics of HTL process. Whilst sufficient reaction time is necessary to ensure a maximum biocrude oil yield, excessive reaction time will reduce the biocrude oil yield by forming biochar or gaseous products.

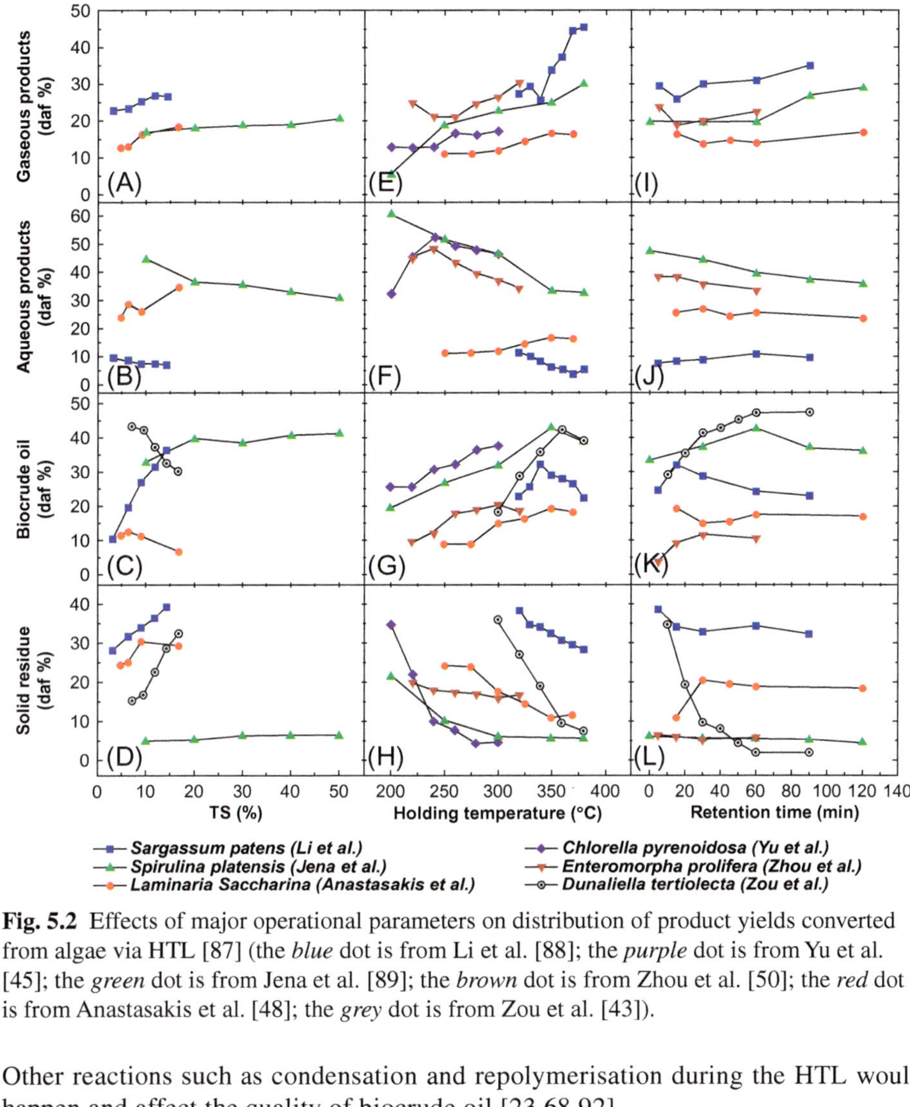

Fig. 5.2 Effects of major operational parameters on distribution of product yields converted from algae via HTL [87] (the *blue* dot is from Li et al. [88]; the *purple* dot is from Yu et al. [45]; the *green* dot is from Jena et al. [89]; the *brown* dot is from Zhou et al. [50]; the *red* dot is from Anastasakis et al. [48]; the *grey* dot is from Zou et al. [43]).

Other reactions such as condensation and repolymerisation during the HTL would happen and affect the quality of biocrude oil [23,68,92].

Heating rates could also significantly affect the completeness of HTL reaction and the biocrude oil conversion. Faeth et al. [71] conducted a fast HTL conversion (heating rates as high as 230 ± 5°C/min) of *Nannochloropsis* and reported that an optimum biocrude oil yield (66 wt% based on dry ash-free biomass) can be obtained within 1 min and the biocrude oil presents a similar carbon content and HHV to those converted from conventional HTL. Jazrawi et al. reached a highest algal biocrude oil yield (41.7 wt%) within 3–5 min at 250–350°C by a continuous HTL reactor [72].

Many HTL studies applied initial pressure by filling up a gas in the reactor headspace prior to heating up. The main purposes of applying initial pressure are to maintain water in liquid phase, reduce the enthalpy of phase change of water, enhance

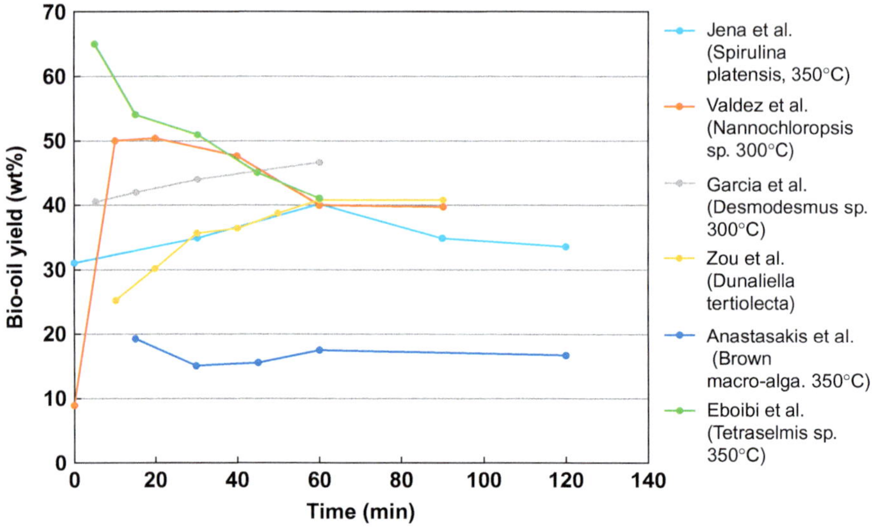

Fig. 5.3 Influence of reaction time on biocrude oil yield under HTL processes [92] (the *sky blue* line is from Jena et al. [91]; the *orange* line is from Valdez et al. [90]; the *grey* line is from Garcia et al. [93]; the *yellow* line is from Zou et al. [43]; the *blue* line is from Anastasakis et al. [48]; the *green* line is from Eboibi et al. [94]).

the solubility of biomass, and improve the energy efficiency [59,87,95]. He et al. used 0–1.30 MPa of nitrogen gas (N_2) as the initial pressure to convert swine manure into biocrude oil via HTL and found that no biocrude oil was formed until the initial pressure reached 0.69 MPa [30]. However, multiple studies have demonstrated that further increasing the initial pressure would not improve the biocrude oil yield [30,32,39,45]. Zhang et al. have extended the initial pressure from 0.69 to 3.45 MPa to covert *Chlorella* into biocrude oil. It was indicated that the initial pressure beyond 0.69 MPa has no significant effect on HTL product yields. Yu et al. showed that increasing the initial pressure from 0 to 0.69 MPa insignificantly impacted the HTL product yields converted from *Chlorella* [45]. On the other hand, Yin et al. converted cattle manure with 0–0.69 MPa of CO and N_2 as the initial pressure. A decreasing biocrude oil yield was observed when the initial pressure increased, possibly due to the self-condensation reaction that converts biocrude oil into solid residues [39]. There are still no absolute conclusions about the effect of the initial pressure on biocrude oil yields when converting protein-containing biomass via HTL.

Reducing gases, such as carbon monoxide (CO) and hydrogen (H_2), has been used as the processing gas to stabilise the fragmented products of HTL; to inhibit the condensation, cyclisation, and/or repolymerisation of free radicals; and thus to prevent char formation [30,39,95]. The following equations give an example for stabilisation of aromatic radicals (Ar) by H_2 to stabilise liquid oil products [96]:

$$Ar + \mathrm{H_2} \rightarrow ArH + \mathrm{H} \tag{4.1}$$

$$Ar + \mathrm{H} \rightarrow ArH \tag{4.2}$$

Yin et al. have converted cattle manure into biocrude oil via HTL with reducing gases such as H_2 and CO [39]. They summarised that using CO and H_2 as HTL processing gases can improve the biocrude oil yield by 5–15 wt%. Similar results were reported by He et al. when converting swine manure into biocrude oil via HTL with CO and H_2 as processing gases [30]. Although H_2 and CO are effective in stabilising the fragmented liquefaction products, they are costly and hazardous options. Alternative processing gases such as synthetic gas (H_2/CO) may be another option to serve the same purpose [95].

Inert gases, such as nitrogen, air, and carbon dioxide, have been used as HTL processing gases [30,39]. However, the role of inert gases in HTL still remains unknown other than keeping reactor headspace pressure and preventing feedstock from gasifying. Yin et al. found that using air as HTL processing gas led to a much lower biocrude oil yield than those converted with N_2, whilst He et al. demonstrated that using compressed air as the processing gas can reach a similar biocrude oil yield as those processed with CO_2 and N_2 [30]. One obvious reason for the difference was the relative mass ratio of the processing gas to feedstock. Too much of air as an initial gas could oxidise the feedstock instead of converting it into biocrude.

Total solid (TS) content of the feedstock is one of the important parameters affecting the biocrude oil yield and the quality of biocrude oil [44,87,95]. The TS of feedstock is highly dependent on the types of biomass and its collection systems. Swine manure collected from a solid floor typically contains a TS of 20–30 wt% whilst from a flushing system with settling contains a TS of 5–10 wt%. Most studies used feedstocks with a TS of 10–30 wt% to achieve positive energy and economic returns [15,31,36,46,48,90,95]. Additional water would result in increased expenses such as heating and wastewater treatment costs, whilst excessively high TS may lead to problems including poor heat transfer and material handling such as pumping. Jena et al. present that 20 wt% of TS is the optimum for HTL of *Spirulina* and greater TS show no significant effect on product yields [91,92]. Similar results were reported by Zhang et al. and Gai et al. when using *Chlorella* as HTL feedstock [32,44]. When TS is lower than 15 wt%, significantly low biocrude oil yields ($<$10 wt%) were obtained [29,32]. He et al. speculated that the formation of biocrude oil may need a micro-organic phase as the media and thus accumulate organic clusters that can be converted into biocrude oil [29,32].

5.2.4 Catalysis

Homogeneous and heterogeneous catalysts have been investigated under HTL processes. Table 5.3 summarises different catalysts used for HTL of protein-containing biomass. The results of employing catalysts in HTL were not all positive [87]. Zhang et al. converted *Chlorella* into biocrude oil under supercritical ethanol (240–300°C) using heterogeneous catalysts (HZSM-5 and Raney Ni) without improvement in biocrude oil yield [73]. Similar results were found when converting microalgae and animal manure into biocrude oil via HTL with the catalysts of Na_2CO_3, Co/Mo, CoMo/Al_2O_3, Ni/Al, Ni/SiO_2-Al_2O_3, Pt/Al, Pd/C, Ru/C, and zeolites [40,69,98].

Table 5.3 Comparison of different catalysts used in HTL of protein-containing biomass

Catalysts	Feedstocks	Optimal biocrude oil yield (dw%)	Optimal biocrude oil HHV (MJ/kg)
Homogeneous			
NaOH [74]	*Chlorella*	45 wt%	N/A
KOH [48]	*Laminaria*	15 wt%	34–38
Na_2CO_3 [40,50,74,88,97]	*Chlorella*	40–45 wt%	N/A
	Dairy manure	N/A[a]	35
	Enteromorpha	25 wt%	30
	Microcystis	30–35 wt%	30–31
	Sargassum	28 wt%	N/A
	Yeast	56.4 wt%	37
Heterogeneous			
Pd/C, Pd/Al_2O_3 [69,74]	*Chlorella*	40–45 wt%	N/A
	Nannochloropsis	45–55 wt%	39
Pt/C, Pt/Al_2O_3 [69,74]	*Chlorella*	43–45 wt%	N/A
	Nannochloropsis	50–52 wt%	40
Pt/Al [98]	*Chlorella*	39 daf% [b]	38
	Nannochloropsis	N/A	N/A
Ru/C [69]	*Nannochloropsis*	45–50 wt%	38
Raney-type alloy [69,74]	*Chlorella*	47–50 wt%	N/A
	Nannochloropsis	40–50 wt%	N/A
Raney Ni [73]	*Chlorella*	50–65 wt%	31–35
HZSM-5 [73]	*Chlorella*	55–65 wt%	32–36
Zeolite [69]	*Nannochloropsis*	40–45 wt%	35–39
Ce/HZSM-5	*Chlorella*	50 wt%	23–28
Co/Mo [69,98]	*Chlorella*	39 daf% [b]	36–38
	Nannochloropsis	42–52 wt%	39
Ni/Al [98]	*Chlorella*	30.0 daf%[b]	35–38
Ni/REHY [99]	*Dunaliella*	72.0 wt%	30
$Fe(CO)_5$-S [100]	*Spirulina*	78.3 wt%	26

[a] Not available.
[b] based on dry ash-free matter.

Phase behaviour (i.e. the interaction among liquid, solid, gas, and sub-/supercritical phases versus heterogeneous catalysts) in HTL of protein-containing biomass is very complex, and thus, catalyst deactivations and strong intraparticle diffusion limitations can also attribute to the ineffectiveness of catalysts [69,73]. Another reason for the catalysts having little effect on the HTL biocrude oil yields could be that the oil yield in the absence of catalyst was already near the upper bound of what is possible, given the constraint of the mass balance and availability of carbon and hydrogen [69]. It was demonstrated that >80% of the C and H atoms in the algal feedstocks were recovered in the HTL biocrude oil without the addition of catalysts [69].

The loadings of catalysts are an influential factor for HTL process. Anastasakis and Ross used 5–100 wt% KOH to study the effect of the loadings of catalysts on HTL of macroalgae. It was reported that the HTL biocrude oil yields decreased by about 10 wt% whilst the heating value of biocrude oil increased by 1.3 MJ/kg, as the amounts of catalysts increased from 5 to 100 wt% [48]. Theegala et al. have confirmed that increasing the catalysts (Na_2CO_3) loading from 5 to 20 wt% did not improve the biocrude oil yield converted from animal manure via HTL [40]. Comparable results were also demonstrated when using CaO to convert olive seeds, in which as the amount of catalysts increased from 5 to 40 wt%, the HTL biocrude oil yield decreased [101].

Homogeneous catalysts seem to have a positive effect on the HTL biocrude oil yield, but the recovery of homogeneous catalysts remains a challenge. Yu et al. investigated the effect of NaOH and Na_2CO_3 on HTL of microalgae at 280°C and reported that the algal biocrude oil yield was improved by 5–10 wt% [74]. Jena et al. converted yeast into biocrude oil via HTL and demonstrated that adding Na_2CO_3 can enhance the biocrude oil yield by about 7 wt% [97].

Regeneration of catalysts has been suggested to improve the technoeconomic feasibility of catalytic HTL conversion of wet biomass [102,103]. Ong has used NaOH to regenerate catalysts such as Raney Ni when hydrothermally treating waste newspapers. Effective and renewable methods to regenerate catalysts under HTL processes are critically needed because the catalysts are typically expensive and their production can impact the environment [102,103].

5.2.5 HTL products separation

After HTL reaction, HTL products can be separated into biocrude oil, solid residue, aqueous products, and gaseous products. Fig. 5.4 presents a typical separation procedure for HTL products [23]. Depending on feedstock and process conditions, HTL products could be naturally separated via gravitation. Organic solvents such as dichloromethane, acetone, and toluene were also used to extract the biocrude oil fraction from HTL products [15,23,40,48,52,53,68,73,91]. Valdez et al. have studied the influence of different organic solvents on HTL product yields and quality. Non-polar solvents (hexadecane, decane, hexane, and cyclohexane) and polar solvents (methoxycyclopentane, chloroform, and dichloromethane) were used to recover the biocrude oil converted from *Nannochloropsis*. It is found that non-polar solvents, such as hexadecane and decane, can provide high gravimetric yields of bio-oil (39 wt%) but the recovered biocrude oil presents a lower carbon content (e.g. 69 wt% for decane) than that recovered with polar solvents such as dichloromethane (76 wt%). The amount of free fatty acids in the biocrude oil is highly dependent on the solvent used, with polar solvents recovering more fatty acids than non-polar solvents [104].

A sequential extraction process was investigated to separate nitrogen-containing compounds from HTL biocrude oil using water and toluene in series to extract HTL biocrude oil converted from wet biowaste [105]. When an ultrasonically assisted extraction by water was conducted, the nitrogen content in algal biocrude oil decreased by 16% with improved carbon and hydrogen contents. However, it was also suggested that alternative extraction techniques should be considered to improve the extractive

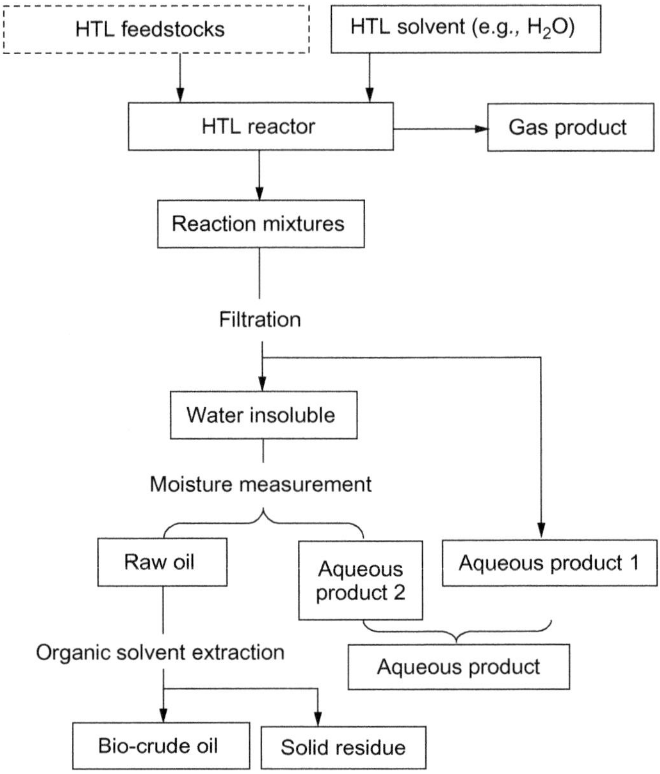

Fig. 5.4 A typical separation procedure for HTL biocrude oil.

denitrogenation efficiency and reduce the amount of biocrude oil lost into the water extract for HTL biocrude oil [105].

Although solvents are usually used to recover HTL biocrude oil, some studies, particularly for those focusing on continuous HTL reactor development, separate HTL biocrude oil by decanting the solid products from aqueous products. For example, the continuous hydrothermal treatment of microalgae in the PNNL studies did not use any organic solvent to recover the algal biocrude oil [70,83]. Similar separation method is also used for continuous HTL of swine manure [84,85].

5.3 Characterization of HTL products

5.3.1 Biocrude oil

5.3.1.1 Elemental analyses and heating value

Elemental analyses including the measurement of carbon, hydrogen, nitrogen (CHN), metal, and inorganic contents are commonly used to understand and evaluate the quality of HTL biocrude oil. Carbon, hydrogen, nitrogen, and sulphur contents of the HTL biocrude oil were measured and used to calculate the higher heating value (HHV). Compared with petroleum crudes, HTL biocrude has a much higher oxygen and nitrogen contents. Further upgrading processes to lower heteroatom contents in HTL biocrude oil are recommended in order to be used as transportation fuel [53]. Similar measurement and suggestions were also reported by multiple studies converting biowaste and/or algal biomass into biocrude oil via HTL [31,38,70,73,91,104]. With the elemental compositions, Van Krevelen diagrams can be constructed based on the H/C, O/C, and N/C molar ratios of HTL biocrude oil (Fig. 5.5). From a fuel perspective, biocrude oil with a high H/C and low O/C and N/C molar ratios is preferred, so that their properties would be close to those of petroleum-derived fuels [23,68,87]. In addition, nutrient recoveries (i.e. nitrogen and carbon) can be computed with the information of elemental compositions [23,44,68].

Although the HHV of HTL biocrude oil is usually calculated by Dulong formula, other equations or methods considering the effect of high-ash contents have been attempted. Anastasakis and Ross evaluate the HHV of biocrude oil converted from macroalgae based on the elemental compositions of carbon, hydrogen, sulphur, oxygen, nitrogen, and ash contents [48]. Oxygen bomb calorimeter is commonly used to directly measure the HHV of HTL biocrude oil [47,52]. The energy recovery (defined as the energy content in the HTL oil to the energy content in the feedstock), energy consumption ratio (defined as the energy required for HTL against the energy that can be recovered from biocrude oil combustion), and fossil energy balance (FEB, defined as the energy in biofuel-fossil fuel input (1 MJ biofuel produced)) of HTL biocrude oil can be further estimated with the measurement of HHV [15,31,46]. To perform the technoeconomic analysis for HTL (or any other bioenergy production), exergy could

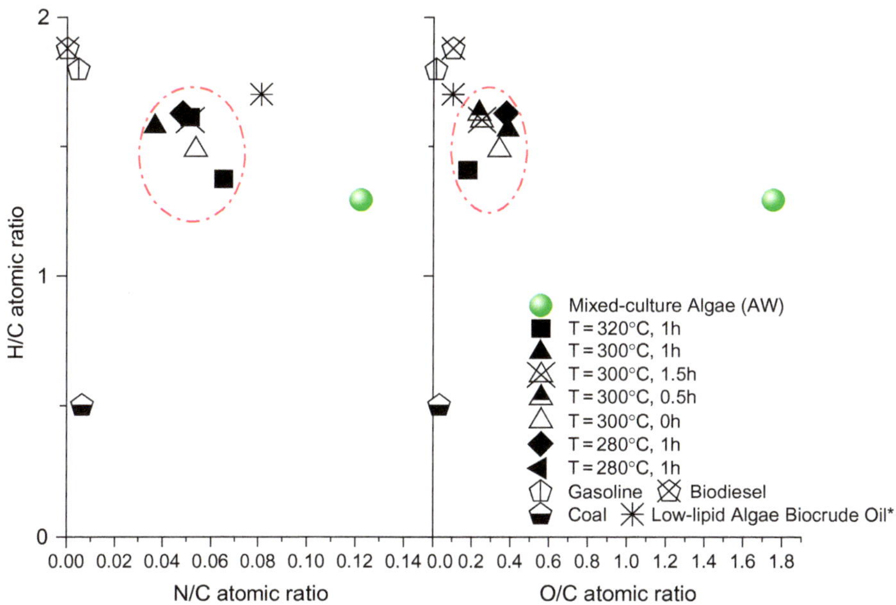

Fig. 5.5 Van Krevelen diagrams of HTL biocrude oil [23].

be a more direct indicator, which is defined as the ratio of energy in the HTL biocrude oil to the sum of energy in the feedstock and the energy required for HTL process.

Metal compositions such as Ca, Mg, K, Na, Al, and Fe of HTL biocrude oil have also been characterised, particularly for the biocrude oil converted from high-ash biomass such as macroalgae [23,48,52,106]. Understanding the fate of the metal in feedstocks during HTL is important to assess whether they would affect HTL conversion efficiency and the quality of resulting biocrude oil or recover the metal as a byproduct. Several studies have pointed out that the metal compositions such as alkaline earth metals and calcium carbonates could act as catalysts during HTL of algal biomass [23,47,48,52]. The existence of calcium carbonate can promote denitrogenation of biocrude oil during HTL of mixed-culture algae [23]. However, these metals may also poison catalyst used for upgrading processes such as desulphurisation and catalytic cracking [106]. Trade-offs between the catalytic effect and poison of the metals in biocrude oil need further investigation.

Inorganic compositions, such as phosphorus (P), in HTL biocrude oil are also of interest, probably because phosphorus is an important element for algal cultivation and its limited quantities emphasise the need to recycle it for sustainable production of algal biofuels [74,90,107]. A number of studies showed that majority of phosphorus in algal biomass were retained in HTL aqueous product [90,93]. The phosphorus content in the HTL aqueous products could be highly affected by the metal compositions of the feedstock, because some metal elements could bond with phosphorus to form precipitates and to be recovered by solid residue [36,74].

5.3.1.2 TGA, Sim-Dist, GPC, and SEC analyses

Thermal gravimetric analysis (TGA) and simulated distillation (Sim-Dist) are two common tools to evaluate the boiling point distribution of HTL biocrude oil. Boiling point distribution is one of the most commonly referenced characteristics of crude oil for fuel applications [15,23,31,44,46,48,53,73]. Boiling points of the biocrude oil converted from microalgae and swine manure predominantly fell in the range of heavy vacuum gas oil (343–538°C). Similar results were also demonstrated in other HTL biocrude converted from different types of biowaste and algal biomass [23,31,44,46,73].

Gel permeation chromatography (GPC) and size-exclusion chromatography (SEC) are two typical methods to assess the molecular weight distribution of HTL biocrude oil [15,53,93,108]. Barreiro et al. have conducted GPC analysis on biocrude oil converted from algal biomass at 250°C and 375°C and found that high temperatures (375°C) and pressures (250–270 bar) seem to be severe enough to degrade the microalgae molecules into similar types of oil molecular weight distributions whilst HTL at mild temperature results in an "improved" extraction, rather than a proper thermochemical conversion [108]. On the other hand, SEC analysis can provide additional information including the weight-average molecular weights (Mw), the number-average molecular weights (Mn), and polydispersity values (PDI). In general, HTL biocrude oil presents a higher Mw, Mn, and PDI than petroleum crude. The higher PDI indicates that there is a greater difference in the minimum and maximum molecular weights of biocrude oil, which in turn is due to the high oxygen content derived from polymer-linking groups such as esters and ethers [15,53].

5.3.1.3 Chemical composition

Chemical compositions of HTL biocrude oil are typically characterised using gas chromatography-mass spectrometer (GC–MS), nuclear magnetic resonance (NMR), and Fourier transform infrared spectroscopy (FTIR) [15,23,31,36,44,46,47,53,73,74, 105]. GC–MS analysis identifies compounds, and NMR and FTIR reveal functionalities. HTL biocrude oil includes a variety of compounds such as hydrocarbons (e.g. tridecene), fatty acids (e.g. hexadecanoic acid), straight amide derivatives (e.g. hexadecanamide), n-heterocyclic compounds (e.g. indole), and oxygenates (e.g. esters, alcohols, aldehydes, and phenols). Compositions of HTL biocrude oil are extremely complex and vary widely from its original feedstocks. Only a fraction of HTL biocrude oil can be identified by GC–MS due to two reasons: (1) A large fraction of the biocrude has high-molecular-weight molecules, and (2) it is limited by the temperature of the instrument and thus boiling point distribution [53]. On the other hand, some low-boiling-point compounds may be masked by the solvent peak or evaporated during evaporation of organic solvents used to recover biocrude oil [105]. Therefore, complementary analyses such as NMR and FTIR are usually conduced to verify the results obtained from GC–MS analysis.

FTIR analysis allows for a more comprehensive comparison of "whole" oil functional group characteristics compared with GC–MS, with spectral band assignments and interpretation based on previous studies [39,53,73,106]. It is believed that HTL biocrude oil presents a prominent C–H stretching, –CH2 bending, and CH3 bending characteristics, which suggests the existence of alkyl C–H [15,44,53,73]. On the other hand, NMR analysis gives complementary functional group information to FTIR spectra and the ability to quantify and compare integration areas between spectra [53]. To date, ^{1}H NMR, ^{13}C NMR, and ^{31}P NMR have been exploited to characterise HTL biocrude oil [15,44,47,53,73,109,110]. One ^{1}H NMR analysis of HTL biocrude oil converted from *Chlorella* observed a high percentage of aliphatic functionality (59%–67%), which is also found in other biocrude oil products converted from *Spirulina* (53%) and duckweed (60%) [44,53,111]. Other techniques for quantifying biocrude oil functional groups are also employed. For example, high-resolution Fourier transform ion cyclotron resonance mass spectrometry (FTICR-MS) has been applied in a systematic way to fully categorise the HTL biocrude oil, and relatively, quantitative results of different functional groups were reported [112–114].

5.3.1.4 Potential value-added products

In addition to transportation biofuels, HTL biocrude oil can be further processed to obtain other value-added chemicals. Chemical and rheological techniques have been used to categorise the applicability of the HTL biocrude oil converted from *Spirulina* and swine manure as raw materials for pavement construction [115]. It was demonstrated that HTL biocrude oil had higher asphaltene content than that of commercial binders. When HTL biocrude oil was mixed with commercial binders at a ratio of 1:8, the component structure resembled that of conventional petroleum binder, indicating its possible use as a softening agent for mixing with old asphalt materials and a potential alternative surrogate for conventional asphalt.

5.3.2 Aqueous product

The post-HTL aqueous product, also refers to as posthydrothermal wastewater (PHWW), contains a significant amount of nutrients (e.g. ammonia and phosphates) and toxic compounds. The PHWW of protein-containing feedstocks is quite different from other feedstocks such as crop residue. First, it contains most nutrients (primarily nitrogen) in the feedstock that can be reused for biomass production; on the other hand, it contains many n-heterocyclic compounds that can be inhibitive to the growth of microorganism, algae, and plants. Therefore, PHWW must be treated prior to being used as a fertilizer or discharge to waterways.

5.3.2.1 Chemical composition

The chemical composition of PHWW can be characterised in terms of chemical oxygen demand (COD), total organic carbon (TOC), total ammonia nitrogen (TAN), total nitrogen (TN), total phosphorus (TP), and total suspended solids (TS), as well as the elemental and GC–MS analyses [79,90,107,116–123]. Table 5.4 summarises the chemical composition of PHWW converted from different feedstocks via HTL. Response surface methodology (RSM) was used to investigate the interactions of reaction temperature, retention time, and total solid ratio of feedstocks on the water quality characteristics of PHWW converted from microalgae in a batch mode [109]. The authors reported that the total solid ratio of feedstocks was the dominant factor affecting the nutrient recovery efficiencies of PHWW. Appleford et al. have characterised PHWW converted from swine manure in a continuous-stirring reactor in terms of chemical compositions, biological oxygen demand (BOD), TS, pH, TAN, chloride, nitrate, phosphate, sulphate, metal ions (e.g. Cr, Fe, Mg, and Zn), and total faecal coliforms [121]. Although bacterial pathogens in swine manure were completely killed during the HTL process, the level of BOD, pH, and elemental concentrations (e.g. Fr, Mn, and Zn) was higher than the USEPA standard for discharge to waterways [121] [79,107]. This is because an HTL process essentially separates a large fraction of elements (except hydrogen and carbon) in the feedstocks into aqueous phase. For that reason, the cytotoxicity of PHWW should be thoroughly characterised. Pham et al. utilised a Chinese hamster ovary (CHO) cell assay to evaluate the cytotoxicity of nine identified nitrogenous organic compounds (NOCs) in the PHWW from HTL of *Spirulina.* The authors reported that the organic mixture extracted from PHWW expressed potent CHO cell cytotoxic activity, with an LC_{50} at 7.5% of PHWW. It is also found that 30% of the toxicity can be removed biologically by recycling the PHWW back into an algal cultivation system. The remaining toxicity of the PHWW could be mostly eliminated by subsequent treatment with granular activated carbon (GAC) [118].

Chemicals of emerging concerns (CECs) that potentially cause undesirable ecological or health effects have also been characterised in the PHWW converted from the algal-bacterial biomass grown in the liquid portion of swine manure [126]. ELISA tests and GC/MS analysis were used to measure the concentrations of the CECs in the algal-bacterial feedstocks and HTL products. It is highlighted that a 99.9% of

Table 5.4 Comparison of average values of physicochemical characteristics of PHWW converted from swine manure and different algal feedstocks

Characteristic	Swine manure [79,121]	Mixed-culture algae [23,116]	*Chlorella* [107,124]	*Spirulina* [79,125]	High-ash algae [52]	USEPA standard[a]	Illinois standard[b]	Filtered municipal wastewater [79]
Biological oxygen demand (mg/L)	35,240	N/A	N/A	N/A	N/A	120	10	N/A
Chemical oxygen demand (mg/L)	52,030	30,000–70,000	62,700–104,000	89,039 ± 3321	N/A	N/A	N/A	54
Total suspended solids (g/L)	33.0	N/A	2.3–21.5	N/A	N/A	183	12	N/A
pH	5.52	7.4–8.6	7.8–8.3	7.5	N/A	6–9	6–9	N/A
Ions (mg/L)								
Ammonia	3413	2000–4500	3000–14,800	10,117 ± 1167	N/A	N/A	N/A	N/A
Total nitrogen (TN)	5355	3500–8000	11,000–31,700	22,981 ± 2482	1973–3941	N/A	N/A	33
Chloride	667	N/A	N/A	N/A	N/A	N/A	N/A	N/A
Nitrate	0.87	N/A	N/A	N/A	N/A	N/A	N/A	N/A
Phosphate	921	N/A	N/A	N/A	N/A	N/A	N/A	N/A
Total phosphorus (TP)	1499	N/A	5440–18,900	4400 ± 198	N/A	N/A	N/A	8
Sulphate	427	N/A	N/A	N/A	N/A	N/A	N/A	N/A
C/N ratio	N/A	N/A	N/A	2.4	N/A	N/A	N/A	N/A

Continued

Table 5.4 Continued

Characteristic	Swine manure [79,121]	Mixed-culture algae [23,116]	*Chlorella* [107,124]	*Spirulina* [79,125]	High-ash algae [52]	USEPA standard[a]	Illinois standard [b]	Filtered municipal wastewater [79]
Elements (mg/L)								
Aluminium (Al)	N/A	N/A	N/A	3.9	N.D.	N/A	N/A	N/A
Boron (B)	N/A	N/A	N/A	22.5	N/A	N/A	N/A	N/A
Calcium (Ca)	N/A	N/A	N/A	13.5	193	N/A	N/A	N/A
Chromium (Cr)	0.27	N/A	N/A	N/A	N/A	2.77	1	N/A
Cobalt (Co)	N/A	N/A	N/A	0.9	N/A	N/A	N/A	N/A
Copper (Cu)	N/A	N/A	N/A	0.6	N/A	N/A	N/A	N/A
Iron (Fe)	28	N/A	N/A	4.7	<15.7	N/A	2	N/A
Magnesium (Mg)	242	N/A	N/A	1.4	2.9	N/A	N/A	N/A
Manganese (Mn)	2	N/A	N/A	0.3	N/A	N/A	1	N/A
Phosphorus (P)	434	N/A	N/A	N/A	N/A	N/A	N/A	N/A
Potassium (K)	1482	N/A	N/A	3400	155	N/A	N/A	N/A
Rubidium (Rb)	0.58	N/A	N/A	N/A	N/A	N/A	N/A	N/A
Sodium (Na)	N/A	N/A	N/A	2254	115	N/A	N/A	N/A
Selenium (Se)	N/A	N/A	N/A	11.7	N/A	N/A	N/A	N/A
Sulphur (S)	9651	N/A	N/A	N/A	N/A	N/A	N/A	N/A
Zinc (Zn)	1.67	N/A	N/A	1.3	N/A	N/A	1	N/A
Total faecal coliforms (#/mL)	<2	N/A	N/A	N/A	N/A	N/A	N/A	N/A
IC_{50} (%)	N/A	N/A	0.35–1.90	N/A	N/A	N/A	N/A	N/A

N/A, Not applied.

[a] Standard is 1 day maximum concentration for the specialty organic chemical industrial category.

[b] Standard is categorical effluent limitation for the State of Illinois.

florfenicol (a common environmental hormone) was removed under the HTL process at 300°C for a 35 min reaction time [126].

5.3.2.2 Treatments on PHWW

PHWW treatment has been investigated using adsorption [124,125], anaerobic digestion [116,127,128], algal cultivation [117,118,126], recirculation to the HTL processes [122,129,130], and catalytic hydrothermal gasification [70,77,120,131]. A photobioreactor (PBR) was used to monitor the long-term performance when treating a combination of municipal wastewater and PHWW for an about 800-day operation [124]. Different operating conditions (retention time, various PHWW loading rate, and with and without the addition of GAC) were performed, and the results show that the PBR can achieve an efficient removal of organic carbon (79%–93% COD) and nitrogen (50%–99% NH_4^+ and 27%–30% total nitrogen) [124]. Increasing the PHWW loading rates reduced the removal efficiency of COD and NH_4^+. Adding GAC facilitated a stable PBR performance and allowed higher PHWW loading rate. GAC addition also resulted in shorter retention times with an improved removal efficiency of COD (from 70% to over 90%), NH_4^+ (from 26% to 100%), and cytotoxicity (from 40% to 60%). Similarly, zeolite, GAC, and polyurethane matrices (PM) were used for detoxification during a two-round anaerobic batch test with PHWW converted from cyanobacteria (e.g. *Spirulina*). GAC was considered promising due to its highest methane yield of 124 mL/g COD at the second feeding, indicating a good recovery of adsorption capacity [125].

Anaerobic digestion (AD) could be also used to treat PHWW. Zhou et al. have investigated the feasibility of using AD to treat PHWW converted from swine manure [124,127]. It is demonstrated that successful AD occurred at relatively low concentrations of PHWW (≤6.7%) and led to a biogas yield of 0.5 mL/mg COD removed. On the other hand, higher concentrations of PHWW (e.g. ≥13.3%) present an inhibitory effect on the AD process [124,127]. Tommaso et al. have also examined the anaerobic degradability of HTL-AP converted from mixed-culture algae grown in a wastewater treatment system [116]. The effect of HTL reaction temperature (260–320°C) and reaction time (0–90 min) on the anaerobic degradability of PHWW was studied. The highest cumulative methane production (with a moderate lag phase) was realised when PHWW was obtained at 320°C. The second highest accumulated methane production (obtained with the PHWW converted at 300°C) was achieved after overcoming the longest lag phase in this study. Further, the authors have identified acetogenesis as a possible rate-limiting pathway in AD of PHWW [116].

Cultivating algal biomass in PHWW can help develop a sustainable system that couples bioenergy production with nutrient recovery, wastewater treatment, and biomass augmentation. Multiple studies have utilised diluted PHWW to grow algal biomass [91,117,124,132]. Biller et al. cultured microalgae strains of *Chlorella* and *Scenedesmus* and the cyanobacteria of *Spirulina* and *Chlorogloeopsis* with diluted PHWW converted from the same microalgae and cyanobacteria strains. It is found that the optimum dilution is strain-dependent but ranges between 200 and 400 times [117]. Zhang et al. investigated the nutrient recovery efficiency (C, N, and P) from PHWW prepared with two biocrude-aqueous separation methods (vacuum filtration and ethyl ether extraction). With ethyl ether extraction as a biocrude-aqueous separation method, a higher algal biomass growth rate was reached. In contrast, a more advantageous nutrient recovery efficiency (about 1.4 times nitrogen recovery, 1.5 times phosphorus recovery, and 2.3 times carbon recovery) was realised when the vacuum filtration was utilised to separate the PHWW converted from *Chlorella* [132].

Alternatively, PHWW still contains 10%–40% of original carbon in the feedstock than can be added back to HTL processes as a reaction substrate. PHWW converted from microalgae was recirculated to HTL processes to recover the organic content and to develop a solvent-free process [129]. It was demonstrated that the biocrude oil yield increased to a constant level after six recycles at 240°C. The authors also summarised that recycling the PHWW not only increased the biocrude oil yield but also reduced the operating temperature of HTL (e.g. from 300°C to 240°C) to achieve an optimal biocrude oil yield.

PHWW can also be converted into hydrogen via catalytic hydrothermal gasification (CHG) [120,131]. Breinl and Zhang has hydrothermally treated the PHWW converted from swine manure with a number of catalytic and noncatalytic batch tests at 350–600°C with a reaction time of 1 h. Activated carbon (AC), α-Al_2O_3, TiO_2, diatomaceous earth (DE), MnO_2, and Raney Ni were employed as catalysts for CHG processes. Raney Ni and AC can lead to the highest catalytic activities, whilst the highest COD removal efficiency (91.3%–98.1%) was reached at 600°C [131]. PHWW converted from microalgae was upgraded into hydrogen via CHG with sodium hydroxide. The high yields of hydrogen can be produced (30 mol H_2/kg algae) with near-complete gasification of the organics (98%) in PHWW [120]. Compared with HTL, the energy recovery of CHG for PHWW is quite low due to its low energy content in the PHWW and high energy input of the CHG process.

5.3.3 Gas products

The chemical compositions of HTL gas products are mainly composed of carbon dioxide (CO_2) and small fraction of carbon monoxide (CO), hydrogen (H_2), methane (CH_4), ethane (C_2H_6), ethene (C_2H_4), benzene, toluene, and styrene [23,68,72,73,114,121]. Gas products converted from swine manure in a continuous-stirring HTL reactor were reported at concentrations (167–288 ppm) of total hydrocarbons as gas (C_4–C_{12}). Faeth et al. also characterised the compositions of gaseous products converted from microalgae in a batch reactor. Carbon dioxide is found to be the primary compound (0.301–0.857 mmol/g dry algae), along with some CO (0.002–0.120 mmol/g dry algae), H_2 (0.0002–0.035 mmol/g dry algae), CH_4 (0.001–0.046 mmol/g dry algae), C_2H_6 (0.004–0.013 mmol/g dry algae), and C_2H_4 (0.009–0.043 mmol/g dry algae).

Since the HTL gaseous products contain mostly carbon dioxide (>95%), it is suggested to recycle the gas product to grow biomass [68,79]. The University of Illinois group has proposed a synergistic system, namely, environment-enhancing energy (E^2-energy), which can reutilise the nutrients in PHWW and the carbon source in HTL gaseous products to achieve biomass amplification by growing more algal-bacterial biomass [68,79].Several compounds (e.g. C_2H_4) in the HTL gaseous products have been identified as USEPA hazardous air pollutant (HAP) [121].

Although these levels are exempt from the regulation of HAP because of low concentrations, health and safety precautions, such as gas masks, for those working around the process should be used [121]. Moreover, abatement facilities such as activated carbon filters are highly suggested to treat the HTL gaseous products before they are vented out to the atmosphere [72].

5.3.4 Solid residue

Elemental, TG, FTIR, and surface analyses are conducted on HTL solid residues to understand their CHN and mineral contents, boiling point distribution, chemical compositions, and surface area, so their potential utilisation can be elucidated. For HTL of mixed-culture algae from wastewater treatment systems, 12%–18% of the carbon remained in the solid residue, whilst TG analysis suggests that the solid residues can be possibly used as asphalt [23]. Most of the potassium and sodium could be recovered in the PHWW, whilst most of the calcium and magnesium are distributed in the solid residue. It is possible that the solid residue can be potentially combustible without problems associated with slagging and fouling, based on the low concentration of alkalis in the solid residue.

Ash contents, surface area (S_{BET}), bulk density, and helium density of the solid residue converted from swine carcasses via HTL [133] are close to the biochar converted from plant biomass via pyrolysis and can be potentially used as soil amendments. The solid residue can also be used as an alternative of activated carbon for metal adsorption or as a renewable solid fuel, based on their CHN contents and HHV [133]. Aliphatic, carboxylic, primary and secondary amides and primary amine functional groups were identified in the solid residues [134]. Similar results were also reported when macroalgae were used as HTL feedstocks [135]. Lignocellulosic fraction in feedstocks could not be completely reacted and remain as solid residue such as biochar [136].

5.4 HTL Reaction mechanism

Though many different feedstocks have been converted into biocrude oil via HTL, the reaction mechanism of HTL remain unclear largely due to the complexity of the feedstock biochemical property. Several studies have attempted [23,36,46,73,109,137] to elucidate the reaction mechanisms, and it is generally believed that there are three basic HTL reaction pathways: (1) depolymerisation of various biomolecules (e.g. lipids, proteins, and carbohydrates); (2) decomposition of different biomass monomers (e.g. fatty acids, amino acids, and glucoses) by cleavage, decarboxylation, and deamination; and (3) recombination and repolymerisation of reactive compounds (e.g. radicals) [37,53,57,59,61].

To elucidate reaction mechanisms, HTL of model compounds such as albumin (representative of proteins), butter, fatty acids (representatives of crude fats), starch, and glucose (representatives of carbohydrates) was conducted [36,56,137]. Lipid can be further converted into hydrocarbons, possibly via decarboxylation [23,36,46,59,73,109,138]. Proteins can experience either deamination or decarboxylation and are potentially converted into amines, amides, nitriles, ammonia, and carbon dioxide [23,31,59,61,73,105,109]. It is generally believed that lignocellulosic materials are first hydrolyzed and then decomposed into char during HTL processes [61,139,140]. Further, the monomers of various biomolecules (e.g. proteins) can have interreactions. For example, amino acids could react with reduced glucose and then produce melanoidins [59]. The reaction kinetics of HTL has been investigated by using model compounds (e.g. soy protein) and different HTL products (e.g. aqueous products and solid residue from HTL reactions) [141,142]. HTL tests with both model compounds and real feedstock (e.g. animal manure) were conducted to understand reaction pathways regarding the major interreactions under HTL reactions [36,137]. It is concluded that an excessive amount of protein in the feedstock would deteriorate the biocrude oil quality whilst the addition of carbohydrates (such as glucose) would promote the biocrude oil formation.

A modified reaction network for HTL of protein-containing feedstocks is present in Fig. 5.6. The reaction network of HTL of microalgae has also been established, whilst the reaction constants and the activation energy of respective reaction paths have been reported [141,143].

The reaction mechanism of HTL of protein-containing biomass mostly relies on the speculation from characterisation results of GC–MS, NMR, FTIR, etc. In order to validate this speculation, insightful techniques such as isotopes (deuterium oxide or isotopic labelling proteins) with representative model compounds in HTL process will greatly enhance the understanding of the reaction kinetics and network in this biomass conversion system. Considering HTL reactions that involve complex free-radical reactions between hundreds of reaction intermediates, it is extremely difficult to elucidate its mechanism via experiment. Computational studies using quantum chemistry could be an effective tool to provide fundamental information on HTL reaction mechanisms.

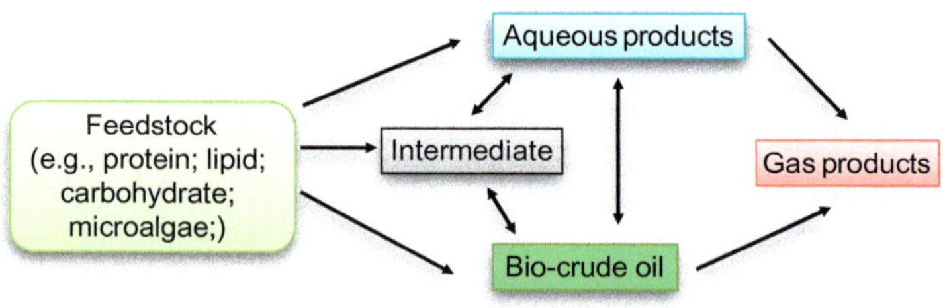

Fig. 5.6 A generic reaction network for HTL of protein-containing feedstocks.
Modified from Valdez P.J.; Savage P.E., A reaction network for the hydrothermal liquefaction of Nannochloropsis sp. Algal Res 2 (2013) 416–425, Luo L.; Sheehan J.D.; Dai L.; Savage P.E., Products and kinetics for isothermal hydrothermal liquefaction of soy protein concentrate. ACS Sustain Chem Eng 4 (2016) 2725–2733, Valdez P.J.; Tocco V.J.; Savage P.E., A general kinetic model for the hydrothermal liquefaction of microalgae. Bioresour Technol 163 (2014) 123–127 and Sheehan J.D.; Savage P.E., Products, pathways, and kinetics for the fast hydrothermal liquefaction of soy protein isolate. ACS Sustain Chem Eng (2016).

5.5 Upgrading of HTL biocrude

Available upgrading techniques for HTL biocrude include steam reforming [144–148], sub-/supercritical fluid (SCF) treatment [3,59,73,76,149–152], hydrocracking [153–155], zeolite cracking [145,154,156], thermal cracking [153,157], hydrotreating [158–162], solvent addition [163–166], chemical extraction from HTL biocrude oil [41,105,167–170], and emulsification [81,171–173]. SCF treatments can inhibit the char formation [3,59] and significantly improve the HTL biocrude oil quality such as reducing oxygen and nitrogen contents and viscosity. SCFs such as water can serve as a hydrogen donor. However, the upgrading mechanism (e.g. how SCFs participate in the reaction) under SCFs still remains unclear.

Traditionally, cracking reactions (e.g. zeolite cracking) and hydrotreating are two major approaches to upgrade petroleum crude because they can reform undesired chemicals into hydrocarbons through chemical reactions. Cracking reactions mainly fragment one heavy molecule into two light molecules, whilst hydrotreating can achieve saturation of olefins, hydrodeoxygenation (remove O as H_2O), hydrodenitrogenation (remove N as NH_3), and hydrodesulphurisation (remove S as H_2S).

Steam reforming of pyrolysis bio-oil has been extensively studied but not of HTL biocrude oil [174]. Steam reforming could be a promising method to upgrade HTL biocrude oil because it can produce a synthesis gas as a renewable fuel. Further, steam reforming is a relatively well-established technique (e.g. reactor development), and the reaction mechanisms are better understood [144,145]. However, the catalysts involved in steam reforming can be significantly deactivated by coking, which is typically caused by oligomerisation of carbonaceous materials such as tar. Steam reforming of HTL biocrude oil can be inherently complex due to the need of different decomposition reactions in order to obtain single carbon species to react with the water on the surface of catalysts [174,175].

Less severe upgrading, such as solvent addition, improves the viscosity and acidity of pyrolysis bio-oil through esterification and alcoholysis. The removal of nitrogen-containing compounds presents a challenge when solvent-addition technique is adapted to upgrading HTL biocrude oil. In contrast, chemical extractions/separations and emulsification involve no chemical reactions during upgrading processes. This feature makes these two methods relatively simple and economic. However, they cannot reform undesired compounds such as phenols and pyridines.

Most upgrading techniques have been exclusively developed for petroleum that nitrogen is not a concern. HTL biocrude oil converted from protein-containing feedstocks typically contains a relatively high nitrogen content (3%–7%) [31,46,176]. The high nitrogen content in HTL biocrude oil would cause fouling of conventional oil-upgrading catalysts (e.g. zeolites) because the high basicity of the heterocyclic nitrogen (N-ring) compounds can result in adhesion to acidic active catalytic sites and poison the catalysts [15,105,159]. If the upgraded HTL biocrude oil is used as drop-in transportation fuel (i.e. blending 5%–10% upgraded HTL biocrude with petroleum fuel), the nitrogen content could be diluted to an acceptable level. Identifying what types of nitrogen-containing compounds that would deteriorate combustion processes is essential such as nitrogen-heterocyclic compounds versus nitrogen-containing alkyl compounds. Engine tests are highly needed to examine the effect of nitrogen-containing compounds on combustion performance and emissions.

Most of the upgrading techniques such as hydrotreating need catalysts to enhance the heteroatom removal efficacy. Deactivation of catalysts is commonly observed in steam reforming, cracking, and hydrotreating of pyrolysis bio-oil. Deactivation of catalysts has also been found when hydrothermally treating HTL algal biocrude oil under supercritical water in batch reactor systems [149]. Duan et al. have recycled the spent Pt/γ-Al_2O_3 to hydrothermally upgrade HTL algal biocrude oil at 400°C for 1 h with an added 6 MPa hydrogen under supercritical water. The spent catalyst presented modest activity loss on the performance of hydrodeoxygenation but not hydrodenitrogenation. On the contrary, catalyst deactivation was not observed when hydrotreating HTL algal biocrude oil in a continuous reactor system during a short-term testing [70,87].

5.6 Continuous HTL reactor development

Efforts have been made to scale-up the HTL process of protein-containing biomass by industries and industry associations. The major findings are summarised in Table 5.5.

The University of Sydney group developed a plug-flow reactor system with a total volume of 2 L for converting algal biomass into biocrude oil under HTL. In their system, high-pressure triplex piston pump was used to pump up to 10 wt% microalgae slurries with high viscosities. Heat exchangers were employed to preheat the algal slurries (to about 170°C) before entering the reactor [72]. The flow rates investigated in this study were in the range of 15–30 L/h and led to residence times of 3–5 min. In addition, the authors claimed that the blockage of the reactor coils has never happened in any of their operations (with a maximum duration of 2 h in a single run) [72,83]. Overall, the University of Sydney group reported that the higher temperatures and longer residence times increased the yields, which is in agreement with the results from batch HTL experiments of algal feedstocks [72].

Continuous HTL conversions of algal feedstocks with organic solvents have been conducted by the University of Sydney group recently [75]. Cosolvent HTL with 10 wt% n-heptane, anisole, or toluene was conducted to achieve in situ fractionation and reduce heteroatoms (e.g. N) of the algal biocrude oil. Technoeconomic (TEC) analyses are suggested to evaluate if the cosolvent HTL process is cost-effective, since the organic solvents are more expensive and less accessible than water. LCA analysis of the cosolvent HTL is also needed, because the production of organic solvents from petroleum crude can lead to the emission of greenhouse gases. Further, it is recommended to explore if the organic solvents can be produced from renewable sources, to prevent the additional emission of greenhouse gases.

The PNNL group has developed a continuous HTL process consisting of a high-pressure pump feeding system, a product recovery system, a 1 L continuous-stirred tank reactor (CSTR) for preheating algal feedstocks, and a 1 L tubular catalytic reactor [70]. In order to avoid potential plugging problems, the PNNL group combined CSTR and plug flow in their continuous HTL process. The authors reported

Table 5.5 Summary of plants/implementation of HTL feedstocks containing protein contents

Group	Capacity of the plant to process feedstocks	Biocrude oil yield/conversion efficiency	Process condition and features
University of Sydney group			
He et al. [75]	Up to 90 L/h of feed microalgae and macroalgae slurries	Up to 25 wt% (dry ash-free basis)	Coprocessed with 10 wt% n-heptane, toluene, or anisole at 300–350°C for 3–5 min with 2–5 wt% loadings of algal feedstocks to achieve in situ fractionation of HTL biocrude oil
Jazrawi et al. [72]	15–90 L/h of feed algal feedstocks (total reaction volume, 2 L)	Up to 41.7 wt% (dry ash-free basis)	Processed at 250–350°C for 3–5 min with 1–10 wt% loadings of microalgae under 150–200 bar
PNNL group			
Elliott et al. [70]	1.5 L/h of microalgae (total reaction volume, 1 L tubular reactor)	38.0–63.6 wt% (dry ash-free basis)	Processed at 350°C for 3–5 min with 1–10 wt% loadings of microalgae under 200 bar
Elliott et al. [77]	1.5 L/h of macroalgae (total reaction volume, 1 L tubular reactor)	8.70–27.7 wt% (dry ash-free basis)	Processed at 350°C with 5–22 wt% loadings of macroalgae under 200 bar
UIUC group			
Ocfemia et al. [84,85]	Up to 3 kg/h of swine manure (total reaction volume, 2 L continuous-stirred reactor)	62.0–70.4 wt% (dry ash-free basis)	Processed at 285–305°C for 40–80 min hydraulic residence time with 20 wt% loadings of swine manure under 90–121 bar

that as the temperature in the CSTR preheater maintained below 200°C, no plugging with algal feedstocks was observed. Trace components (e.g. minerals) were also removed by processing steps so that they would not cause process difficulties during HTL conversions and/or catalytic hydroprocessing stages [70]. For example, before the algal biocrude enters the fixed catalyst bed for catalytic hydroprocessing, the algal biocrude would be pretreated in the CSTR and then passed through the solid separator and a sulphur stripping bed.

In addition to algal biocrude oil production, the PNNL group is also interested in recovering the nutrients from HTL byproducts. The PNNL group studied the compositions of the HTL gaseous products and pointed out that the depressurisation of the high-pressure liquid collecting system reveals possibility to recover unaccounted ammonia. Development of an alternate collection method to allow the ammonia to be collected as dissolved ammonium in the HTL aqueous byproduct is therefore recommended in their studies [70]. Further, improving the recovery system of the solid residues is also suggested, so that a better recovery efficiency of trace elements (e.g. P) can be achieved [77].

Different from the studies reported by the University of Sydney group, the PNNL group not only produces biocrude in a continuous HTL reactor but also upgrades the algal biocrude oil through catalytic hydroprocessing. By using molybdenum sulfide as catalysts, the nitrogen content in the HTL algal biocrude can be effectively reduced from 4.0–4.7 dw% to <0.05–0.25 dw%, depending on the types of algal biocrude oil. Nevertheless, the catalyst activity for a long-term operation would need further investigation, since the PNNL group typically operates the continuous HTL reaction over a range of days. Additionally, the PNNL group also noted that the mineral residues in the HTL biocrude oil may remain as a concern for a long-term operation. Further removal of the trace elements is thus proposed [83].

Compared with the continuous HTL of microalgae in the PNNL studies, the process design and reactor set-up of the University of Sydney group's unit exhibit some differences. The reactor employed by the University of Sydney group is unstirred, and the reactants flow through coiled stainless steel tubes submerged into a fluidised sand bath [83]. Moreover, the University of Sydney group utilised solvents to recover biocrude, whilst PNNL studies recovered the biocrude by gravity separation without the requirement of solvent handling [70].

In contrast to the previously mentioned studies in the University of Sydney and PNNL, the University of Illinois at Urbana-Champaign (UIUC) group focuses on using agricultural biowaste—swine manure and food processing waste—as their HTL feedstocks. The UIUC group has developed a CSTR system to process 20 wt% loadings of swine manure under HTL condition. Different from algal slurry, the swine manure slurry was very viscous (~32,000 cp) and contains solid particles and other compositions such as pig hair that cannot be well homogenised before entering the CSTR system. Therefore, a special high-end metering pump, the valveless rotary piston pump, was utilised in their study, because the valveless design would avoid blockage and undesired backflow caused by flaky particles in the swine manure [84].

The UIUC group reported that no clogging or accumulation was experienced in using the continuous-stirred tank reactor [84]. However, there was a pump durability issue as the piston and sleeve wore out quickly, which was possibly caused by the abrasive nature of swine manure slurries. Similar to the PNNL group's unit, the UIUC group separated the HTL products by gravity without any organic solvents. A decanting method was used to separate the HTL biocrude oil and aqueous products. [84]. Different from the results reported by the University of Sydney and PNNL groups, a longer residence time (60 min vs. 3–5 min) was employed by the UIUC group. The authors stated that the continuous mode operation of their reactor showed that steady state could be achieved in about 60 min [84].

In summary, technical challenges associated with HTL of protein-containing biomass are highly dependent on the feedstock compositions, optimisation of the liquefaction/upgrading process variables, and effective separation techniques to recover biocrude oil or remove precipitated nutrients [70]. The physiochemical properties of the oil could be different due to the separation methods alone (e.g. solvent extraction, distillation, or centrifugation) [84,85]. Further studies to investigate different types of protein-containing feedstocks, various separation techniques, and HTL operation strategies are needed. In addition, high-pressure feeding systems for protein-containing biomass slurries have been recognised as a process development challenge for a continuous HTL system [83,84]. Moreover, the above-mentioned studies mostly focus on pilot-scale implementations (the total reactor volume is generally 1–2 L); a demonstration scale (e.g. 1–2 t processing capacity) would be highly recommended to more practically realise the potential of using HTL to produce biofuel from protein-containing biomass.

References

[1] Narayan S, Muldoon J, Finn M, Fokin VV, Kolb HC, Sharpless KB. On water: unique reactivity of organic compounds in aqueous suspension. Angew Chem Int Ed 2005;44:3275–9.

[2] Calvo L, Vallejo D. Formation of organic acids during the hydrolysis and oxidation of several wastes in sub-and supercritical water. Ind Eng Chem Res 2002;41:6503–9.

[3] Savage PE. A perspective on catalysis in sub-and supercritical water. J Supercrit Fluids 2009;47:407–14.

[4] Miao C, Chakraborty M, Dong T, Yu X, Chi Z, Chen S. Sequential hydrothermal fractionation of yeast Cryptococcus curvatus biomass. Bioresour Technol 2014;164:106–12.

[5] Eboibi B, Lewis DM, Ashman PJ, Chinnasamy S. Influence of process conditions on pretreatment of microalgae for protein extraction and production of biocrude during hydrothermal liquefaction of pretreated Tetraselmis sp. RSC Adv 2015;5:20193–207.

[6] Chen W-T, Ma J, Zhang Y, Gai C, Qian W. Physical pretreatments of wastewater algae to reduce ash content and improve thermal decomposition characteristics. Bioresour Technol 2014;169:816–20.

[7] Minarick M, Zhang Y, Schideman L, Wang Z, Yu G, Funk T, et al. Product and economic analysis of direct liquefaction of swine manure. BioEnergy Res 2011;4:324–33.

[8] Biller P, Friedman C, Ross AB. Hydrothermal microwave processing of microalgae as a pre-treatment and extraction technique for bio-fuels and bio-products. Bioresour Technol 2013;136:188–95.

[9] Chakraborty M, Miao C, McDonald A, Chen S. Concomitant extraction of bio-oil and value added polysaccharides from *Chlorella sorokiniana* using a unique sequential hydrothermal extraction technology. Fuel 2012;95:63–70.

[10] Cheng J, Huang R, Yu T, Li T, Zhou J, Cen K. Biodiesel production from lipids in wet microalgae with microwave irradiation and bio-crude production from algal residue through hydrothermal liquefaction. Bioresour Technol 2014;151:415–8.

[11] Chakraborty M, AG MD, Nindo C, Chen S. An α-glucan isolated as a co-product of biofuel by hydrothermal liquefaction of *Chlorella sorokiniana* biomass. Algal Res 2013;2:230–6.

[12] Miao C, Chakraborty M, Chen S. Impact of reaction conditions on the simultaneous production of polysaccharides and bio-oil from heterotrophically grown *Chlorella sorokiniana* by a unique sequential hydrothermal liquefaction process. Bioresour Technol 2012;110:617–27.

[13] Costanzo W, Jena U, Hilten R, Das K, Kastner JR. Low temperature hydrothermal pretreatment of algae to reduce nitrogen heteroatoms and generate nutrient recycle streams. Algal Res 2015;12:377–87.
[14] Jazrawi C, Biller P, He Y, Montoya A, Ross AB, Maschmeyer T, et al. Two-stage hydrothermal liquefaction of a high-protein microalga. Algal Res 2015;8:15–22.
[15] Vardon DR, Sharma BK, Blazina GV, Rajagopalan K, Strathmann TJ. Thermochemical conversion of raw and defatted algal biomass via hydrothermal liquefaction and slow pyrolysis. Bioresour Technol 2012;109:178–87.
[16] Leng L, Yuan X, Shao J, Huang H, Wang H, Li H, et al. Study on demetalization of sewage sludge by sequential extraction before liquefaction for the production of cleaner bio-oil and bio-char. Bioresour Technol 2016;200:320–7.
[17] Barreiro DL, Samorì C, Terranella G, Hornung U, Kruse A, Prins W. Assessing microalgae biorefinery routes for the production of biofuels via hydrothermal liquefaction. Bioresour Technol 2014;174:256–65.
[18] Kaushik R, Parshetti GK, Liu Z, Balasubramanian R. Enzyme-assisted hydrothermal treatment of food waste for co-production of hydrochar and bio-oil. Bioresour Technol 2014;168:267–74.
[19] Pei Y, Wang J, Wu K, Xuan X, Lu X. Ionic liquid-based aqueous two-phase extraction of selected proteins. Sep Purif Technol 2009;64:288–95.
[20] Leema JM, Kirubagaran R, Vinithkumar N, Dheenan P, Karthikayulu S. High value pigment production from Arthrospira (Spirulina) platensis cultured in seawater. Bioresour Technol 2010;101:9221–7.
[21] Kumar S, Hablot E, JLG M, Obeid W, Hatcher PG, BM DQ, et al. Polyurethanes preparation using proteins obtained from microalgae. J Mater Sci 2014;49:7824–33.
[22] Spolaore P, Joannis-Cassan C, Duran E, Isambert A. Commercial applications of microalgae. J Biosci Bioeng 2006;101:87–96.
[23] Chen W-T, Zhang Y, Zhang J, Yu G, Schideman LC, Zhang P, et al. Hydrothermal liquefaction of mixed-culture algal biomass from wastewater treatment system into bio-crude oil. Bioresour Technol 2014;152:130–9.
[24] Chen J-H, Huang C-E. Selective separation of Cu and Zn in the citric acid leachate of industrial printed wiring board sludge by D2EHPA-modified Amberlite XAD-4 resin. Ind Eng Chem Res 2007;46:7231–8.
[25] Sprynskyy M. Solid–liquid–solid extraction of heavy metals (Cr, Cu, Cd, Ni and Pb) in aqueous systems of zeolite–sewage sludge. J Hazard Mater 2009;161:1377–83.
[26] Chen W-T, Lin C-W, Shih P-K, Chang W-L. Adsorption of phosphate into waste oyster shell: thermodynamic parameters and reaction kinetics. Desalin Water Treat 2012;47:86–95.
[27] He B, Zhang Y, Funk TL, Riskowski GL, Yin Y. Thermochemical conversion of swine manure: an alternative process for waste treatment and renewable energy production. Trans ASAE 2000;43:1827–33.
[28] He B, Zhang Y, Yin Y, Funk TL, Riskowski GL. Operating temperature and retention time effects on the thermochemical conversion process of swine manure. Trans ASAE 2000;43:1821–5.
[29] He B, Zhang Y, Yin Y, Funk TL, Riskowski GL. Effects of feedstock pH, initial CO addition, and total solids content on the thermochemical conversion process of swine manure. Trans ASAE 2001;44:697–704.
[30] He B, Zhang Y, Yin Y, Funk TL, Riskowski GL. Effects of alternative process gases on the thermochemical conversion process of swine manure. Trans ASAE 2001;44:1873–80.

[31] Chen W-T, Zhang Y, Zhang J, Schideman L, Yu G, Zhang P, et al. Co-liquefaction of swine manure and mixed-culture algal biomass from a wastewater treatment system to produce bio-crude oil. Appl Energy 2014;128:209–16.
[32] Zhang J, Luo Z, Zhang Y. Hydrothermal liquefaction of *Chlorella pyrenoidosa* in water and ethanol. Trans ASABE 2013;56:253–9.
[33] Quitain AT, Sato N, Daimon H, Fujie K. Production of valuable materials by hydrothermal treatment of shrimp shells. Ind Eng Chem Res 2001;40:5885–8.
[34] Yoshida H, Terashima M, Takahashi Y. Production of organic acids and amino acids from fish meat by sub-critical water hydrolysis. Biotechnol Prog 1999;15:1090–4.
[35] Dong R, Zhang Y, Christianson LL, Funk TL, Wang X, Wang Z, et al. Product distribution and implication of hydrothermal conversion of swine manure at low temperatures. Trans ASABE 2009;52:1239–48.
[36] Wang Z. Reaction mechanisms of hydrothermal liquefaction of model compounds and biowaste feedstocks. PhD Dissertation, Urbana, IL: University of Illinois at Urbana-Champaign; 2011.
[37] Zhang Y. Hydrothermal liquefaction to convert biomass into crude oil. In: Biofuels from agricultural wastes and byproducts. Hoboken, NJ: Wiley-Blackwell; 2010. p. 201–32.
[38] Xiu S, Shahbazi A, Wallace CW, Wang L, Cheng D. Enhanced bio-oil production from swine manure co-liquefaction with crude glycerol. Energy Convers Manag 2011;52:1004–9.
[39] Yin SD, Dolan R, Harris M, Tan ZC. Subcritical hydrothermal liquefaction of cattle manure to bio-oil: Effects of conversion parameters on bio-oil yield and characterization of bio-oil. Bioresour Technol 2010;101:3657–64.
[40] Theegala CS, Midgett JS. Hydrothermal liquefaction of separated dairy manure for production of bio-oils with simultaneous waste treatment. Bioresour Technol 2012;107:456–63.
[41] Eboibi BE-O, Lewis DM, Ashman PJ, Chinnasamy S. Hydrothermal liquefaction of microalgae for biocrude production: Improving the biocrude properties with vacuum distillation. Bioresour Technol 2014;174:212–21.
[42] Brown TM, Duan P, Savage PE. Hydrothermal liquefaction and gasification of Nannochloropsis sp. Energy Fuel 2010;24:3639–46.
[43] Zou S, Wu Y, Yang M, Li C, Tong J. Bio-oil production from sub-and supercritical water liquefaction of microalgae *Dunaliella tertiolecta* and related properties. Energy Environ Sci 2010;3:1073–8.
[44] Gai C, Zhang Y, Chen W-T, Zhang P, Dong Y. Energy and nutrient recovery efficiencies in biocrude oil produced via hydrothermal liquefaction of *Chlorella pyrenoidosa*. RSC Adv 2014;4:16958–67.
[45] Yu G, Zhang Y, Schideman L, Funk TL, Wang Z. Hydrothermal liquefaction of low lipid content microalgae into bio-crude oil. Trans ASABE 2011;54:239–46.
[46] Yu G. Hydrothermal liquefaction of low-lipid microalgae to produce bio-crude oil. PhD Dissertation Urbana, IL: University of Illinois at Urbana-Champaign; 2012.
[47] Zhang P. Characterization of biofuel from diatoms via hydrothermal liquefaction [Master Thesis]. Urbana: University of Illinois at Urbana-Champaign; 2014.
[48] Anastasakis K, Ross AB. Hydrothermal liquefaction of the brown macro-alga *Laminaria saccharina*: Effect of reaction conditions on product distribution and composition. Bioresour Technol 2011;102:4876–83.
[49] López Barreiro D, Beck M, Hornung U, Ronsse F, Kruse A, Prins W. Suitability of hydrothermal liquefaction as a conversion route to produce biofuels from macroalgae. Algal Res 2015;11:234–41.

[50] Zhou D, Zhang L, Zhang S, Fu H, Chen J. Hydrothermal liquefaction of macroalgae *Enteromorpha prolifera* to bio-oil. Energy Fuel 2010;24:4054–61.
[51] Roberts GW, Fortier M-OP, BSM S, Stagg-Williams SM. Promising pathway for algal biofuels through wastewater cultivation and hydrothermal conversion. Energy Fuel 2013;27:857–67.
[52] Tian C, Liu Z, Zhang Y, Li B, Cao W, Lu H, et al. Hydrothermal liquefaction of harvested high-ash low-lipid algal biomass from Dianchi Lake: effects of operational parameters and relations of products. Bioresour Technol 2015;184:336–43.
[53] Vardon DR, Sharma BK, Scott J, Yu G, Wang Z, Schideman L, et al. Chemical properties of biocrude oil from the hydrothermal liquefaction of Spirulina algae, swine manure, and digested anaerobic sludge. Bioresour Technol 2011;102:8295–303.
[54] Li H, Yuan X, Zeng G, Huang D, Huang H, Tong J, et al. The formation of bio-oil from sludge by deoxy-liquefaction in supercritical ethanol. Bioresour Technol 2010;101:2860–6.
[55] Pavlovič I, Ze K, Škerget M. Hydrothermal reactions of agricultural and food processing wastes in sub-and supercritical water: a review of fundamentals, mechanisms, and state of research. J Agric Food Chem 2013;61:8003–25.
[56] Biller P, Ross A. Potential yields and properties of oil from the hydrothermal liquefaction of microalgae with different biochemical content. Bioresour Technol 2011;102:215–25.
[57] Demirbaş A. Effect of lignin content on aqueous liquefaction products of biomass. Energy Convers Manag 2000;41:1601–7.
[58] Zhong C, Wei X. A comparative experimental study on the liquefaction of wood. Energy 2004;29:1731–41.
[59] Peterson AA, Vogel F, Lachance RP, Fröling M, Antal Jr. MJ, Tester JW. Thermochemical biofuel production in hydrothermal media: a review of sub-and supercritical water technologies. Energy Environ Sci 2008;1:32–65.
[60] Kabyemela BM, Adschiri T, Malaluan RM, Arai K. Glucose and fructose decomposition in subcritical and supercritical water: detailed reaction pathway, mechanisms, and kinetics. Ind Eng Chem Res 1999;38:2888–95.
[61] Toor SS, Rosendahl L, Rudolf A. Hydrothermal liquefaction of biomass: a review of subcritical water technologies. Energy 2011;36:2328–42.
[62] Rogalinski T, Herrmann S, Brunner G. Production of amino acids from bovine serum albumin by continuous sub-critical water hydrolysis. J Supercrit Fluids 2005;36:49–58.
[63] Adams T, Appel B, Samson P, Roberts M. Converting turkey offal into bio-derived hydrocarbon oil with the CWT thermal process. Power-Gen Renewable Energy Conference 2004.
[64] Tsukahara K, Sawayama S. Liquid fuel production using microalgae. J Japan Petrol Inst 2005;48:251.
[65] Dote Y, Sawayama S, Inoue S, Minowa T, Yokoyama S-y. Recovery of liquid fuel from hydrocarbon-rich microalgae by thermochemical liquefaction. Fuel 1994;73:1855–7.
[66] Chen Y-Q, Wang N, Zhang P, Zhou H, Qu L-H. Molecular evidence identifies bloom-forming Phaeocystis (Prymnesiophyta) from coastal waters of southeast China as *Phaeocystis globosa*. Biochem Syst Ecol 2002;30:15–22.
[67] Sheehan J, Dunahay T, Benemann J, Roessler P. A look back at the us department of energy's aquatic species program: biodiesel from algae; Close-Out Report. 1998.
[68] Yu G, Zhang YH, Schideman L, Funk T, Wang ZC. Distributions of carbon and nitrogen in the products from hydrothermal liquefaction of low-lipid microalgae. Energy Environ Sci 2011;4:4587–95.

[69] Duan P, Savage PE. Hydrothermal liquefaction of a microalga with heterogeneous catalysts. Ind Eng Chem Res 2010;50:52–61.

[70] Elliott DC, Hart TR, Schmidt AJ, Neuenschwander GG, Rotness LJ, Olarte MV, et al. Process development for hydrothermal liquefaction of algae feedstocks in a continuous-flow reactor. Algal Res 2013;2:445–54.

[71] Faeth JL, Valdez PJ, Savage PE. Fast hydrothermal liquefaction of Nannochloropsis sp. to produce biocrude. Energy Fuel 2013;27:1391–8.

[72] Jazrawi C, Biller P, Ross AB, Montoya A, Maschmeyer T, Haynes BS. Pilot plant testing of continuous hydrothermal liquefaction of microalgae. Algal Res 2013;2:268–77.

[73] Zhang J, Chen W-T, Zhang P, Luo Z, Zhang Y. Hydrothermal liquefaction of *Chlorella pyrenoidosa* in sub- and supercritical ethanol with heterogeneous catalysts. Bioresour Technol 2013;133:389–97.

[74] Yu G, Zhang Y, Guo B, Funk T, Schideman L. Nutrient flows and quality of bio-crude oil produced via catalytic hydrothermal liquefaction of low-lipid microalgae. BioEnergy Res 2014;1–12.

[75] He Y, Liang X, Jazrawi C, Montoya A, Yuen A, Cole AJ, et al. Continuous hydrothermal liquefaction of macroalgae in the presence of organic co-solvents. Algal Res 2016;17:185–95.

[76] Cole A, Dinburg Y, Haynes BS, He Y, Herskowitz M, Jazrawi C, et al. From macroalgae to liquid fuel via waste-water remediation, hydrothermal upgrading, carbon dioxide hydrogenation and hydrotreating. Energy Environ Sci 2016;9:1828–40.

[77] Elliott DC, Hart TR, Neuenschwander GG, Rotness LJ, Roesijadi G, Zacher AH, et al. Hydrothermal processing of macroalgal feedstocks in continuous-flow reactors. ACS Sustain Chem Eng 2014;2:207–15.

[78] Clarens AF, Resurreccion EP, White MA, Colosi LM. Environmental life cycle comparison of algae to other bioenergy feedstocks. Environ Sci Technol 2010;44:1813–9.

[79] Zhou Y, Schideman L, Yu G, Zhang Y. A synergistic combination of algal wastewater treatment and hydrothermal biofuel production maximized by nutrient and carbon recycling. Energy Environ Sci 2013;6:3765–79.

[80] Bolin K. In: The SlurryCarb process: turning municipal wastewater solids into a profitable renewable fuel. Proceedings of the Water Environment Federation; 2001. p. 1456–70. 2001.

[81] Leng L, Yuan X, Chen X, Huang H, Wang H, Li H, et al. Characterization of liquefaction bio-oil from sewage sludge and its solubilization in diesel microemulsion. Energy 2015;82:218–28.

[82] Zhao P, Shen Y, Ge S, Chen Z, Yoshikawa K. Clean solid biofuel production from high moisture content waste biomass employing hydrothermal treatment. Appl Energy 2014;131:345–67.

[83] Elliott DC, Biller P, Ross AB, Schmidt AJ, Jones SB. Hydrothermal liquefaction of biomass: developments from batch to continuous process. Bioresour Technol 2015;178:147–56.

[84] Ocfemia K, Zhang Y, Funk T. Hydrothermal processing of swine manure into oil using a continuous reactor system: Development and testing. Trans ASAE 2006;49:533.

[85] Ocfemia K, Zhang Y, Funk T. Hydrothermal processing of swine manure to oil using a continuous reactor system: effects of operating parameters on oil yield and quality. Trans ASABE. 2006;49:1897–904.

[86] Fogler HS. Elements of chemical reaction engineering. Pearson Education; 2016.

[87] Tian C, Li B, Liu Z, Zhang Y, Lu H. Hydrothermal liquefaction for algal biorefinery: a critical review. Renew Sust Energ Rev 2014;38:933–50.

[88] Li D, Chen L, Xu D, Zhang X, Ye N, Chen F, et al. Preparation and characteristics of bio-oil from the marine brown alga Sargassum patens C. Agardh. Bioresour Technol 2012;104:737–42.

[89] Jena U, Das K. Comparative evaluation of thermochemical liquefaction and pyrolysis for bio-oil production from microalgae. Energy Fuel 2011;25:5472–82.

[90] Valdez PJ, Nelson MC, Wang HY, Lin XN, Savage PE. Hydrothermal liquefaction of Nannochloropsis sp.: systematic study of process variables and analysis of the product fractions. Biomass Bioenergy 2012;46:317–31.

[91] Jena U, Das K, Kastner J. Effect of operating conditions of thermochemical liquefaction on biocrude production from Spirulina platensis. Bioresour Technol 2011;102: 6221–9.

[92] Guo Y, Yeh T, Song W, Xu D, Wang S. A review of bio-oil production from hydrothermal liquefaction of algae. Renew Sust Energ Rev 2015;48:776–90.

[93] Garcia Alba L, Torri C, Samorì C, van der Spek J, Fabbri D, Kersten SR, et al. Hydrothermal treatment (HTT) of microalgae: evaluation of the process as conversion method in an algae biorefinery concept. Energy Fuel 2011;26:642–57.

[94] Eboibi B, Lewis D, Ashman P, Chinnasamy S. Effect of operating conditions on yield and quality of biocrude during hydrothermal liquefaction of halophytic microalga Tetraselmis sp. Bioresour Technol 2014;170:20–9.

[95] Akhtar J, Amin NAS. A review on process conditions for optimum bio-oil yield in hydrothermal liquefaction of biomass. Renew Sust Energ Rev 2011;15: 1615–24.

[96] Shuping Z, Yulong W, Mingde Y, Kaleem I, Chun L, Tong J. Production and characterization of bio-oil from hydrothermal liquefaction of microalgae *Dunaliella tertiolecta* cake. Energy 2010;35:5406–11.

[97] Jena U, AT MC, Warren A, Summers H, Ledbetter RN, Hoekman SK, et al. Oleaginous yeast platform for producing biofuels via co-solvent hydrothermal liquefaction. Biotechnol biofuels 2015;8(1):167.

[98] Biller P, Riley R, Ross A. Catalytic hydrothermal processing of microalgae: decomposition and upgrading of lipids. Bioresour Technol 2011;102:4841–8.

[99] Yang C, Jia L, Chen C, Liu G, Fang W. Bio-oil from hydro-liquefaction of *Dunaliella salina* over Ni/REHY catalyst. Bioresour Technol 2011;102:4580–4.

[100] Matsui T-O, Nishihara A, Ueda C, Ohtsuki M, Ikenaga N-O, Suzuki T. Liquefaction of micro-algae with iron catalyst. Fuel 1997;76:1043–8.

[101] Tekin K. Hydrothermal conversion of russian olive seeds into crude bio-oil using a CaO catalyst derived from waste mussel shells. Energy Fuel 2015;29:4382–92.

[102] Ong M. Evaluation of anaerobic membrane bioreactors and hydrothermal catalytic gasification for enhanced conversion of organic wastes to renewable fuels. Urbana, IL: University of Illinois at Urbana-Champaign; 2013.

[103] Biddy M, Davis R, Jones S, Zhu Y. Whole algae hydrothermal liquefaction technology pathway. Golden, CO: National Renewable Energy Laboratory Golden; 2013.

[104] Valdez PJ, Dickinson JG, Savage PE. Characterization of product fractions from hydrothermal liquefaction of Nannochloropsis sp. and the influence of solvents. Energy Fuel 2011;25:3235–43.

[105] Chen W-T, Tang L, Qian W, Scheppe K, Nair K, Wu Z, et al. Extract nitrogen-containing compounds in biocrude oil converted from wet biowaste via hydrothermal liquefaction. ACS Sustain Chem Eng 2016;4:2182–90.

[106] Xiu S, Shahbazi A, Shirley VB, Wang L. Swine manure/crude glycerol co-liquefaction: physical properties and chemical analysis of bio-oil product. Bioresour Technol 2011;102:1928–32.

[107] Gai C, Zhang Y, Chen W-T, Zhou Y, Schideman L, Zhang P, et al. Characterization of aqueous phase from the hydrothermal liquefaction of *Chlorella pyrenoidosa*. Bioresour Technol 2014;150:139–48.

[108] Barreiro DL, Zamalloa C, Boon N, Vyverman W, Ronsse F, Brilman W, et al. Influence of strain-specific parameters on hydrothermal liquefaction of microalgae. Bioresour Technol 2013;146:463–71.

[109] Gai C, Zhang Y, Chen W-T, Zhang P, Dong Y. An investigation of reaction pathways of hydrothermal liquefaction using *Chlorella pyrenoidosa* and Spirulina platensis. Energy Convers Manag 2015;96:330–9.

[110] Patel B, Hellgardt K. Hydrothermal upgrading of algae paste: application of 31P–NMR. Environ Prog Sustain Energy 2013;32:1002–12.

[111] Duan P, Chang Z, Xu Y, Bai X, Wang F, Zhang L. Hydrothermal processing of duckweed: effect of reaction conditions on product distribution and composition. Bioresour Technol 2013;135:710–9.

[112] Leonardis I, Chiaberge S, Fiorani T, Spera S, Battistel E, Bosetti A, et al. Characterization of Bio-oil from hydrothermal liquefaction of organic waste by NMR spectroscopy and FTICR mass spectrometry. ChemSusChem 2013;6:160–7.

[113] Chiaberge S, Leonardis I, Fiorani T, Cesti P, Reale S, Angelis FD. Bio-oil from waste: a comprehensive analytical study by soft-ionization FTICR mass spectrometry. Energy Fuel 2014;28:2019–26.

[114] Faeth JL, Savage PE, Jarvis JM, McKenna AM. Characterization of products from fast and isothermal hydrothermal liquefaction of microalgae. AIChE J 2016;62:815–28.

[115] Dhasmana H, Ozer H, Al-Qadi IL, Zhang Y, Schideman L, Sharma BK, et al. Rheological and chemical characterization of biobinders from different biomass resources. Transport Res Record 2015;121–9.

[116] Tommaso G, Chen W-T, Li P, Schideman L, Zhang Y. Chemical characterization and anaerobic biodegradability of hydrothermal liquefaction aqueous products from mixed-culture wastewater algae. Bioresour Technol 2015;178:139–46.

[117] Biller P, Ross AB, Skill S, Lea-Langton A, Balasundaram B, Hall C, et al. Nutrient recycling of aqueous phase for microalgae cultivation from the hydrothermal liquefaction process. Algal Res 2012;1:70–6.

[118] Pham M, Schideman L, Scott J, Rajagopalan N, Plewa MJ. Chemical and biological characterization of wastewater generated from hydrothermal liquefaction of Spirulina. Environ Sci Technol 2013;47:2131–8.

[119] Pham M, Schideman L, Sharma BK, Zhang Y, Chen WT. Effects of hydrothermal liquefaction on the fate of bioactive contaminants in manure and algal feedstocks. Bioresour Technol 2013;149:126–35.

[120] Cherad R, Onwudili J, Biller P, Williams P, Ross A. Hydrogen production from the catalytic supercritical water gasification of process water generated from hydrothermal liquefaction of microalgae. Fuel 2016;166:24–8.

[121] Appleford J, Ocfemia K, Zhang Y, Christianson L, Funk T, Dong R. Analysis and characterization of the product oil and other products of hydrothermal conversion of swine manure. 2005 ASAE Annual Meeting: American Society of Agricultural and Biological Engineers; 2005. p. 1.

[122] Biller P, Madsen RB, Klemmer M, Becker J, Iversen BB, Glasius M. Effect of hydrothermal liquefaction aqueous phase recycling on bio-crude yields and composition. Bioresour Technol 2016;220:190–9.

[123] Albrecht KO, Zhu Y, Schmidt AJ, Billing JM, Hart TR, Jones SB, et al. Impact of heterotrophically stressed algae for biofuel production via hydrothermal liquefaction and catalytic hydrotreating in continuous-flow reactors. Algal Res 2016;14:17–27.

[124] Zhou Y. Improving integrated systems for algal biofuels and wastewater treatment [PhD Dissertation]. Urbana: University of Illinois at Urbana-Champaign; 2015.
[125] LCS MZ, Tommaso G, Chen W-T, Zhou Y, Nair K, Qian W, Zhang Y, Wang K. Anaerobic digestion of wastewater generated from the hydrothermal liquefaction of Spirulina: toxicity assessment and minimization. Energy Convers Manag 2017;141:420–8.
[126] Shin YH, Schideman L. Characterizing the fate and transport of Chemicals of Emerging Concerns (CECs) from integrated bioenergy and manure management system. 2015 ASABE Annual International Meeting: American Society of Agricultural and Biological Engineers; 2015. p. 1.
[127] Zhou Y, Schideman L, Zheng M, Martin-Ryals A, Li P, Tommaso G, et al. Anaerobic digestion of post-hydrothermal liquefaction wastewater for improved energy efficiency of hydrothermal bioenergy processes. Water Sci Technol 2015;72:2139–47.
[128] Chen H, Wan J, Chen K, Luo G, Fan J, Clark J, et al. Biogas production from hydrothermal liquefaction wastewater (HTLWW): focusing on the microbial communities as revealed by high-throughput sequencing of full-length 16S rRNA genes. Water Res 2016;106:98–107.
[129] Ramos-Tercero EA, Bertucco A, Brilman D. Process water recycle in hydrothermal liquefaction of microalgae to enhance bio-oil yield. Energy Fuel 2015;29:2422–30.
[130] Zhu Z, Rosendahl L, Toor SS, Yu D, Chen G. Hydrothermal liquefaction of barley straw to bio-crude oil: Effects of reaction temperature and aqueous phase recirculation. Appl Energy 2015;137:183–92.
[131] Breinl J, Zhang Y. Hydrothermal gasification of HTL wastewater. 2015.
[132] Zhang L, Lu H, Zhang Y, Li B, Liu Z, Duan N, et al. Nutrient recovery and biomass production by cultivating *Chlorella vulgaris* 1067 from four types of post-hydrothermal liquefaction wastewater. J Appl Phycol 2016;28:1031–9.
[133] Zheng J-L, Zhu M-Q, Wu H-T. Alkaline hydrothermal liquefaction of swine carcasses to bio-oil. Waste Manag 2015;43:230–8.
[134] Li R, Zhang B, Xiu S, Wang H, Wang L, Shahbazi A. Characterization of solid residues obtained from supercritical ethanol liquefaction of swine manure. Am J Eng Appl Sci 2015;8:465.
[135] Singh R, Balagurumurthy B, Bhaskar T. Hydrothermal liquefaction of macro algae: effect of feedstock composition. Fuel 2015;146:69–74.
[136] Kruse A, Funke A, Titirici M-M. Hydrothermal conversion of biomass to fuels and energetic materials. Curr Opin Chem Biol 2013;17:515–21.
[137] Zhang C, Tang X, Sheng L, Yang X. Enhancing the performance of Co-hydrothermal liquefaction for mixed algae strains by the Maillard reaction. Green Chem 2016;18:2542–53.
[138] Vardon DR, Sharma BK, Jaramillo H, Kim D, Choe JK, Ciesielski PN, et al. Hydrothermal catalytic processing of saturated and unsaturated fatty acids to hydrocarbons with glycerol for in situ hydrogen production. Green Chem 2014;16:1507–20.
[139] Guo B, Zhang Y, Ha S-J, Jin Y-S, Morgenroth E. Combined biomimetic and inorganic acids hydrolysis of hemicellulose in Miscanthus for bioethanol production. Bioresour Technol 2012;110:278–87.
[140] Guo B, Zhang Y, Yu G, Lee W-H, Jin Y-S, Morgenroth E. two-stage acidic–alkaline hydrothermal pretreatment of lignocellulose for the high recovery of cellulose and hemicellulose sugars. Appl Biochem Biotechnol 2013;169:1069–87.
[141] Valdez PJ, Savage PE. A reaction network for the hydrothermal liquefaction of Nannochloropsis sp. Algal Res 2013;2:416–25.
[142] Luo L, Sheehan JD, Dai L, Savage PE. Products and kinetics for isothermal hydrothermal liquefaction of soy protein concentrate. ACS Sustain Chem Eng 2016;4:2725–33.

[143] Valdez PJ, Tocco VJ, Savage PE. A general kinetic model for the hydrothermal liquefaction of microalgae. Bioresour Technol 2014;163:123–7.
[144] Xiu S, Shahbazi A. Bio-oil production and upgrading research: a review. Renew Sust Energ Rev 2012;16:4406–14.
[145] Saber M, Nakhshiniev B, Yoshikawa K. A review of production and upgrading of algal bio-oil. Renew Sust Energ Rev 2016;58:918–30.
[146] Vagia EC, Lemonidou AA. Hydrogen production via steam reforming of bio-oil components over calcium aluminate supported nickel and noble metal catalysts. Appl Catal A Gen 2008;351:111–21.
[147] Xu Q, Lan P, Zhang B, Ren Z, Yan Y. Hydrogen production via catalytic steam reforming of fast pyrolysis bio-oil in a fluidized-bed reactor. Energy Fuel 2010;24:6456–62.
[148] Trane-Restrup R, Jensen AD. Steam reforming of cyclic model compounds of bio-oil over Ni-based catalysts: product distribution and carbon formation. Appl Catal B Environ 2015;165:117–27.
[149] Duan P, Bai X, Xu Y, Zhang A, Wang F, Zhang L, et al. Catalytic upgrading of crude algal oil using platinum/gamma alumina in supercritical water. Fuel 2013;109:225–33.
[150] Duan P, Savage PE. Catalytic treatment of crude algal bio-oil in supercritical water: optimization studies. Energy Environ Sci 2011;4:1447–56.
[151] Li W, Pan C, Zhang Q, Liu Z, Peng J, Chen P, et al. Upgrading of low-boiling fraction of bio-oil in supercritical methanol and reaction network. Bioresour Technol 2011;102:4884–9.
[152] Xu X, Zhang C, Zhai Y, Liu Y, Zhang R, Tang X. Upgrading of bio-oil using supercritical 1-butanol over a Ru/C heterogeneous catalyst: role of the solvent. Energy Fuel 2014;28:4611–21.
[153] Li Z, Savage PE. Feedstocks for fuels and chemicals from algae: treatment of crude bio-oil over HZSM-5. Algal Res 2013;2:154–63.
[154] Mortensen PM, Grunwaldt J-D, Jensen PA, Knudsen K, Jensen AD. A review of catalytic upgrading of bio-oil to engine fuels. Appl Catal A Gen 2011;407:1–19.
[155] Zhu Y, Albrecht KO, Elliott DC, Hallen RT, Jones SB. Development of hydrothermal liquefaction and upgrading technologies for lipid-extracted algae conversion to liquid fuels. Algal Res 2013;2:455–64.
[156] Cheng D, Wang L, Shahbazi A, Xiu S, Zhang B. Catalytic cracking of crude bio-oil from glycerol-assisted liquefaction of swine manure. Energy Convers Manag 2014;87:378–84.
[157] Roussis SG, Cranford R, Sytkovetskiy N. Thermal treatment of crude algae oils prepared under hydrothermal extraction conditions. Energy Fuel 2012;26:5294–9.
[158] Zacher AH, Olarte MV, Santosa DM, Elliott DC, Jones SB. A review and perspective of recent bio-oil hydrotreating research. Green Chem 2014;16:491–515.
[159] Furimsky E, Massoth FE. Hydrodenitrogenation of petroleum. Catal Rev 2005;47:297–489.
[160] Huber GW, Iborra S, Corma A. Synthesis of transportation fuels from biomass: chemistry, catalysts, and engineering. Chem Rev 2006;106:4044–98.
[161] Mullen CA, Boateng AA, Reichenbach SE. Hydrotreating of fast pyrolysis oils from protein-rich pennycress seed presscake. Fuel 2013;111:797–804.
[162] Biller P, Sharma BK, Kunwar B, Ross AB. Hydroprocessing of bio-crude from continuous hydrothermal liquefaction of microalgae. Fuel 2015;159:197–205.
[163] Hilten RN, Bibens BP, Kastner JR, Das KC. In-line esterification of pyrolysis vapor with ethanol improves bio-oil quality. Energy Fuel 2010;24:673–82.
[164] Xiong W-M, Zhu M-Z, Deng L, Fu Y, Guo Q-X. Esterification of organic acid in bio-oil using acidic ionic liquid catalysts. Energy Fuel 2009;23:2278–83.

[165] Oasmaa A, Kuoppala E, Selin J-F, Gust S, Solantausta Y. Fast pyrolysis of forestry residue and pine. 4. Improvement of the product quality by solvent addition. Energy Fuel 2004;18:1578–83.
[166] Xu J, Jiang J, Dai W, Zhang T, Xu Y. Bio-oil upgrading by means of ozone oxidation and esterification to remove water and to improve fuel characteristics. Energy Fuel 2011;25:1798–801.
[167] Cheng D, Wang L, Shahbazi A, Xiu S, Zhang B. Characterization of the physical and chemical properties of the distillate fractions of crude bio-oil produced by the glycerol-assisted liquefaction of swine manure. Fuel 2014;130:251–6.
[168] Hoffmann J, Jensen CU, Rosendahl LA. Co-processing potential of HTL bio-crude at petroleum refineries—Part 1: fractional distillation and characterization. Fuel 2016;165:526–35.
[169] Ott LS, Smith BL, Bruno TJ. Advanced distillation curve measurement: application to a bio-derived crude oil prepared from swine manure. Fuel 2008;87:3379–87.
[170] Cao J-P, Zhao X-Y, Morishita K, Li L-Y, Xiao X-B, Obara R, et al. Triacetonamine formation in a bio-oil from fast pyrolysis of sewage sludge using acetone as the absorption solvent. Bioresour Technol 2010;101:4242–5.
[171] Chiaramonti D, Bonini M, Fratini E, Tondi G, Gartner K, Bridgwater A, et al. Development of emulsions from biomass pyrolysis liquid and diesel and their use in engines—Part 1: emulsion production. Biomass Bioenergy 2003;25:85–99.
[172] Li Y, Wang T, Liang W, Wu C, Ma L, Zhang Q, et al. Ultrasonic preparation of emulsions derived from aqueous bio-oil fraction and 0# diesel and combustion characteristics in diesel generator. Energy Fuel 2010;24:1987–95.
[173] Lavanya MR, Meenakshisundaram A, Renganathan S, Chinnasamy S, Lewis DM, Nallasivam J, et al. Hydrothermal liquefaction of freshwater and marine algal biomass: A novel approach to produce distillate fuel fractions through blending and co-processing of biocrude with petrocrude. Bioresour Technol 2015.
[174] Ortiz-Toral PJ. Steam reforming of bio-oil: effect of bio-oil composition and stability. 2008 Iowa State University, Ames, Iowa.
[175] Takanabe K, Aika K-i, Inazu K, Baba T, Seshan K, Lefferts L. Steam reforming of acetic acid as a biomass derived oxygenate: Bifunctional pathway for hydrogen formation over Pt/ZrO2 catalysts. J Catal 2006;243:263–9.
[176] Audo M, Paraschiv M, Cm Q, Louvet I, Hémez J, Fayon F, et al. Subcritical hydrothermal liquefaction of microalgae residues as a green route to alternative road binders. ACS Sustain Chem Eng 2015;3:583–90.

Further reading

[177] Sheehan JD, Savage PE. Products, pathways, and kinetics for the fast hydrothermal liquefaction of soy protein isolate. ACS Sustain Chem Eng 2016;24:6459–6462.

Recipe-based co-HTL of biomass and organic waste

6

S.S. Toor, L. Rosendahl, I. Sintamarean
Aalborg University, Aalborg, Denmark

Hydrothermal liquefaction (HTL) has been investigated for several decades as a means to convert a wide range of biomasses and residues efficiently and with high yield into biofuel intermediates. As a wet processing technology, HTL is generally carried out in aqueous media of at least 50%–60% water making it very suitable for most natural biomasses and organic residues [1–4]. Processed at temperatures around 280–370°C, even up to approximately 400°C, and pressures between 10 and 30 MPa, the original biomass breaks down into a bio-crude phase (typically the main and desired product), an aqueous phase with water-soluble organics, and minor fractions of solid residue and gas. The yield and composition of these products depends not only on reactor type, process conditions, and product workup but also to a very large extent on the biochemical composition of the biomass.

Over the past few years, some efforts have been devoted by researchers to examine the coliquefaction of lignocellulosic biomass with other feedstocks such as algae, manure, glycerol, and synthetic plastics [5–16]. Exploring what has been termed 'synergistic effects', these studies suggest that coliquefaction of various feedstocks forms an interesting operational space in which many parameters can be influenced, for example, yield and product characteristics and partial replacement expensive feedstocks with cheaper ones or using coliquefaction to increase organic loading capacity of continuous processing systems.

These effects will be explored in the following.

6.1 Synergistic biomass mixing effects for continuous liquefaction

Feedstock processability can benefit from a biomass coprocessing strategy. Biomass such as lignocellulosic materials typically form heterogeneous slurries with water, extremely challenging to pump in continuous HTL systems. Other biomass such as algae develop stronger interactions with water being able to form aqueous slurries that can resist mechanical dewatering when the feedstock is pumped into the HTL system. Combining the two biomass streams, lignocellulosics and algae, is a successful strategy to improve the processability of lignocellulosic feeds compared with the pure wood slurries. Microalgae and brown seaweeds were found as suitable cosubstrates to improve the wood slurry processability. When added into the wood-water mixture,

Direct Thermochemical Liquefaction for Energy Applications. https://doi.org/10.1016/B978-0-08-101029-7.00005-9

the algae chemically bind the water and thicken the solution reducing the mechanical dewatering of wood slurries and solid particle settling [17]. Microalgae have several advantages over the brown seaweeds: (a) Due to the lower porosity of the particles, the wood slurries prepared with microalgae have higher biomass loading, and (b) due to the lower ash content of microalgae compared with seaweeds, the organic fraction of the wood-microalgae mixtures is high.

Not only solid cosubstrates but also liquid streams can be mixed with the biomass slurries. An example of a liquid stream is the recycled HTL bio-crude. In this case, a fraction of the water phase is replaced with recycle bio-crude to obtain a more stable biomass slurry. The recycled bio-crude phase serves as vehicle fluid for the wood particles, while the water acts as pore-filling material. To avoid the phase separation between the water phase and the bio-crude, the biomass needs to be mixed with water first followed by the addition of the bio-crude [18]. The lubricating properties of the bio-crude facilitate the flow of the high-viscosity slurry and overcome the bridging of wood particles when the slurry is passed through an orifice. The presence of the bio-crude in the slurry also ensures that the water absorbed by the biomass is not removed when the feed is pressurised, thus reducing the risk of clogging.

Another liquid cosubstrate that can be used in the HTL feedstock is a predissolved biomass solution. The dissolution of the biomass prior to the HTL allows the processing of coarse and mixed biomass substrates. To the biomass solution obtained, raw and finely milled biomass can be added to further increase the biomass loading of the slurry. This method was applied in a continuous HTL system where willow chips were previously dissolved under an alkaline hydrothermal pretreatment at 180°C. The wood paste obtained was further mixed with raw wood powder to obtain a pumpable slurry containing 25% mass fraction of wood [19]. The final mixture was pumped at 30 MPa for several hours in a continuous HTL plant (Table 6.1).

6.2 Coliquefaction with algae

6.2.1 Microalgae and lignocellulosic biomass

Lignocellulosic biomass is probably the most widely used feedstock for bio-oil production via HTL such as wood, barley straw, and miscanthus [20,21]. Previous studies indicate that the bio-oil yield of hydrothermal processing of algal biomass is usually higher than that of lignocellulosic biomass; however, limited information is available about the coliquefaction of these two biomass feedstocks. Gai et al. [5] aimed to explore the coliquefaction of microalgae (*Chlorella pyronoidosa*) and lignocellulosic biomass (rice husk) for bio-crude oil production. The effects of liquefaction temperature (200–350°C), residence time (10–90 min), solid concentration (10–30 wt%), and mass ratio of microalgae/rice husk on product distribution were investigated. As shown in Fig. 6.1, the yield does not follow the linearity of the mixing ratio, with a reduction in impact from about 50/50 towards 100% microalgae. Thus, from a yield perspective, it could be argued that a significant amount of microalgae (expensive) could be replaced by rice husk (cheap) without equivalent effect on oil yield, up to 50%.

Table 6.1 Coprocessing strategies to obtain pumpable wood-based slurries for continuous HTL [1–3]

Biomass		Cosubstrate		
Type	Concentration (wt% of slurry)	Type	Concentration (wt% of slurry)	Advantages for feedstock processability
Wood powder	10–15	Microalgae, brown seaweeds	10–15	Prevents mechanical dewatering of the slurry and solid settling and increases the biomass loading of the feedstock
Wood powder	20	Recycled HTL bio-crude	20–50	Reduces biomass particle settling and mechanical dewatering of the slurry and reduces the risk of clogging
Wood powder	12	Liquefied wood chips	13	Increases the biomass loading of the slurry and allows the utilisation of coarse and mixed biomass substrates

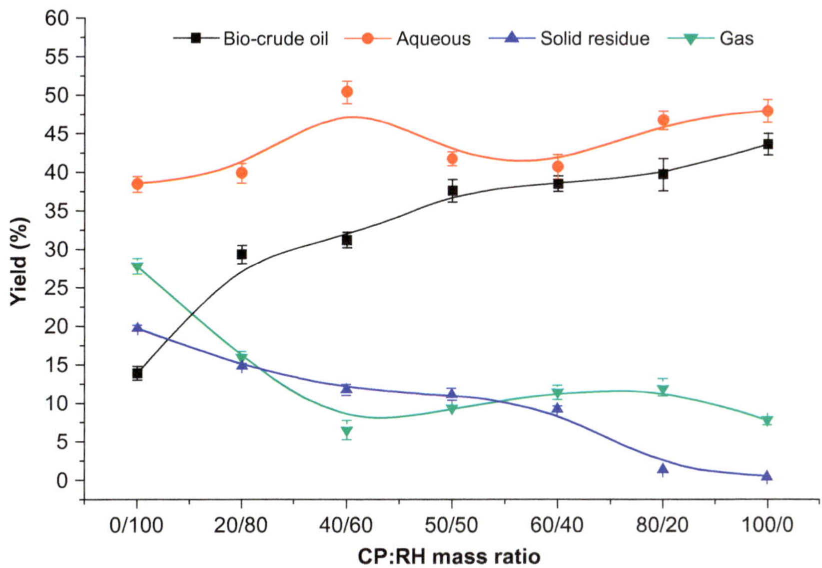

Fig. 6.1 Effect of mass ratio of microalgae/rice husk on product distribution (at 300°C temperature, 60 min residence time, and 20 wt% solid concentration). *CP*, *Chlorella pyronoidosa*; *RH*, rice husk [5].

However, such a conclusion should be supported by inclusion also of oil composition and energy recovery, particularly for fuel applications, as rice husk and microalgae give rise to significantly different oils from HTL.

With the increment of mass ratio of microalgae/rice husk, the bio-crude oil yield increased constantly from 13.9% at the combination of 0 microalgae with 100% rice husk to 43.6% at the combination of 100% microalgae with 0 rice husk. The yields for solid residue gradually decreased from 19.8% to 1.2% with increased mass ratio of microalgae/rice husk from the combination of 0 microalgae with 100% rice husk to the combination of 100% microalgae with 0 rice husk. The solid residue is mainly composed of ash and biochar, and the ash content for microalgae and rice husk was 5.7% and 18.91%, respectively [5]. The yield of solid residue decreased gradually due to the increased proportion of microalgae with low ash content in the feedstock. Another reason is that in hydrothermal environment, the composition of lignin contributes more to the formation of solid residue compared with other constituents. The variations of yields for gas and aqueous fractions were both not monotonic with the increased mass ratio of microalgae/rice husk. A possible reason may be that during hydrothermal processing, the interactions between microalgae and rice husk may alter the product distribution, and it indicates that a synergistic interaction may exist during the coliquefaction process.

Characterisation of the bio-crude oils is necessary in order to evaluate the coliquefaction process. As shown in Fig. 6.2 the seven samples of bio-crude oils obtained at different mass ratios of microalgae/rice husk were analysed using GC/MS to identify the major components.

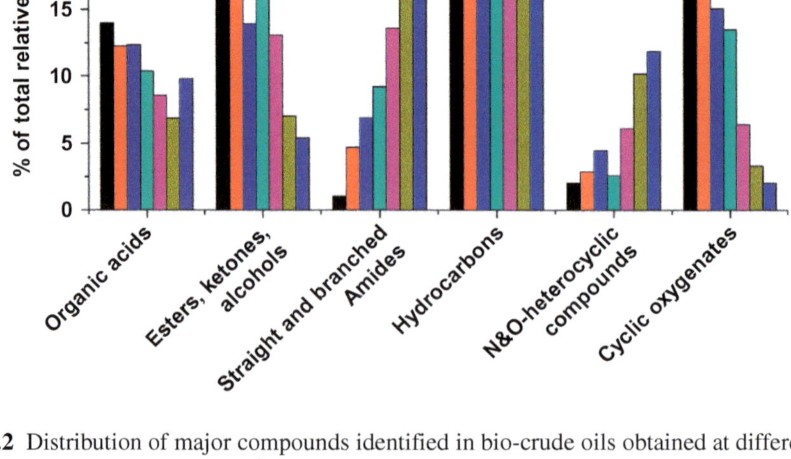

Fig. 6.2 Distribution of major compounds identified in bio-crude oils obtained at different mass ratios of microalgae/rice husk [5].

The mass ratio of microalgae/rice husk showed different effects on chemical composition of the bio-crude oils. With the increasing percentage of microalgae in mixed feedstock from 0% to 100% (sample 1–7), the content of straight and branched amides and N&O-heterocyclic compounds gradually increased, while the concentration of cyclic oxygenates and esters, ketones, and alcohols constantly decreased. It can also be observed in Fig. 6.2 that the variations of organic acids and hydrocarbons were opposite with respect to the increase of the microalgae/rice husk mass ratio. This is because the organic acids are mainly formed from the hydrolysis of cellulose and hemicellulose in lignocellulosic biomass or the hydrolysis of proteins and lipids in microalgae. Hydrocarbons are partially produced from organic acids via decarboxylation. Interestingly, with the increase of the microalgae/rice husk mass ratio, the content of organic acids declined gradually, while the hydrocarbons increased correspondingly.

Sintamarean et al. [21] studied HTL of wood-algae slurries at 400°C and 15 min in 10 mL tubular batch reactors. They concluded that macroalgae slightly decrease the bio-crude yield from 28.5 to 21.6–25.5 wt%, while microalgae increase the bio-crude yield by more than 40% relative to pure wood liquefaction. The negative impact of macroalgae on the bio-crude yield was attributed to the lower conversion of macroalgae carbohydrates into bio-crude compounds, while the higher fractions of lipids and proteins in microalgae led to the increase of the bio-crude yield. Another benefit of microalgae addition is that the total biomass loading and the organic mass fraction of the slurry can be increased by 100% and 90%, respectively. Therefore, when coprocessed with wood, microalgae can improve the wood feedstock pumpability, biomass loading, organic matter, and bio-crude yield. The main disadvantage of using algae (microalgae and seaweeds) as cosubstrate is the introduction of sulphur and nitrogen heteroatoms into the bio-crude. The high ash content of seaweeds is also a major drawback because it reduces the fraction of organic matter in the feedstock.

6.2.2 Microalgae and macroalgae

Microalgae are also employed in the coliquefaction with macroalgae. Compared with microalgae, macroalgae are lower yielding due to their high carbohydrate content [6], which typically results in water-soluble organic products. Therefore, one would expect that a positive synergistic effect might also exist during the coliquefaction of microalgae and macroalgae. Jin et al. [6] examined the coliquefaction of microalgae (*Spirulina platensis*) and macroalgae (*Enteromorpha prolifera*, EP) at 250–370°C, time varied from 5 to 120 min, microalgae/macroalgae mass ratio varied from 0 to 1.0, and water/algae mass ratio varied from 1:1 to 6:1. They explored how these parameters affect the coliquefaction behaviour and the possible synergistic effects between microalgae and macroalgae. The results suggest that a positive synergetic effect existed during the coliquefaction of microalgae and macroalgae, and this synergetic effect was dependent on reaction conditions. Coliquefaction alleviated the severe reaction conditions compared with the separate liquefaction of microalgae and macroalgae. It was concluded that 340°C is sufficient for the codecomposition of microalgae and macroalgae to form an oily compound mixture. Fig. 6.3 showed the effect of

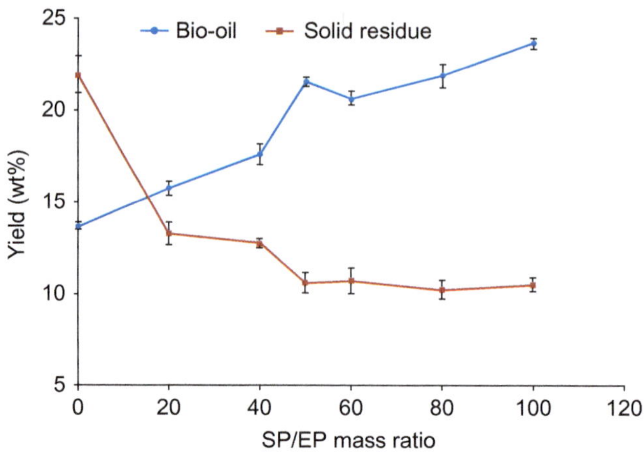

Fig. 6.3 Effects of microalgae/macroalgae mass ratio on the yields of product fractions at 340°C, 40 min, and total algae loading of 2.5 g. *EP*, *Enteromorpha prolifera*, macroalgae; SP, *Spirulina platensis*, microalgae [6].

microalgae/macroalgae mass ratio on the yields of product fractions at 340°C, 40 min, and algal biomass loading of 2.5 g with varying the microalgae/macroalgae mass ratio from 0% to 100%.

The higher bio-oil yield was achieved at a microalgae/macroalgae ratio of 50%. The bio-oil yields obtained at microalgae/macroalgae ratio of 0% and 100% present the individual liquefaction of macroalgae and microalgae, respectively. The bio-oil yield from the microalgae was always higher than that from macroalgae due to the higher organic content especially the lipid in microalgae. The fatty acids derived from the microalgae promoted the conversion of macroalgae, while the inorganic salts in macroalgae suppressed the oil production. As inferred from Fig. 6.10, the synergetic effects during coliquefaction of microalgae and macroalgae were all positive, suggesting that the enhancement from the microalgae was always larger than the suppression from macroalgae during the coliquefaction. The SE value increased with increasing the microalgae/macroalgae ratio, and it hit the highest at a microalgae/macroalgae ratio of 50%. Further increasing the microalgae/macroalgae ratio gradually weakened the synergetic effect. Therefore, the best microalgae/macroalgae mass ratio for the coliquefaction of microalgae and macroalgae is 50% where the largest synergetic effect was achieved. The solid residue yield decreased with increasing the microalgae/macroalgae mass ratio. Recall, the ash contents for microalgae and macroalgae are 17.5 and 38.0 wt%, respectively [6]. Therefore, increasing the microalgae/macroalgae ratio decreased the total ash content in the starting material and thus resulted in decreased solid residue yield. The lowest solid residue yield of 10.6 wt% was achieved at a microalgae/macroalgae mass ratio of 50%. Further increasing the microalgae/macroalgae mass ratio slightly increased the solid residue yield.

The major differences of the bio-oil produced from the HTL of microalgae are that it contained larger amount of unsaturated fatty acids and smaller amount of fatty acid amides and phenols. The bio-oil from coliquefaction of microalgae and macroalgae, which contained both in oils from pure microalgae and pure macroalgae, was more similar in chemical composition to the oil obtained from pure macroalgae. There was also a group of compounds present only in the oil from the coliquefaction, suggesting some secondary reactions occurred between the reaction products. This explains the synergistic effect observed in the coliquefaction of the microalgae and macroalgae. The mutual interaction between microalgae and macroalgae not only promoted the conversion but also improved the energy recovery.

6.2.3 Microalgae and swine manure

Swine manure and algae are each individually viewed as promising feedstocks for bio-crude oil production via HTL [8,22–24]. Chen et al. [23] explored the product distribution, physicochemical properties, nutrient recovery, and energy balance in the coliquefaction of these two feedstocks. HTL was processed at 300°C and 1 h reaction time, which are previously determined optimum conditions for converting mixed-culture algae into bio-crude oil [23].

The distribution of liquefaction products is shown in Fig. 6.4. The highest bio-crude oil yield (35.7% dw) of the coliquefaction process was obtained from the combination of 25% mixed-culture algae (AW) and 75% swine manure (SW). Although the bio-crude oil yield of pure SW was higher than pure AW biomass, the bio-crude oil yields did not increase as the SW fraction of the feedstock increased from 0% to 50%. Instead, the average bio-crude oil yields decreased slightly from 25.8% to 22.0% as the addition of SW increased from 25% to 50%. When the SW fraction was increased to 75%, the bio-crude oil yield increased to 35.7%. This reveals that when

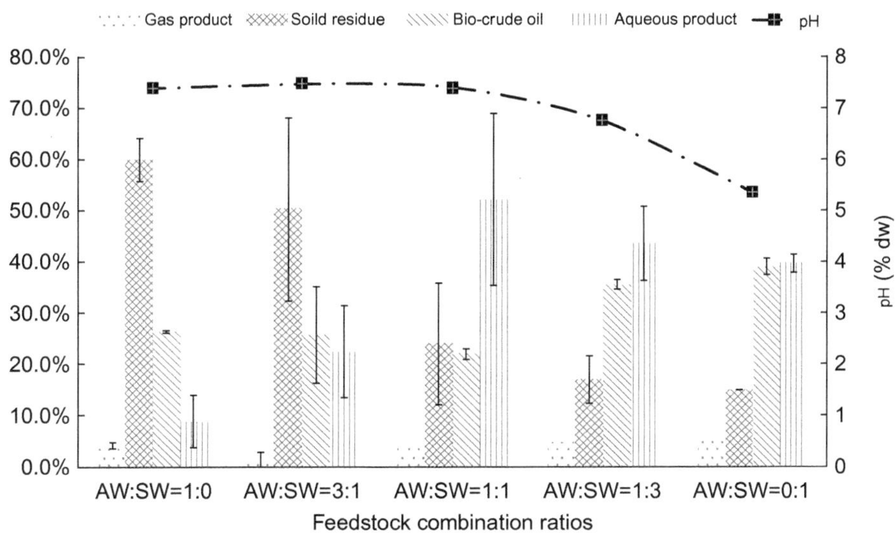

Fig. 6.4 The effect of feedstock combination ratios on yields of liquefaction products based on dry feedstock; *AW*, mixed-culture algae from wastewater; *SW*, swine manure [7].

the ash content of feedstock was too high (in this case, above 30% dw), the feedstock combinations may be preferentially converted into aqueous or solid products rather than bio-crude oils.

As the SW fraction was increased from 25% to 75%, the solid residue yield dramatically decreased from 50.3% to 24.1% and then to 17.1%, which implied that the interaction between AW and SW led liquefaction products to be more favourably converted into aqueous products rather than solid residue. This may also be affected by the pH of the coliquefaction system. As the addition of SW increased from 0% to 75%, the pH of aqueous product was decreased from 7.4 to 6.8, but the pH was relatively stable from 0% to 50% SW. The dilute acidic environment with 75% SW may facilitate the hydrolysis of hemicellulose and the disruption of lignin structure [25], followed by some water-soluble products such as phenols.

Fig. 6.5A shows the GC–MS results for the bio-crude oils. Notably, as the crude fat increased with the addition of SW, the amount of hydrocarbons in the bio-crude oil decreased or remained constant. Typically, fatty acids would undergo decarboxylation to produce hydrocarbons and carbon dioxide under HTL conditions. However, this reaction may be suppressed under hydrothermal conditions.

Fig. 6.5B provides the GC–MS results for the aqueous HTL products. Notably, the organic acids (e.g. acetic acid) were greatly reduced after SW was added into the coliquefaction system, suggesting that the degradation of either lignocellulose or proteins was suppressed. The reduction of cyclic oxygenates in the bio-crude oil obtained with 75% SW and pure SW also supports this argument. Cellulose from different biological sources has different structure and physicochemical properties that can affect its reactivity [26]. Surprisingly, the alkaloids group (mainly pyrazine derivatives) increased as the fraction of SW was increased. This might be due to the fact that chloride was naturally present in manure and the chloride might facilitate the synthesis of pyrazines via self-condensation [27,28].

The carbon recovery (CR) of the bio-crude oils Fig. 6.6A did not increase as the crude fat increased in the feedstock, indicating that there may be antagonisms between SW and AW in terms of oil formation. In contrast, when the cellulose content increased in the coliquefaction system, CR of the solid residue increased as well. This suggests that there is a positive correlation between the amount of cellulose and solid residue. Fig. 6.6B shows that as the addition of SW increased, the nitrogen recovery (NR) of bio-crude oil increased from 16.8% to 43.9%, implying that protein derivatives may tend to react with crude fat derivatives when the crude fat was drastically increased in the coliquefaction system. GC–MS results also demonstrated the straight amide derivatives increased as the crude fat increased.

Fig. 6.7 shows that the energy consumption ratio (ECR), the energy required for thermochemical conversion against the energy that can be recovered from bio-crude oil combustion, of the coliquefaction system with 25% SW plus 75% AW was lower than that of pure AW.

The above observation suggests that more net energy gain was obtained in the coliquefaction system. ECRs suggest that coliquefaction of AW with SW was a competitive alternative system for bio-crude oil production.

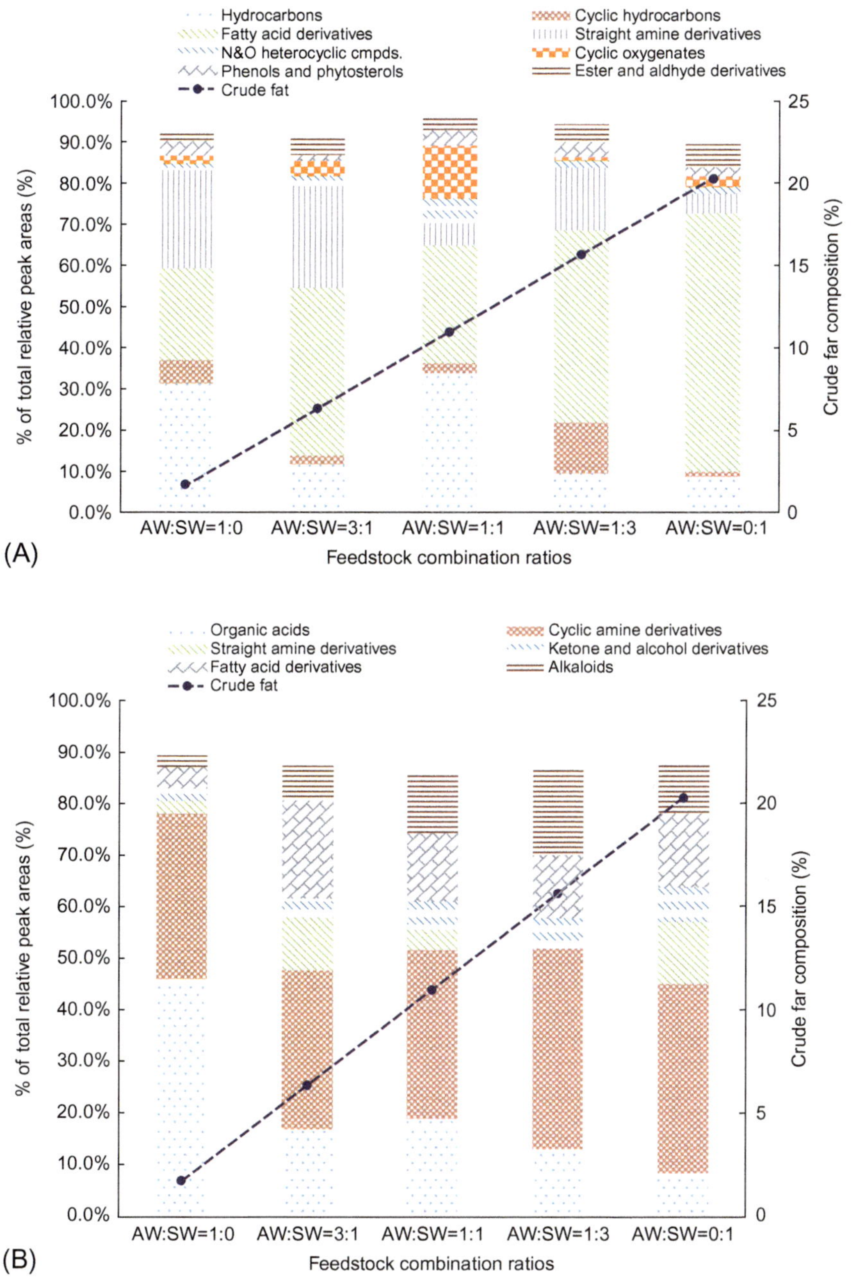

Fig. 6.5 Effect of feedstock combination ratios on the composition of (A) bio-crude oil and (B) aqueous products at 300°C and 1 h reaction time [7].

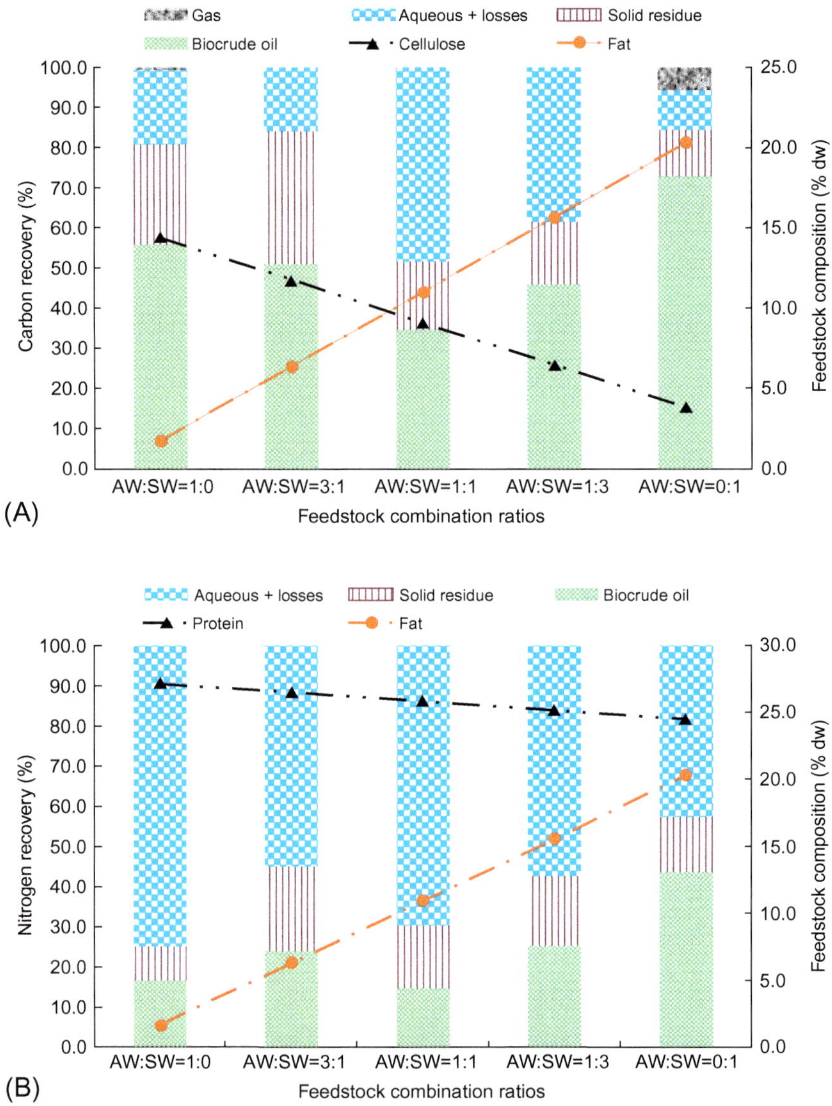

Fig. 6.6 The effect of AW/SW feedstock combination ratios on (A) carbon recovery and (B) nitrogen recovery distributed in various HTL products at the operating condition of 300°C and 1 h reaction time and protein and cellulose contents in the feedstock [7].

6.3 Coliquefaction with crude glycerol

6.3.1 Crude glycerol and aspen wood

In a lignocellulosic context, char formation results primarily from two counteracting mechanisms: (1) dehydration reactions of free sugars favoured at intermediate temperatures, high feed concentrations, and acidic conditions and (2) radical formation

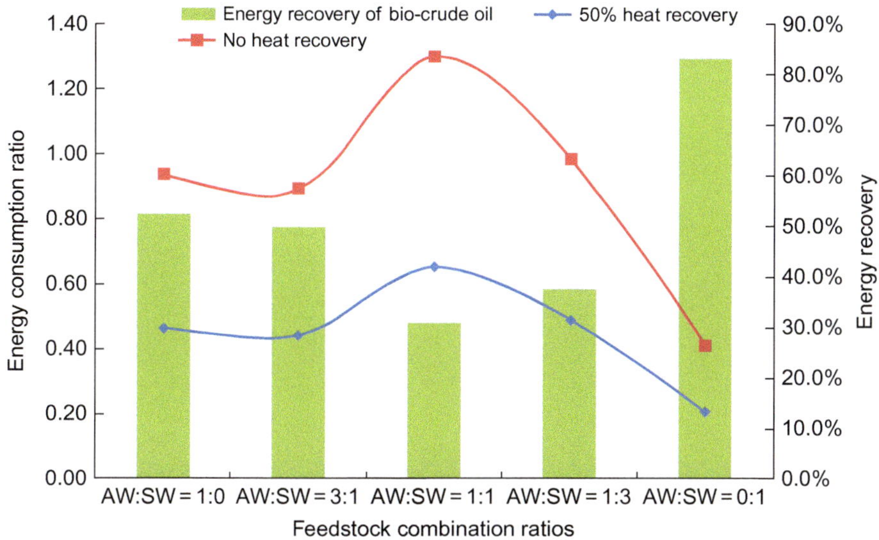

Fig. 6.7 The energy recovery (ER) of bio-crude oils and energy consumption ratios (ECR) for the coliquefaction system (operated at 300°C and 1 h reaction time) with different feedstock combination ratios [7].

during lignin degradation leading to retrogressive char formation. Char formation from carbohydrates can be suppressed by applying high heating rates to reach high reaction temperatures, preferably above supercritical conditions, and alkaline conditions. Lignin radical formation is purely a thermal effect; hence, supercritical conditions tend to enhance lignin char formation. Means of overcoming these counteracting thermal effects includes the addition of a reducing agent such as hydrogen or carbon monoxide or the addition of a stabilising cosolvent acting as a radical scavenger through hydrogen donation. Low-molecular-weight cosolvents such as phenol, propanol, ethanol, methanol, and glycerol have been applied and are preferable in order to obtain low-molecular-weight product compounds, when transport fuel precursors are targeted [10].

Among cosolvents, glycerol is of particular interest since it is already produced in large quantities. Today, glycerol is mainly a refined bio-based by-product from biodiesel production. Pedersen et al. [11] studied glycerol-assisted hydrothermal coliquefaction of aspen wood in batch microreactors. Coprocessing aspen wood and neat glycerol led to a significant reduction in the char yield, and glycerol is hypothesised to act as a radical scavenger, alleviating repolymerisation of especially lignin-derived fragments. In the temperature range of 380–420°C, it was found that bio-crude and char yield and bio-crude quality were all invariant to the reaction temperature. Fig. 6.8 shows that by increasing the crude-glycerol-to-aspen-wood mass ratio from 0:1 to 3:1, char yield was decreased from 18.3 (only aspen wood) to 3.4%. Furthermore, the bio-crude quality in terms of the effective hydrogen-to-carbon ratio (H/C_{eff}) was significantly affected by the crude-glycerol-to-aspen-wood ratio. At a crude-glycerol-to-aspen-wood ratio of 3:1, the H/C_{eff} was 1.6 compared with 0.96 in the case of only aspen wood.

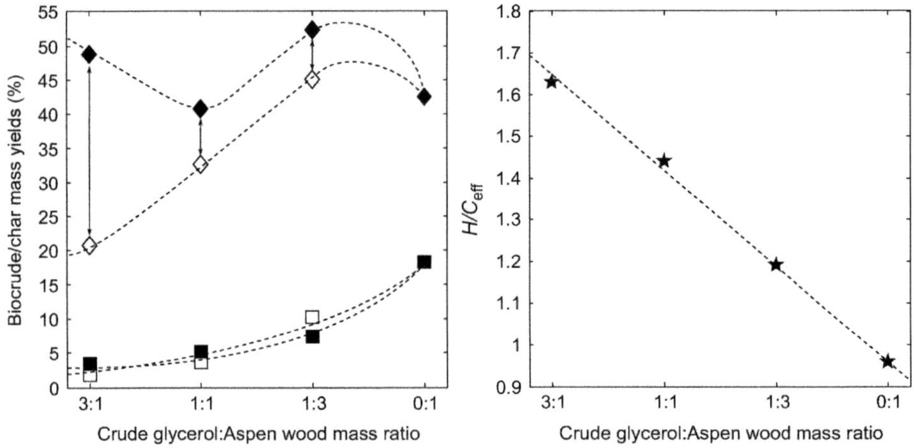

Fig. 6.8 Left: bio-crude (◆) and char (■) yields as a function of crude-glycerol-to-aspen-wood mass ratio in the feedstock. Overlay of bio-crude (◇) and char (□) yields from this figure for comparison between neat and crude glycerol coprocessing with aspen wood. Right: H/C_{eff} (★) as a function of crude-glycerol-to-aspen-wood mass ratio in the feedstock [11].

The composition of all bio-crudes was found to be dominated by oxygenates from wood degradation, unconverted free fatty acids (FFA), and fatty acid methyl esters (FAME) brought by the crude glycerol. Wood-derived organic acids provided a benign recovery of FFAs and FAMEs in the crude glycerol, by breaking the saponified emulsion. FFAs passed through the process unchanged, whereas FAMEs were hydrolysed into corresponding FFAs at high aspen wood shares in the feedstock. Bio-crude and char yield and bio-crude composition and quality were found invariant to reaction temperature.

6.3.2 Crude glycerol and straw

Cao et al. [12] examined the effect of glycerol used as a cosolvent on yields of bio-oil derived from rice straw with different volume ratios of glycerol to water through HTL. The optimal conditions were obtained: 1:1 volume ratio of glycerol to water with 5 wt% of Na_2CO_3 at 260°C for 1 h. Fig. 6.9 showed the effect of glycerol-to-water ratio (v/v) on the product yield. In all of the cases, the yield of oil remarkably increased when glycerol was added as a cosolvent.

The compound contents of the glycerol-water and water-derived bio-oils are obviously different. Classifying the compounds in bio-oils is difficult because they basically contained two or more functional groups. Therefore, in this study, the different organic compounds were classified in terms of the amount of carbon atoms (e.g. C7 represents all the compounds whose molecules contain seven carbon atoms), and distribution of C1–C6, C7–C10, and C11+ compounds in the bio-oils obtained from rice straw with glycerol at different ratios and with or without Na_2CO_3 is presented in Fig. 6.10. C7–C10 compounds, regarded as completely chain-breaking decomposed products of lignin in biomass, were the main categories in the bio-oils of both derivations, which occupied a large proportion in all the three bio-oils. The C11+ compounds, the incompletely chain-breaking decomposed

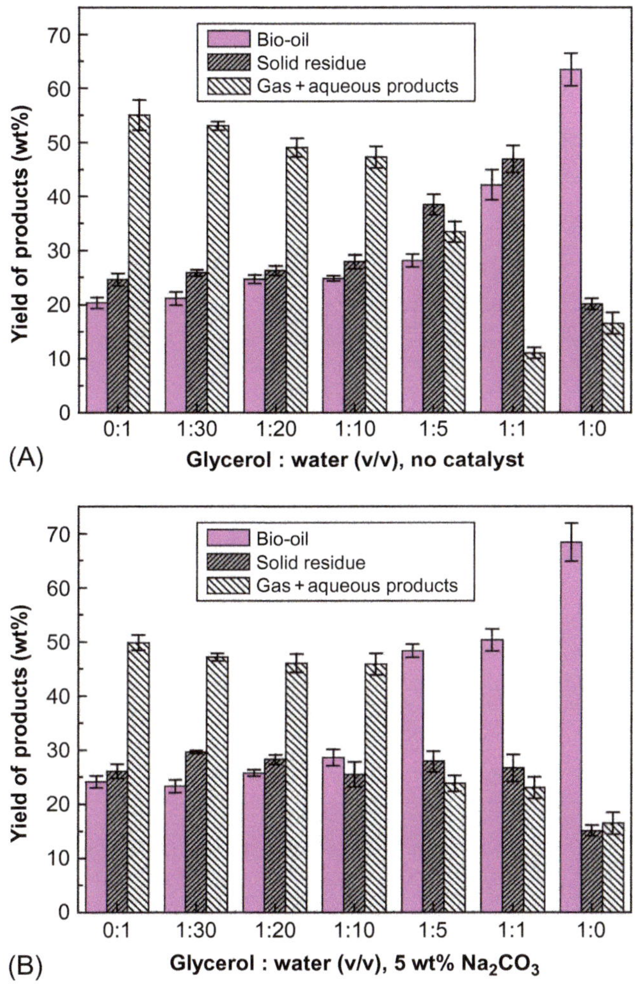

Fig. 6.9 Effect of glycerol-to-water ratio (v/v) on the product yield in HTL of rice straw. (A) No catalyst and (B) with catalyst. Temperature, 260°C; solid/liquid, 15 g/150 mL; initial air pressure, 0.1 MPa; and reaction pressure, 5.5–5.8 MPa [12].

products of lignin, accounted for only a small proportion in the bio-oils. As it can be seen in Fig. 6.10, the compound distribution changed slightly. Since yield of bio-oils increased steadily in considerable amount, it can be concluded that glycerol and Na_2CO_3 promoted the decomposition of lignin, cellulose, and hemicelluloses simultaneously.

6.3.3 Crude glycerol and swine manure

Xiu et al. [8,9] also evaluated the use of crude glycerol as a cosubstrate to improve the bio-oil production during the liquefaction processing of swine manure over a temperature range of 260–360°C and a retention time of 5–90 min. The highest oil yield was

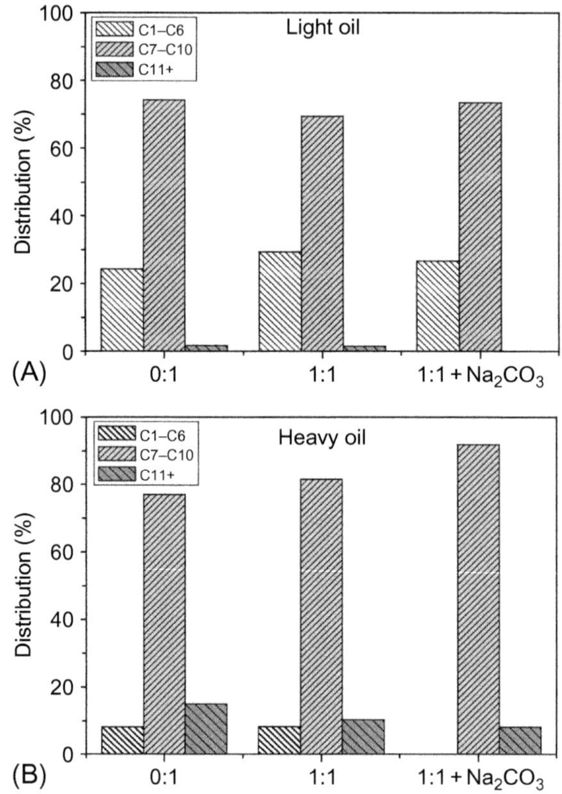

Fig. 6.10 Distribution of C1–C6, C7–C10, and C11+ compounds in the bio-oils obtained from rice straw with glycerol at different ratios and with or without Na_2CO_3 (glycerol/water, v/v) [12].

obtained at 24% on a dry matter basis at reaction temperature of 340°C and retention time of 15 min. At the optimum conditions, crude glycerol was added in the swine manure slurry at different ratios in order to evaluate the use of glycerol as a cosubstrate. Fig. 6.11 shows the effect of crude glycerol on the oil yield at different ratios. In all the cases, the oil yield increased dramatically by adding crude glycerol as the cosubstrate. The maximum oil yield was obtained at 68% with swine-manure-to-crude-glycerol ratio 1:3. At the optimum conditions, compared with the oil production without glycerol addition, the addition of glycerol enhanced the oil production by approximately 1.83 times.

Bio-oil from swine manure, swine manure/crude glycerol, and swine manure/acid phase were thoroughly identified using different analysis procedures. Coliquefaction of swine manure with crude glycerol/acid phase produced oil with desirable physical properties like lower viscosity, lower pH, lower solid content, and higher H contents. Chemical analysis showed that the entire phenol derivative and most of the organic acids were eliminated, while more methyl esters were produced from the addition of crude glycerol or acid phase to swine manure. High concentration of potassium was found in bio-oils obtained from coliquefaction of swine manure/crude glycerol.

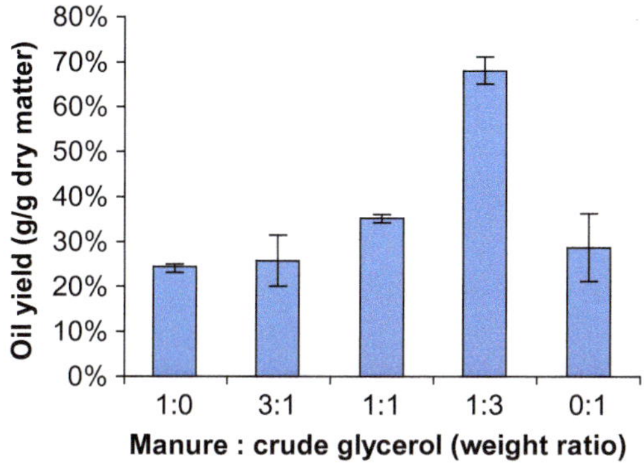

Fig. 6.11 Effect of swine-manure-to-crude-glycerol ratio (weight ratio) on oil yield in the liquefaction of swine manure. Temperature, 340°C; solid content, 20%; initial N_2 pressure, 0.69 MPa; and corresponding pressure, 12.0–15.9 MPa [8].

6.4 Coliquefaction of lignocellulosic biomass with synthetic plastics

Yuan et al. [16] studied the coliquefaction of biomass and high-density polyethylene (HDPE) mixtures in sub- and supercritical water. They investigated the effects of reaction temperature (340–440°C), holding time (0–80 min), water-to-biomass/HDPE ratio (4–10), water filling ratio (6–18 vol%), and biomass plastic composition (100/0, 80/20, 50/50, 20/80, and 0/100) on the liquefaction of lignocellulosic biomass and plastic to oils in 150–270 bar. At 380°C, the most important parameter for the oil yield was biomass/HDPE ratio in the feedstock, and oil yield close to 60 wt% was obtained for the 20/80 weight ratio of biomass/HDPE mixture. From Fig. 6.12, it can be seen that the yield of oil increases with the HDPE concentration for both sawdust/HDPE and rice straw/HDPE mixtures. Meanwhile, at 380°C, the yields of oil obtained in mixture runs are nearly two times higher than the expected ones (dashed lines, calculated as the sum of oil fractions produced by liquefaction of sole component). In some cases, experimental values are lower than those calculated by additive rule. This effect, generally, took place at biomass/plastic ratio lower than 50/50 (weight ratio) and become weaker at 400°C. These experimental results indicate a significant synergistic effect during plastic and biomass coliquefaction at the high-temperature region. Compared with the oils obtained from liquefaction of sawdust, the oils from liquefaction of the sawdust/HDPE mixture have higher contents of carbon and hydrogen and lower concentrations of nitrogen and oxygen, leading to a significantly increasing of higher heating value.

According to GC–MS characterisation, the oil fractions strongly differ depending on the composition of the sawdust/HDPE mixture. The oils from pure sawdust run contain mostly oxygen-containing compounds including aldehydes, carbonyl groups, and carboxylic acids. Both saturated and unsaturated linear hydrocarbons dominate in the oils from pure polyethylene run, while oxygen-containing compounds are in much smaller quantities. The proportion of hydrocarbons increases with a decrease of the sawdust/HDPE ratio. The chemical composition of oils from sawdust/HDPE mixture

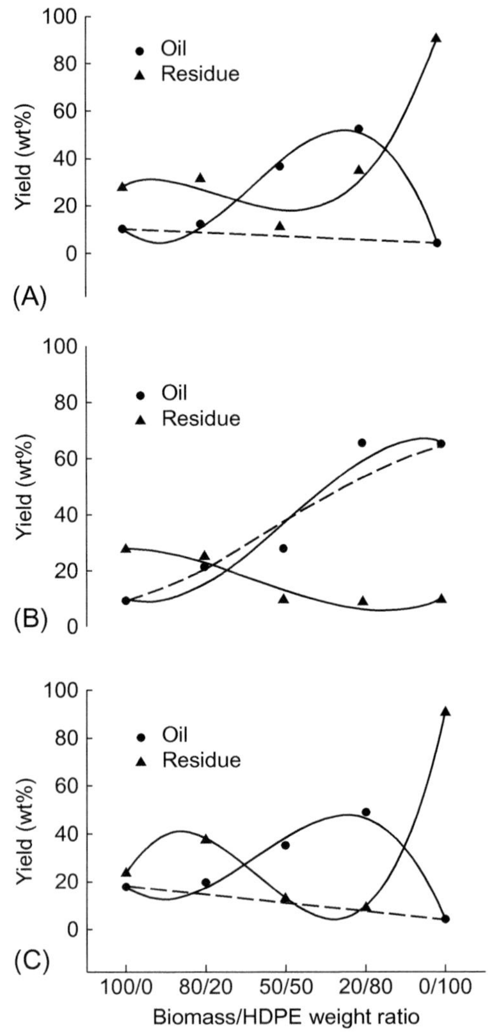

Fig. 6.12 Effect of wood-plastic composition on product yields. (A) Temperature 380°C, reaction time 20 min, H_2O/(sawdust + HDPE) = 6/1, and water filling ratio 12 vol%. (B) Temperature 400°C, reaction time 20 min, H_2O/(sawdust + HDPE) = 6/1, and water filling ratio 12 vol%. (C) Temperature 380°C, reaction time 20 min, H_2O/(rice straw + HDPE) = 6/1, and water filling ratio 12 vol%. The *dashed lines* in figures represent the theoretical additive evolutions of the oil fraction [16].

is more similar to the oil obtained from pure polyethylene. The product of sawdust/HDPE 50/50 run contains compounds that contain both in oils from pure sawdust and pure polyethylene. Moreover, there is also a group of compounds present only in the oil from sawdust/HDPE mixture. Indicatively, during liquefaction, some secondary reactions between thermal decomposition products take place. This explains the synergistic effect observed in the coliquefaction of the biomass/plastic mixture.

6.5 Coliquefaction of various biomasses

6.5.1 Pulp/paper-mill sludge and waste newspaper

Zhang et al. [14] conducted the coliquefaction of secondary pulp/paper-mill sludge (solid concentration, 1.6 wt%) and waste newspaper with a total solid concentration of 11.3 wt% with and without the addition of catalysts at temperatures of 250–380°C for 20 min. In order to investigate the synergistic effects between secondary pulp/paper-mill sludge (SPP) and waste newspaper, feedstock mixtures with the same total solid content (11.3 wt%) but varying SPP-to-newspaper ratios (1:0, 2:1, 1:1, 1:2, 1:6.7, and 0:1 wt/wt) were prepared and liquefied at 300°C for 20 min without catalyst. As shown in Fig. 6.13, the heavy oil (HO) yields for SPP or waste newspaper alone were 18.0 and 21.0 wt%, respectively. The higher potential of waste newspaper to yield HO could be related to its higher volatile organic matter (90.0 wt% in waste newspaper vs 58.7 wt% in SPP) and lignin (21.1 wt% in waste newspaper vs 16.9 wt% in SPP) contents. On the other hand, the gas yield for SPP (12.1 wt%) was higher in comparison with that of newspaper (9.7 wt%). This could possibly be due to the catalytic effects of the high initial pH value (e.g. 8) and the presence of alkali metals, particularly Na, naturally present in the SPP ash.

6.5.2 Spent coffee grounds with lignocellulosic biomass

Yang et al. [15] examined the coliquefaction of spent coffee grounds (SCG) with paper filter (PF), corn stalk (CS), and white pine bark (WPB), respectively, in subcritical

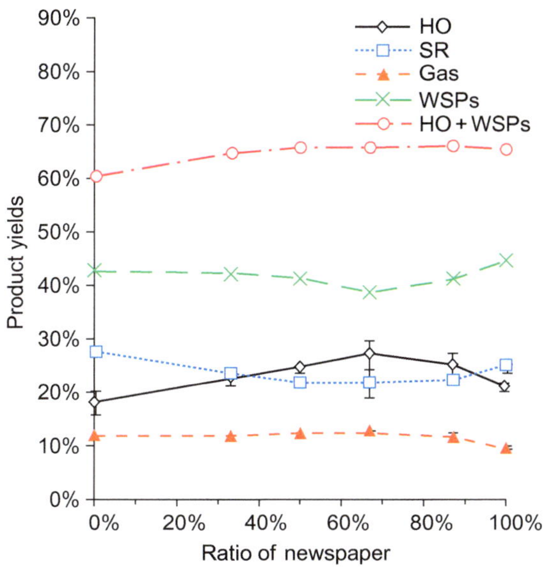

Fig. 6.13 Effects of SPP-to-waste-newspaper ratio on the product yields in the coliquefaction of SPP/waste newspaper mixture at 300°C for 20 min without catalyst. Standard deviations of the yields of HO, SR, and gas appear as error bars; however, for most tests, the errors were too small to be seen [14].

water for bio-crude oil production. The optimum reaction temperature was 250°C, and the mixing biomass ratio was 1:1. SCG and CS were identified as the best feedstock combination with a significant positive synergetic effect in the coliquefaction process with 5% NaOH as a catalyst.

As shown in Fig. 6.14, when the temperature increased from 225°C to 250°C, the yield of bio-crude oil increased from 13.8%, 10.2%, and 15.5% to 18.3%, 21.6%, and 15.6% for the coliquefaction of SCG with PF, CS, and WPB, respectively. However, a further increase of temperature (reached 325°C) led to a slight decrease in oil yield to 15.2%, 16.7%, and 15.4%, for three feedstock combinations, respectively. It is well established that biomass degrades through hydrolysis, dehydration, condensation, decarboxylation and decarbonylation, etc., all of which occur in sequence and/or in parallel. Generally, biomass decomposition was accelerated by increasing temperature. Lignin is a rigid component, and a high temperature promotes its complete decomposition. However, too high temperature might further decompose the oil compounds/intermediates to form water-soluble short carbon-chain compounds, which reduces the oil yield.

Fig. 6.15 illustrates the chemical components of the oil products obtained from the coliquefaction of SCG, PF, CS, and WPB. Interestingly, the major compounds in the oils derived from the coliquefaction of SCG and PF were similar to the SCG oil, containing mainly long-chain carboxylic acids (94.64%), namely, *n*-hexadecanoic acid (45.40%) and octadecadienoic acid (*Z*,*Z*) (46.82%), and a small amount of ester/aldehydes contributed by PF. This result indicates that the volatile compounds in the oil obtained from coliquefaction of SCG and PF were primarily derived from the

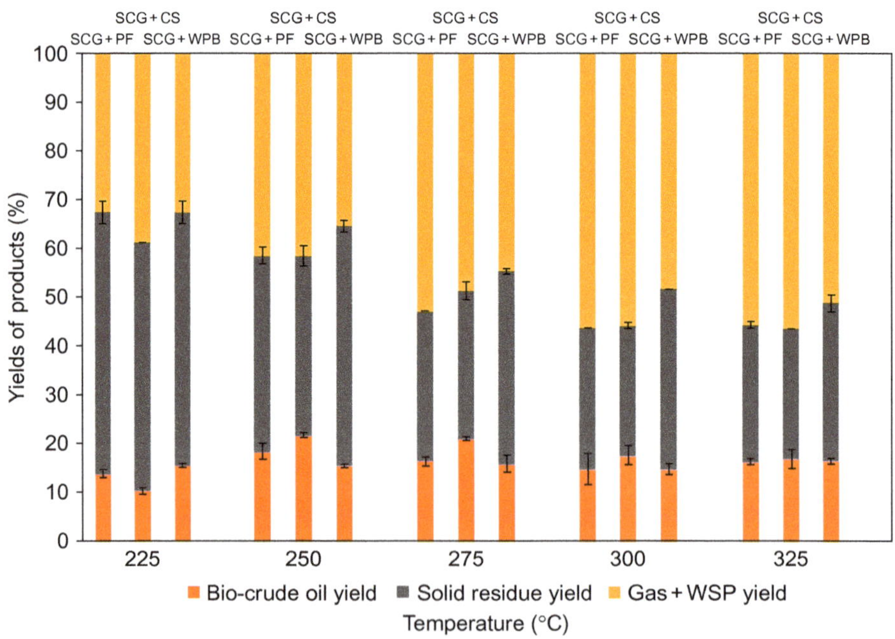

Fig. 6.14 Yields of bio-crude oil, solid residue, gas, and WSP from coliquefaction of SCG with PF, CS, and WPB at various reaction temperatures [15].

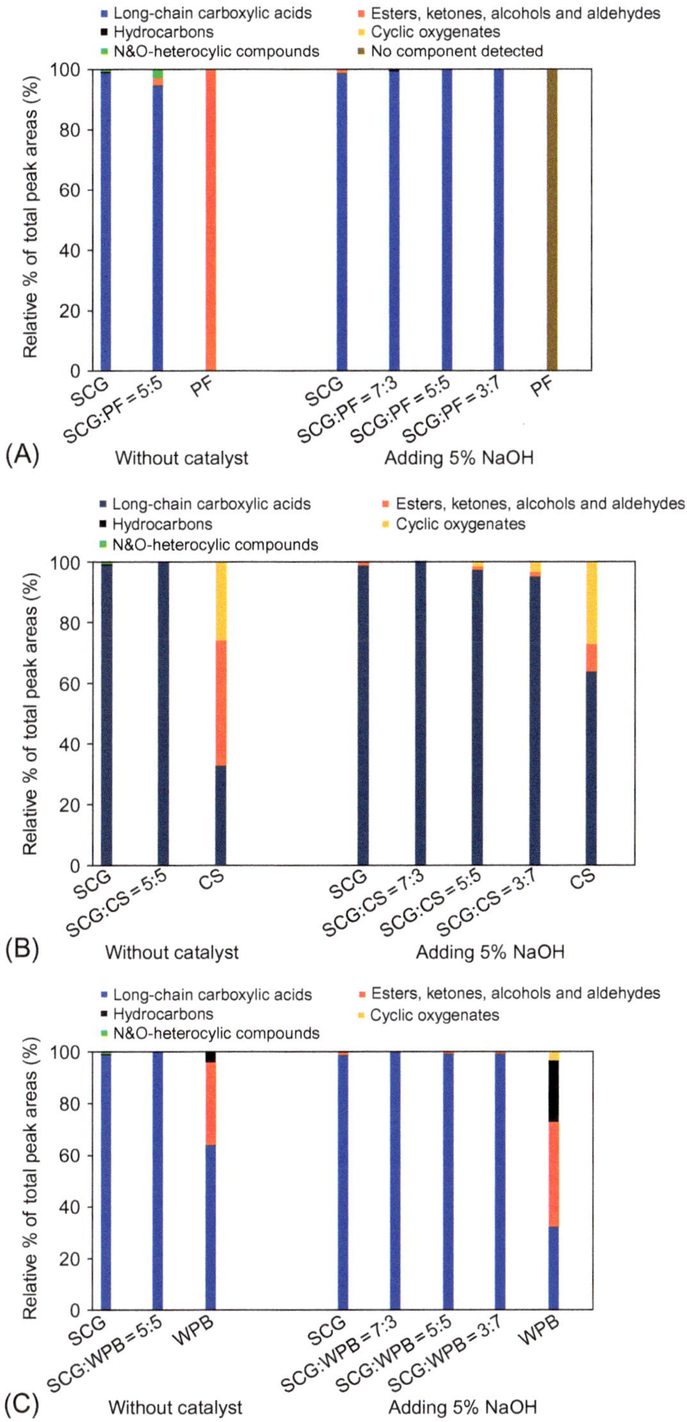

Fig. 6.15 Effect of catalyst and feedstock combination ratios on the composition of bio-crude oil derived from coliquefaction. (A) SCG and PF, (B) SCG and CS, and (C) SCG and WPB [15].

decomposition/depolymerisation of lipid content in the SCG when no catalyst was added. As for the coliquefaction of SCG and CS is concerned, the major compounds in the oil product detectable by GC was carboxylic acids, similar to the coliquefaction of SCG and PF. Similarly, a large amount of carboxylic acids was also detected in the oil product from coliquefaction of SCG and WPB, which were mainly derived from the lipids in the SCG, while the contribution of WPB to the volatile-compound formation was negligible.

References

[1] Vardon DR, Sharma BK, Scott J, Yu G, Wang Z, Schideman L, et al. Chemical properties of biocrude oil from the hydrothermal liquefaction of *Spirulina* algae, swine manure, and digested anaerobic sludge. Bioresour Technol 2011;102:8295–303.

[2] Karagöz S, Bhaskar T, Muto A, Sakata Y. Comparative studies of oil compositions produced from sawdust, rice husk, lignin and cellulose by hydrothermal treatment. Fuel 2005;84:875–84.

[3] Biller P, Ross AB. Potential yields and properties of oil from the hydrothermal liquefaction of microalgae with different biochemical content. Bioresour Technol 2011;102:215–25.

[4] Anastasakis K, Ross AB. Hydrothermal liquefaction of the brown macro-alga *Laminaria Saccharina*: Effect of reaction conditions on product distribution and composition. Bioresour Technol 2011;102:4876–83.

[5] Gai C, Li Y, Peng N, Fan A, Liu Z. Co-liquefaction of microalgae and lignocellulosic biomass in subcritical water. Bioresour Technol 2015;185:240–5.

[6] Jin B, Duan P, Xu Y, Wang F, Fan Y. Co-liquefaction of micro- and macroalgae in subcritical water. Bioresour Technol 2013;149:103–10.

[7] Chen WT, Zhang Y, Zhang J, Schideman L, Yu G, Zhang P, et al. Co-liquefaction of swine manure and mixed-culture algal biomass from a wastewater treatment system to produce bio-crude oil. Appl Energy 2014;128:209–16.

[8] Xiu S, Shahbazi A, Wallace CW, Wang L, Cheng D. Enhanced bio-oil production from swine manure co-liquefaction with crude glycerol. Energ Conver Manage 2011;52:1004–9.

[9] Xiu S, Shahbazi A, Shirley VB, Wang L. Swine manure/crude glycerol co-liquefaction: physical properties and chemical analysis of bio-oil product. Bioresour Technol 2011;102:1928–32.

[10] Pedersen TH, Grigoras IF, Hoffmann J, Toor SS, Daraban IM, Jensen CU, et al. Continuous hydrothermal co-liquefaction of aspen wood and glycerol with water phase recirculation. Appl Energy 2016;162:1034–41.

[11] Pedersen TH, Jasiũnas L, Casamassima L, Singh S, Jensen T, Rosendahl LA. Synergetic hydrothermal co-liquefaction of crude glycerol and aspen wood. Energ Conver Manage 2015;106:881–91.

[12] Cao L, Zhang C, Hao S, Luo G, Zhang S, Chen J. Effect of glycerol as co-solvent on yields of bio-oil from rice straw through hydrothermal liquefaction. Bioresour Technol 2016;220:471–8.

[13] Sintamarean IM, Pedersen TH, Zhao X, Kruse A, Rosendahl LA. Application of algae as cosubstrate to enhance the processability of willow wood for continuous hydrothermal liquefaction. I & EC Research 2017;56:4562–71.

[14] Zhang L, Champagne P, Xu C. Bio-crude production from secondary pulp/paper-mill sludge and waste newspaper via co-liquefaction in hot-compressed water. Energy 2011;36:2142–50.

[15] Yang L, He Q, Havard P, Corscadden K, Xu C, Wang X. Co-liquefaction of spent coffee grounds and lignocellulosic feedstocks. Bioresour Technol 2017;237:108–21.
[16] Yuan X, Cao H, Li H, Zeng G, Tong J, Wang L. Quantitative and qualitative analysis of products formed during co-liquefaction of biomass and synthetic polymer mixtures in sub- and supercritical water. Fuel Process Technol 2009;90:428–34.
[17] Sintamarean IM, Pedersen TH, Zhao X, Kruse A, Rosendahl LA. Application of algae as cosubstrate to enhance the processability of willow wood for continuous hydrothermal liquefaction. Ind Eng Chem Res 2017;56:4562–71. https://doi.org/10.1021/acs.iecr.7b00327.
[18] Daraban IM, Rosendahl LA, Pedersen TH, Iversen SB. Pretreatment methods to obtain pumpable high solid loading wood-water slurries for continuous hydrothermal liquefaction systems. Biomass Bioenergy 2015;81:437–43. https://doi.org/10.1016/j.biombioe.2015.07.004.
[19] Sintamarean IM, Grigoras IF, Jensen CU, Toor SS, Pedersen, Thomas H, et al. Two-stage alkaline hydrothermal liquefaction of wood to biocrude in a continuous bench scale system. Biomass Convers Biorefin 2017; 1–11. http://dx.doi.org/10.1007/s13399-017-0247-9.
[20] Zhu Z, Toor SS, Rosendahl LA, Yu D, Chen G. Influence of alkali catalyst on product yield and properties via hydrothermal liquefaction of barley straw. Energy 2015;80:284–92.
[21] Hafez I, Hassan EB. Rapid liquefaction of giant miscanthus feedstock in ethanol-water system for production of biofuels. Energ Conver Manage 2015;91:219–24.
[22] Zhou Y, Schideman L, Yu G, Zhang Y. A synergistic combination of algal wastewater treatment and hydrothermal biofuel production maximized by nutrient and carbon recycling. Energ Environ Sci 2013;6:3765–79.
[23] Chen W-T, Zhang Y, Zhang J, Yu G, Schideman LC, Zhang P, et al. Hydrothermal liquefaction of mixed-culture algal biomass from wastewater treatment system into bio-crude oil. Bioresour Technol 2014;152:130–9.
[24] He BJ, Zhang Y, Funk TL, Riskowski GL, Yin Y. Thermochemical conversion of swine manure: an alternative process for waste treatment and renewable energy production. Trans ASAE 2000;43:1827–33.
[25] Guo B, Zhang Y, Yu G, Lee W-H, Jin Y-S, Morgenroth E. Two-stage acidic alkaline hydrothermal pretreatment of lignocellulose for the high recovery of cellulose and hemicellulose sugars. Appl Biochem Biotechnol 2013;169:1069–87.
[26] Peterson AA, Vogel F, Lachance RP, Froling M, Antal Jr MJ, Tester JW. Thermochemical biofuel production in hydrothermal media: a review of sub-and supercritical water technologies. Energ Environ Sci 2008;1:32–65.
[27] Pelletier S, Tremblay GF, Belanger G, Chantigny MH, Seguin P, Drapeau R, et al. Nutritive value of timothy fertilized with chloride or chloride-containing liquid swine manure. J Dairy Sci 2008;91:713–21.
[28] Barlin GB. The chemistry of heterocyclic compounds: the Pyrazines. Wiley, Canada; 2009.

Part Three

Downstream processing

Stabilisation of pyrolysis oils

7

R. Lødeng, H. Bergem
SINTEF Materials and Chemistry, Trondheim, Norway

7.1 Introduction

7.1.1 Motivation

The major objective of this chapter is to provide a representative overview of state of the art on chemical and technological aspects related to instability of bio-oils from lignocellulose pyrolysis. This includes stability dedicated scientific and applied research and closely related topics along the whole liquid value chain from biomass to fuels. The interest sphere of this chapter is indicated by red colour in Fig. 7.1. Stability is a critical aspect for providing required freedom in time and being determined for applicability, that is, representing quality and value. It is de facto one of the most challenging problems for extending the practical usage of bio-oil, particularly for enabling ultimate refinement, that is, carbon and energy efficient production of highly refined advanced transport fuels. Bio-oil stability is dependent upon a number of factors, ranging from type and properties of feedstock, characteristics of the actual liquefaction process and its operating conditions, and position in the processing chain, and requires that challenging procedures of handling and use are closely followed. Bio-oil stability, quality, and levels of application potential with increasing demands are close linked, as indicated below:

- Storage and transportability
- Heat and power applications
- (Advanced) transport fuels

Future developments need to evolve from a more comprehensive fundamental understanding of chemical and physical ageing mechanisms, thus bringing forward new solutions enabling production of bio-oils with sufficient stability for handling by conventional procedures and constituting a basis for higher efficiency of further upgrading. A widened spectrum of commercial renewable applications can hopefully be established.

Fig. 7.1 has been adapted from Bridgewater [16].

7.1.2 Background

Academic and commercial interest in biomass as a sustainable multipurpose resource for producing heat, electricity, fuels, chemicals, and materials is at a steadily increasing trend. According to Briens et al. [1], biomass can provide about 25% of the world energy demand, that is, meaning that there is a substantial future renewable potential in this field. Europe aims at increasing the share of renewables in the energy sector by

Direct Thermochemical Liquefaction for Energy Applications. https://doi.org/10.1016/B978-0-08-101029-7.00006-0

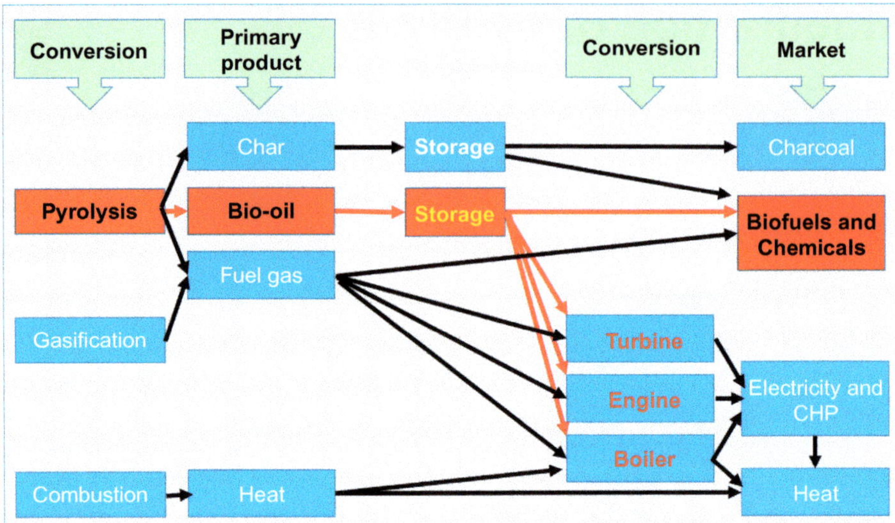

Fig. 7.1 Thermochemical conversion of biomass to different products. Major chapter focus indicated by *red* colour.

20% within 2020 and 27% within 2030, that is, the '20–20-20' and '40–27-27' targets [2]. New technology for replacing conventional energy liquids in the transport sector is an ultimate challenge. Pyrolysis oil instability is a critical hindrance to achieve this, and this chapter's focus is therefore obviously worthy of more and dedicated attention. Table 7.1 provides a summary of reviews related to pyrolysis, bio-oil properties, and stability.

Distributed energy densification is a key requirement for biomass utilisation. Bio-oils obtained by pyrolysis represent such energy. However, liquid density does not imply that crude product stability is obtained. Table 7.2 shows a comparison of volumetric and gravimetric energy densities of common energy carriers. While wood, particularly as pellets, has very low volumetric energy density in the range <10% relative to diesel, it can increase to a level above 50% relative to diesel when liquefied, which is moderate but acceptable. In a densified and preferably stabilised and storable state, the oil provides a better starting point for distribution, further processing, and upgrading.

7.1.3 Status

7.1.3.1 Bio-oil history

Biomass conversion by hydrothermal liquefaction (HTL) and fast pyrolysis (FP) has already been studied for decades. This was triggered by the rising crude oil prices in the 1970s, awakening the concern for alternative strategies for fuels and energy. The interest in renewable fuels is again increasing, now with an emerging understanding that coproduction of chemicals and materials might be required for the resource

Table 7.1 Literature reviews focussing on biomass liquefaction and bio-oil properties

First author	Year	Feedstock	Upstream	Liquefaction	Product bio-oil aspects	Upgrading/use	References
Anjani R.K. Gollakota	2016	Pyrolysis oils			Properties	General approaches	[3]
Tao Kan	2016	Lignocellulosic	Pretreatment	General	Properties	Parameter optimisation	[4]
Anke Krutof	2016	Bio-oils/fossil oils			Properties	Blending for fuels	[5]
Xun Hu	2017	Bio-oils				Acidic catalysis in alcohols	[6]
Zixu Yang	2015	Bio-oils			Stabilisation developments		[7]
Joo-Sik Kim	2015	Lignocellulose		General	Properties	Applications	[8]
Huiju Yang	2014	Pyrolysis oil				General approaches	[9]
Pankaj K. Kanaujia	2014	Lignocellulose			Analytic strategies		[10]
Jani Lehto	2014	Lignocellulose		Fast pyrolysis	Fuel oil quality	Combustion	[11]
Abnisa, F.	2014	Biomass		Co-pyrolysis			[12]
Soo-Young No	2014	Lignocellulose				Application	[13]
Pankaj K. Kanaujia	2013	Biomass		General	Analytic approaches		[14]
Shuangning Xiu	2012	Biomass		Flash pyrolysis/HTL	Properties	General upgrading and use	[15]
Bridgwater, A.V.	2012	Biomass		Fast pyrolysis	Properties	General approaches	[16]

Continued

Table 7.1 Continued

First author	Year	Feedstock	Upstream	Liquefaction	Product bio-oil aspects	Upgrading/use	References
Javaid Akhtar	2011	Biomass		HTL		Parameter optimisation	[17]
Butler, E.	2011	Biomass		Fast pyrolysis		General/ blending/ combustion	[18]
Joseph Zakzeski	2010	Lignin	Pretreatment		Chemistry	Lignin to chemicals	[19]
Elliott, D.C.	2007	Biomass		General	Properties	HT	[20]
Zhang Qi	2007	Biomass		General	Bio-oil properties	General approaches	[21]
Dinesh Mohan	2006	Wood/biomass		General	Physical and chemical properties		[22]
Diebold, J.P.	2000	Fast pyrolysis oils		Fast pyrolysis	Storage stability/ physical/ chemistry		[23]
Bridgwater, A.V.	2000	Biomass		Fast pyrolysis			[24]
Elliott, D.C.	1991	Biomass		General	Extraction, analysis, etc.	General upgrading/ techno-econom.	[25]

Table 7.2 Comparison of energy densities of wood and pellets to other energy carriers

Material	Volumetric (MJ/L)	Gravimetric (MJ/kg)
Diesel	35.8	48
Gasoline	34.2	46.4
Kerosene	37.4	46
LNG	22.2	53.6
LPG	26	46.4
Propane (liquid)	21.9	61.5
Solid coal (black)	39	30
Black coal (bulk)	26.1	30
Bio-oil	ca. 50% of diesel	
Fat (animal/vegetable)	34	37
DME	19.3	28.8
Ethanol	20.9	26.4
Methanol	15.6	19.7
Ammonia	14.6	13.3
H_2 (liquid)	9.4	160
H_2 (700 bars)	5.6	142
Wood	13	16.2
Wood pellets	1.4	19

Adapted from https://en.wikipedia.org/wiki/Energy_density.

utilisation, being a broader basis for a successful business strategy. The history of liquefaction has been reviewed by Elliott in a series of reviews [20, 25, 26]. A high number of general and topic-dedicated publications have been published on the fields of biomass conversion to fuels, chemicals, and products. A recommended selection of highly informative literature is [16, 18, 27–30]. State of the art on technology and research on both pyrolysis and upgrading are summarised by for example Butler et al. [18]. A review providing compatible insight into subcritical HTL has been provided by Saqib et al. [31].

7.1.3.2 Processes

An overview of principal pyrolysis processes, typical conditions, and their major target products and yields is given in Table 7.3. Fast and flash pyrolysis technology is regarded as among the most promising for converting lignocellulose biomass into liquids. An overview of fast pyrolysis technology developments is found in Venderbosch and Prins [32].

Liquid yield is not the only interesting parameter. For fuel production, the nature of the liquid is equally important. Identifying the best process is therefore not straightforward. Variants of fast pyrolysis and HTL technologies have reached the semicommercial stage [18]. However, commercial plants with a complete value chain to final products, that is compatible transport fuels, are still not in operation. This is at least partly due to chemical issues of stability and predictability, and is by nature

Table 7.3 **Pyrolysis technology and principal products**

			Yields/%		
Pyrolysis processes	**Residence time**	**Temp. (K)**	**Liquids**	**Char**	**Gas**
Carbonisation	Days	ca. 673		(Charcoal)	
Torrefaction	10–60 min	ca. 563		(Solid)	
Gasification	Long	1073	5	10	85
Slow pyrolysis	Very long	ca. 673	30 (70% H_2O)	35	35
Intermediate pyrolysis	10–20 s	ca. 773	50 (50% H_2O)	25	25
Fast pyrolysis	0.5–5 s	ca. 723	60–70 (25% H_2O)	18	20
Flash pyrolysis	< 1 s	< 723	65–75	12	13

Adapted from Kan T, Strezov V, Evans TJ, Lignocellulosic biomass pyrolysis: a review of product properties and effects of pyrolysis parameters, Renewable and sustainable. Energy Reviews 2016;57:1126–1140; Mohan D, Pittman CU, Jr., Steele PH, Pyrolysis of Wood/Biomass for bio-oil: a critical review, Energy Fuels 2006;20:848–889; Bulushev DA, Ross, JRH, Catalysis for conversion of biomass to fuels via pyrolysis and gasification: a review, Catalysis Today 2011;171:1–13.

constituting high risk. Candidate bio-oil upgrading steps are still to a high degree addressed at an individual basis at laboratory or pilot scale.

Examples of some fast pyrolysis type demonstration plants are shown in Table 7.4. Complementary information is provided in Butler et al. [18]. Insight into the performance of a commercial scale biomass fast pyrolysis plant for bio-oil production is found in a recent publication of Cai and Liu [33].

7.1.3.3 Applications

Commercial use of second-generation transport fuels or chemicals is still strongly limited. Biofuels are adding up to ca. 4% share of the fuel pool, according to 2014 numbers of IEA [34], and expected to rise slowly to 4.3% towards 2020. The present share is dominated by first-generation biofuels. Availability and use of second-generation sustainable biofuels from lignocellulose are still highly limited, which is to a large extent caused by chemical instability of such bio-oils.

Neat bio-oils are unsuitable for use as transportation fuels due to the water and oxygen content and the low heating value. Major problems for diesel engines running on neat oils are inefficient ignition and corrosiveness and clogging of injection nozzles due to coke formed during combustion. Viscosity is higher than for conventional diesel and leading to poor atomisation and consequently to inefficient ignition and combustion.

Bio-oils are applicable as substitutes or as a component added to fuel oil or diesel in boilers, furnaces, static engines, and turbines for electricity. Relevant reviews are [8,13]. In spite of their low heating value, bio-oils have similar combustion characteristics as light fuel oil and can therefore replace fuel oils in industrial boiler applications. In case of electricity generation, engine modifications are required due to the high acidity levels of the bio-oils. Bio-oils can be blended with organic solvents or be transformed into emulsions with conventional fuels to improve overall properties [5,13].

7.1.4 Bio-oil instability and the challenge of stabilisation

A bio-oil stabilisation strategy must evolve from a detailed chemical understanding of the potential chemical pathways, optimising the oil for further handling or upgrading. A strong emphasis will therefore be given to state of the art regarding knowledge on bio-oil chemistry that is particularly relevant for self-transformations and stability.

Bio-oil chemistry is highly complex due to a number of different aspects. >450 compounds have been identified, illustrating that bio-oil is a complex mixture with components of widely different properties. A high number of components present in bio-oils are not structurally determined yet. The chemical and physical properties of raw bio-oils are far from being compatible with those of standard fossil-based fuels. The challenge of thermal stability is due to the high acidity and the levels and types of oxygenates and aromatics that can easily undergo oligomerisation and polymerisation reactions. The chemical composition of raw bio-oil from pyrolysis is dominated by oxygenated organic compounds. Components include not only carboxylic acids, aldehydes, alcohols, ketones, olefins, and esters but also more stable ethers, phenols, vanillins, furans, pyrones, guaiacols, syringols, and terpenes, besides water. The removal

Table 7.4 **Examples of fast pyrolysis demo-plants and their status ([18], wiki; http://pyrowiki.pyroknown.eu/index.php?title=Pilot_plants)**

Host organisation	Country	Technology	kg feed/h	kg bio-oil/h	Applications	Status	Year
BTG-BTL/EMPYRO	The Netherlands	Rotating cone	5	3.25	Fuel	Operational	2014
Fortum-VALMET	Finland	Fluid bed/riser	10		Fuel	Operational	2013
AE Cote-Nord Bioenergy/Ensyn	Canada	Fluid bed/riser	9	6.4	Fuel	Construction	2017
Red Arrows—Ensyn	Canada	Fluid bed/riser	1667		Fuel	Operational	1996
Ensyn	Canada	Fluid bed/riser	3.5		Fuel	Operational	20xx
Genting	Malaysia	Rotating cone	2		Fuel	Dormant	2005
KiOR	The United States	Catalytic FP	21		Transp. fuel	dormant	2014

of the most reactive oxygen in the mixture is therefore the most requested subject in order to transfer the low-quality bio-oil towards a more stable liquid.

A high oxygen and water content of lignocellulosic biomass (O/C ratio ca. 1) cause a number of chemical and physical challenges along the thermochemical liquid chain. Oxygen chemistry is a true cause of instability, high acidity and corrosiveness, immiscibility with crude-oil-based fuels, high viscosity, and low energy content. A minimum treatment with removal of the oxygen located in unsaturated functional groups is required for achieving pyrolysis oil stabilisation. The reactivity span of typical components is very wide, which is a particular challenge for selecting optimised conditions.

Simple, robust, and efficient strategies for improving bio-oil stability are desired. General reviews on common upgrading strategies are, for example, [3, 9, 15, 16, 21]. Principally different strategies, that is, chemical, thermal, and physical approaches, are treated in dedicated subchapters herein.

7.1.5 Reference stability of conventional fuels

An ambitious goal could be to reach a comparable stability for pyrolysis-derived fuels as for conventional fuels, since this is implying a 'no-change' scenario for society. According to BP [35], under normal conditions, diesel fuels can be expected to stay in a useable condition for

- 12 months or longer at ambient of 293 K
- 6–12 months at an ambient temperature higher than 303 K

While saturated hydrocarbons are relatively stable, the presence of olefins brings along reactive properties and storage challenges. Crude petroleum has in general low levels of olefins, but refinery operations like visbreaking, thermal cracking, and catalytic cracking are causing formation of olefinic components. In addition to the olefin content, storage stability is also dependent upon the levels of heteroatoms. Oxygen and nitrogen are acting as olefin polymerisation catalysts. Dissolved oxygen can react with hydrocarbons forming carboxylic acids or with olefins forming peroxides and hydroperoxides. Formation of hydroperoxides is one mechanistic key to instability, causing free-radical reactions. Sulphur components can act as stabilisers, that is, antioxidants, via quenching of free radicals.

Ageing of conventional diesel leads to fine sediments and gum components, stemming from reactions between diesel components and air. Olefinic polymers formed during ageing have low volatility and are not evaporating well in the fuel systems, thus causing deposits called gums. Sediments and gum are potentially blocking filters, leading to fuel starvation and engine stopping. Such solids burn poorly and can lead to carbon and soot deposits on injectors and high-temperature surfaces. Heteroatoms make up a relatively high portion of such deposits. Expected lifetime of diesel fuel is indicated by an oxidation stability test ASTM D2274, exposing the fuel at 369 K in the presence of oxygen for 16 h, by measuring the amount of gum and sediments. A value of <20 mg/L is acceptable for normal diesel. Antioxidants of lubricating oils include zinc dialkyldithiophosphate, aromatic amines, and alkylated phenols [23].

Ageing is affected by contact with metals, such as Zn, Cu, or alloys containing such. These react with diesel forming unstable compounds. The presence of water, also highly relevant for bio-oils, can allow for growth of fungus and bacteria. These again produce natural by-products such as organic acids, which are causing instability. Exposure to higher temperature and exposure to dust and dirt containing trace elements can destabilise the fuel (Zn and Cu). Fuel composition is of importance since some components age faster than other, that is, olefinic components.

Precautions for prolonged storage are as follows:

- Avoid metal surfaces (Zn and Cu).
- Maintenance programme, removing water and dirt. Water should be drained weekly.
- Keep storage tanks full (exclude oxygen from storage tanks).
- Tanks should have defined low point for collecting and draining of water.
- Tanks should be emptied and cleaned every 10 years.
- Always fill fuel that meets winter type conditions, to avoid wax formation.
- Establish a system for filtering through recirculation filter system.

Additives as indicated below can be used to improve storage life:

- Metal deactivators
- Fungicides/biocides
- Antioxidants
- Fuel stability foam (can keep a fuel stable for 10 years, but does not work in the presence of water and fungus)

Stability of emulsions is of relevance for fossil fuel blends. As long as micelles are small enough to pass through, filters are not creating operational problems [36]. Temperature changes and mixing incompatibility can destabilise emulsions, causing micelle flocculation or aggregation and agglomeration into larger droplets, as well as gravitational separation, that is, leading to sludge. Operational problems will then be encountered. Emulsifying agents can be added for stabilisation. Naturally present phenols in low concentrations can act as antioxidants, but if present in too high concentrations, they can participate in coupling and sludge formation.

A good general introduction to stability issues is provided by Zahira et al. [37], emphasising on oxidative stability. An accelerated test of pure biodiesel (B100) [38] indicates that oxidation begins immediately. Induction times indicate that the fuel would go out of specification for either stability or acid value within 4 months. It was further concluded that it was seemingly the biodiesel stability that was the main parameter of blends, affecting the stability of B5 and B20 (5% and 20% biodiesel blends in conventional fuel). However, the actual level of sulphur (S) in this study was higher than ULSD requirements, and since *S* components can act as antioxidants and free-radical traps, the results must therefore be considered with care.

Thus, a reasonable quantified goal for bio-oils should be to ensure a proper time stability from at least a few months, preferably approaching a year.

7.2 Stability relevant properties of pyrolysis oils

7.2.1 General introduction

This subchapter gives an introduction to the general properties of bio-oils and focusses on some major parameters and aspects of interest in the context of bio-oil instability. Nice overviews of bio-oil properties are found in, for example, [3, 21, 22, 39]. As a comment, pyrolysis bio-oils are addressed by many names. Common terminology includes pyrolysis oil, pyrolysis liquid, bio-crude oil, wood liquids, wood oil, liquid wood, liquid smoke, and pyroligneous acid [7]. For consistency, the simple term bio-oil will be used consequently in this chapter.

Pyrolysis of biomass under conditions of rapid heating and short residence times at elevated temperatures, followed by rapid cooling, can produce low-viscosity single-phase liquids (bio-oils). Yield levels above 70% are reported [3, 4]. While the initial properties of the fresh product would be the generally preferred as a starting point for applications and upgrading, the bio-oils tend to become gradually more viscous during storage. A fundamental reason is found in the thermodynamics. The bio-oils are actually complex nonequilibrium mixtures of oxygenates and other organic and inorganic compounds derived by depolymerisation and degradation of the major constituents, cellulose, hemicellulose, and lignin. They appear as dark brown or black organic liquids and mainly consist of highly oxygenated compounds. The odour can be intense. The obtained condensate is not at thermodynamic equilibrium at typical storage temperatures either, since the rapid quenching is trapping many products that would have reacted further if contact time allows. Consequently, the chemical composition tends to change with time in direction of the thermodynamic equilibrium. Table 7.5 gives a comparison of major properties of bio-oils and heavy fuel oils.

Physical properties of relevance for bio-oil instability include gravimetric density, acidity, viscosity, and actual chemical composition. The elemental composition of bio-oil is not very different from the original lignocellulosic feedstock. A typical elemental composition on a weight basis is ca. 50% C, ca. 7% H, and ca. 43% O, according to Mullen [40]. Of the total carbon, ca. 51% was pyrolytic lignin. Ca. 30% was presented as acids, aldehydes, ketones, and furan-based components. The sugar fraction and derivatives are substantial. Ca. 20% was identified as levoglucosan, sorbitol, and xylose. C_5 sugars amounted to almost 12%. Relevant information on quantified composition and distribution of carbon in the pyrolysis oils is, for example, provided by Routray et al. [41].

Pyrolytic lignin is basically oligomers of phenolics with molecular masses in the range anywhere between 50 and 3000. According to [42–44], an average is typically between 400 and 1300.

Pyrolysis bio-oils are actually microemulsions, being less stable than mineral oils. Viscosity is increasing over time, and then, even phase separations may easily happen. This phase instability is probably a result of a breakdown of the stabilised microemulsion due to chemical reactions proceeding in the bio-oil. Bio-oils consist of molecules of different sizes obtained via depolymerisation and fragmentation of

Table 7.5 **Properties of bio-oils relative to heavy fuel oil**

Physical property	Bio-oil	Heavy fuel oil
Moisture content	15–30	0.1
pH	0.	
Specific gravity	1.2	0.9
Elemental composition		
C	54–58	85
H	5.5–7.0	11
O	35–40	1
N	0–0.2	0.3
Ash	0–0.2	0.1
HHV(MJ/kg)	16–19	40
Viscosity (at 50 °C) (cP)	40–100	180
Solids (wt%)	0.2–1	1
Distillation residue (wt%)	up to 50	1

cP, centipoise; 1 cP = 1 mPa s.
Adapted from Qi Z, Jie C, Tiejun W, Ying X, Review of biomass pyrolysis oil properties and upgrading research. Energy Convers Manag 2007;48:87–92.

lignocellulosic biomass. The resulting multicomponent mixtures are quite different from fossil-derived fuels. The presence of oxygen is the main reason for the difference in the properties and behaviour relative to fossil-derived fuels. The high polarity and hydrophilic nature is responsible for immiscibility with liquid hydrocarbon fractions.

The most important driving force with respect to chemical ageing is temperature. Since bio-oils are different from 'conventional' liquid fuels, certain hurdles have to be overcome, both from a technical, scientific, and marketing point of view. The quality and yield are highly dependent on different process conditions, like temperature, type of feedstock, contact time, and moisture content, as well as catalysts applied. The molecular mass distribution depends on the heating rate up to the final pyrolysis temperature, the contact time, and the particle size and type of biomass feedstock applied. In a way, as mentioned earlier, the bio-oil can be considered as a microemulsion in which the continuous phase is an aqueous solution of lignocellulosic decomposition products. The continuous liquid phase stabilises the discontinuous phase that is mainly composed of pyrolytic lignin macromolecules. The microemulsion stabilisation is largely achieved by hydrogen bonding and micelle formation.

7.2.2 Analysis

The water content is typically measured by Karl Fischer titration, whereas the elemental balances (H/C and O/C ratios) are monitored with CHN elemental analysers. Solvent fractionation can be applied for registration of changes with respect to different product groups, and capillary electrophoresis is a powerful tool for the measurement of volatile acids. ^{1}H NMR and ^{13}C NMR spectroscopy are the preferred methods for checking hydrogenation of carbonyls and acids, change in phenolic and aromatic components, and detachment of methoxy groups from the aromatics.

Physical analysis of the bio-oil is covered by the measurements of viscosity, density, pH value (acidity), heating value, flash point, pour point, water (moisture), ash, and solid content, whereas the chemical composition usually is determined by gas chromatography (GC–MS), at least with respect to the volatile compounds. Some oligomeric structures are not detectable by GC–MS; however, high-pressure liquid chromatography in combination with electrospray mass spectrometry (HPLC/ES-MS) can provide some separation and molecular mass determination of nonvolatile compounds. Since bio-oils have a number of special properties, characteristics, and features, a proper application-specific analysis is required before any storage, upgrading, or utilisation can be considered. Some key properties and dedicated analysis required for fulfilling specification and end-user requirements are provided by Oasmaa et al. [45]. A package of complementary tools for an extensive characterisation of bio-oils is suggested by Nadege Charon et al. [46].

7.2.2.1 Chemical composition

Bio-oils are complex mixtures of >450 compounds containing mainly carbon, oxygen, and hydrogen, composed of carboxylic acids, aldehydes, alcohols, ketones, esters, ethers, phenols, vanillins, furans, pyrones, guaiacols, syringols, terpenes, and others, in addition to water. Literature providing useful insight into the chemical detailing is found in, for example, [13, 18, 23, 47–50].

The pyrolysis temperature has large impact on the chemical composition of the obtained bio-oils. In case of pinewood pyrolysis, the concentration of acetic acid was observed to be almost stable upon the change in pyrolysis temperature, whereas the concentrations of propionic and crotonic acid were increasing. On the other hand, a rapid increase in the concentration of phenol derivatives (phenol, cresols, and catechols) has been observed upon increasing pyrolysis temperature from 723 to 748 K. In parallel, the concentration of guaiacol (2-methoxy-phenol) and its derivatives decreased as the pyrolysis temperature increased from 723 to 773 K. A corresponding behaviour has been observed for bio-oils obtained by pyrolysis of beechwood, spruce wood, olive husk, and hazelnut shell. At enhanced pyrolysis temperatures, the guaiacol compounds are further cracked to phenols, cresols, and catechols. These products are easier to hydrogenate than the guaiacol derivatives. Thus, the pyrolysis temperature is a subject for optimisation, but not as a solution to the stabilisation issue.

7.2.2.2 Water content

Water in bio-oils derives from the original moisture content in the feedstock and from the dehydration processes proceeding during pyrolysis and storage. A typical concentration range is 15–30 wt% [21]. Cooling conditions for pyrolysis vapours can affect the final water content. Water has both positive and negative effects on the properties of the liquid product. It reduces the viscosity of the bio-oil and improves the fluidity and increases the pH level; however, the heating value and flame temperatures are lowered. The solubilising effect of hydrophilic components makes it difficult to selectively remove water by separation.

Water measurements can be useful for addressing stability. According to Diebold [23], Karl Fischer was used to follow the water content of an oil fraction during ageing, and it was observed to increase from 16.1% to 17.7% at either 333 or 363 K. The water content of bio-oil is typically analysed using Karl Fischer titrimetric method [7, 45], following the ASTM E 203 standard.

7.2.2.3 Oxygen content

The oxygen contents of bio-oils are being reported at different levels from 35 to 50 wt% [7]. The high number and concentration levels of oxygen compounds lower the energy density to less than half the level of fossil-derived conventional fuel. It causes immiscibility with hydrocarbon-based fuels as well. Due to the complex chemical composition of bio-oils, a wide boiling point range has to be taken into account. However, distillation cannot be applied for separation purposes, since slow heating of bio-oils already induces polymerisation of the most reactive compounds, leaving up to 50 wt% solid residues. A large fraction of the oxygenated species is reactive and leads to instability during storage and transportation. Typically, the oxygen content is measured by performing an elemental analysis of C, H, S, and N in percent of weight, and the difference is considered as oxygen.

7.2.2.4 Viscosity

Viscosity is considered as a direct indicator of the actual bio-oil state and stability. It is a key for practical handling, pumping, and combustor operation, that is, the application potential. Increasing viscosity is often caused by polymerisation type reactions, consequently leading to increasing average molecular weight.

Viscosity tests can be performed using capillary or rotary viscometers at different temperatures, but temperatures of 293 and 313 K are usually recommended [22]. It is recommended by IEA that accelerated measurements (24 h, 353 K) should be performed at well-controlled and standardised conditions to obtain the required repeatability [51]. Mass losses due to volatilisation should be kept below 0.05 wt%. Kinematic viscosity should be measured at 313 K following the ASTM D 445. The percent change in viscosity is a typical way of reporting the test result and is one quantified value of stability.

Depending upon the actual pyrolysis conditions applied and the nature of the biomass-derived feedstocks used, the viscosities of the obtained bio-oils vary largely, 35–1000 cP at 313 K, according to Czernik [52]. In general, the viscosity is higher than that of fossil-derived hydrocarbon fuels. However, the viscosities are lower in bio-oils with higher water content and less water-insoluble components. Upon storage, the viscosity of the bio-oils increases. Addition of alcohols can reduce the density and viscosity and contribute to improved stability, however, with the limitation of a lowered flash point of the mixture. Upon an enhancement of the pyrolysis temperature, the viscosity can be affected. Unlike hydrocarbon fuels, the viscosity decreases rapidly above room temperature [22, 53], and it changes with time. It also decreases with addition of polar solvents, exemplified by methanol or acetone [54].

A pyrolysis oil used by Routray et al. [41] had a measured viscosity of 0.1771 Pas at 298 K and 0.0638 Pas at 313 K.

7.2.2.5 Aromatics and average molecular weight (ave-Mr)

Gel permeation chromatography (GPC) can be used for detecting aromatic compounds selectively and also provides a measure of the increase in average molecular weight, which is typically increasing over time due to condensation and oligo- and polymerisation reactions. Common methods are so-called GPC or size-exclusion chromatography (SEC), but these are not fully applicable to provide generally valid quantitative data. Diebold [23], demonstrated by using GPC on a pyrolysis oil from oak, that the low-molecular-weight (100–300 units) materials decreased and the high-molecular-weight (300–100,000) ones increased.

7.2.2.6 Carbonyls

The water-soluble fraction of the bio-oils are containing significant amounts of aldehydes and ketones, as well as the sugars and acids, containing carbonyl groups. These are primary participants in the chemical changes proceeding during storage. Reactions such as esterification, acetalisation, and aldol condensation then lead to the growth of larger molecular structures. These reactions occur due to the presence of high concentrations of reactive groups, such as the unsaturated carbonyl ($C = O$) that is easily susceptible to chemical attacks.

A linear correlation of carbonyl content and water-insoluble content has been observed during accelerated ageing [55].

The concentrations of carbonyl groups or components can be measured by chemical titration or by other analytical techniques. Oximation is one titration technique in which a hydroxylamine is selectively reacted with aldehydes or ketones in a mixture also containing esters and carboxylic acids to form the oximes. Carbonyls can also be measured by UV spectroscopy, Fourier transform infrared spectroscopy (FTIR), gas chromatography or mass spectrometry (GC/MS), and nuclear magnetic resonance (NMR).

7.2.2.7 Acidity

Bio-oils contain substantial amounts of carboxylic acids, which are causing low pH values, typically in the range 2–3 [7]. Besides being major contributors to instability, the high acidity contributes in addition to a quite corrosive behaviour, especially at elevated temperatures, at which the pH value is increasing. These properties must be taken into account with respect to selection of materials and equipment. Besides carboxylic acids, there are also other acidic compounds contributing to acidity.

Acidity is usually evaluated following ASTM D 664 by measuring the total acid number (TAN, mg of KOH g^{-1}), which is the quantity of KOH in milligrams needed to neutralise the acids in 1 g of the bio-oil [7]. According to Huhnke et al. [7], a stabilised bio-oil should have a total acid number <10.

7.2.2.8 Heating value

The large water content (15–35 wt%) is responsible for a rather low heating value and also flame temperature (flash point), compared with hydrocarbon-derived fractions and fuels. However, bio-oils derived from oil plants usually have a higher heating value compared with those obtained from lignocellulosic biomass or agricultural residues.

7.2.2.9 Pour point

The pour point is the lowest temperature at which the oil can be pumped. For lignocellulosic-derived bio-oils, the pour point is typically about 246 K or even lower.

7.2.2.10 Flash point

The flash point is one important criteria of relevance for oil stability, signifying the lowest temperature at which the fuel sets off to give some vapours, thus mixing with air and igniting [56]. This parameter must comply with insurance and fire regulation with regard to fuel handling and storage. Low viscosity low cetane (LVLC) fuels have lower flash points than diesel, and hence, appropriate measures must be taken for safe storage. On the contrary, pine oil and eucalyptus oil will have higher flash points than ethanol and methanol and thus can be more easily stored.

The flash point of pyrolysis-based bio-oil does not correlate with its ignition properties like for mineral-based oils. A flash is caused by the presence of easily volatile components. However, the water evaporated will immediately distinguish the flame temperature. This results in a low flash point of the bio-oil although it is ignited above 773 K.

7.2.2.11 Ash content

Common minerals present in lignocellulose include K, Na, Ca, Mg, and P [22], are typically entering the value chain already in the pyrolysis step. Ash is of relevance for bio-oil yield and stability. Its presence can catalyse nondesired reactions, for example, causing gas, that is, the loss of liquid yields, and water formation via condensation [18]. This can shift pyrolysis temperatures for maximum liquid yields to lower levels. Maximum recommended ash level in the biomass feedstock is 3%. Higher levels can easily cause phase separation of bio-oil during ageing [57]. Other impurities like Cl^- and S can also lead to deactivation in subsequent catalytic steps. However, there are also other challenges further downstream for the end users.

The ash content is usually not affected by the pyrolysis temperature, but its presence can cause erosion, corrosion, and kicking problems in the engines and the valves and even deterioration when the content is higher than 0.1 wt%. The most problematic components are sodium, potassium, and vanadium, which are responsible for high-temperature corrosion and deposition, whereas calcium is responsible for hard deposits. Severe corrosion of turbines and deposition of surfaces and pipelines can occur during combustion [18].

The ash content is determined by quantifying the noncombustible residues in the bio-oil following the ASTM D482 standard.

It is obviously a potential for general improvements from the removal of ash elements by biomass pretreatment, that is, removing the problem at the root of the value chain. Some interesting papers on demineralisation and its effects on the pyrolysis oil quality are written by Wigley et al. [58], Song Hu et al. [59], Michailof et al. [60], and Stefanidis et al. [61].

7.2.2.12 Char content

According to Huhnke [7], a stabilised bio-oil should have a content of char fines of <0.5 wt%. Solid char residues, consisting mainly of oxides and salts of metals, contribute to changes in bio-oil properties [16]. While char increases the viscosity of bio-oil, it offers plenty of catalytic sites for promoting reactions [21].

7.3 Chemistry of bio-oil self-transformations

7.3.1 General introduction

An overview of relevant literature related to the topic of stability is provided in Table 7.6.

Different terms and meaning of stability is encountered in literature. According to Batts [95], the term 'fuel stability' is typically used for addressing and evaluating oils or fuel resistance to change. The complementary term 'storage stability' [7, 37] is used with the dedicated purpose specified, covering both chemical and physical changes. 'Phase stability' is another variant that occur, with a dominantly physical meaning. In this chapter, the term 'stability' is used in a general context related to self-transformation properties, thus addressing the fundamental cause of both aspects.

Compared with fossil oils, the bio-oils are typically both less chemically and physically stable, due to high levels of chemically unstable oxygen-containing reactive groups, the presence of solids, and low-boiling point volatiles [51, 55]. A number of oxygenated functional groups like the hydroxyaldehydes, hydroxyketones, carboxylic acids, levoglucosan, and also phenolics make pyrolysis oil subject to serious chemical changes at ambient conditions. Aldehydes are considered among the most critical since they easily react with other components, which include water, phenolics, and also alcohols, thus also forming condensed products.

Two different mechanisms targeted in research on bio-oil stability are the resistance to thermally induced changes, and the other is the resistance to oxidation. The thermal stability is defined as a bio-oil or a fuel's ability to withstand stress in terms of temperature over time.

7.3.2 Detailing chemistry of bio-oil instability

7.3.2.1 General behaviour

Pyrolysis processing cracks biomass thermally in the absence of oxygen resulting in solid (char), gaseous (syngas and light hydrocarbons), and liquid (bio-oil) products, according to Table 7.3. This is principally similar to the transformations proceeding during combustion or gasification of biomass. As bio-oils are produced by rapid heating followed by quenching of the product mixture, a thermodynamically stable state is not reached. Species of widely different reactivity are produced, including components that are able to trigger chemistry at room temperature and even below. Some are highly reactive also at ambient conditions tending to undergo chemical changes

Table 7.6 A selection of stability relevant literature

First author	Year	Review	Feedstock	Liquefaction	Bio-oil focus	Upgrading	Reference
Garcia M.,	2017		Bio-oils		Stability	Antioxidants	[62]
Khairuddin Md Isa	2016		Pyrolysis oil		Stability aspects	Supercritical H_2O	[63]
Le Zhang	2016		Crude oil		Ageing, crude oil and ethanol stabilised		[64]
Shinyoung Oh	2016		Bio-oil		Storage performance after HDO	HDO by noble metals	[65]
Renzhan Yin	2016		Bio-oil		Stabilisation by composite additives		[66]
Wenran Gao	2016		Oxygenates, bio-oils		Properties glycerol/ methanol/bio-oil		[67]
Geert Haarlemmer	2016		Beech wood	Fast pyrolysis/ HTL	Analysis		[49]
Xun Hu	2017	Yes	Bio-oils		Stability	Acid catalysis/ esterification	[6]
Anke Krutof	2016	Yes	Bio-oils/fossil oils		Quality	Blending for fuels	[5]
Stedile, T.	2015		Ligno/ triglyceride oils		Physical and chemical properties		[50]
Laleh Nazari	2015		Woody biomass	Cat-HTL	Characterisation of bio-crudes		[68]
Ghodka Praveen	2015		Pyrolysis oils		Stabilisation	Reactive distillation	[69]
Marjorie R. Rover	2015		Bio-oils		Stabilisation	Low-T/P HT	[70]
Xu Ying	2015		Raw oil/models		Stabilisation/methanol as H-donor	In situ HT/Ni	[71]
Etienne Le Roux	2015		Aspen oil		Stability and quality	Pressurised hot H_2O	[72]

Mingming Zhang	2015		Blends		Impurities in glycerol/methanol/bio-oil		[73]
Chiara Boscagli	2015		Straw pyrolysis oil		Light fraction of pyrolysis oil	Mild hydrotreatment/Ni catalysts	[74]
Zixu Yang	2015	Yes	Bio-oils		Stabilisation developments		[7]
Hari P. Reddy	2015		p-Cresol/furfural		Stabilisation of oxygenates	Catalytic H-transfer	[75]
Hao Li	2015		Rice straw	Fast pyrolysis	Stability evaluation		[76]
Jani Lehto	2014	Yes	Lignocellulose	Fast pyrolysis	Fuel oil quality		[11]
Xuefei Cao	2014		Lignin, hemi-, cellulose		Characterisation/comparison		[77]
Wen Chen	2014		Bio-oil		Stabilisation	One-pot esterification/catalytic	[78]
Rounghou Liu	2014		Bio-oils		Physicochemical properties	Acetone addition	[79]
Dengyu Chen	2014		Bio-oils		Storage stability methods evaluation		[80]
Mauro Cordella	2014		Switchgrass, corn	Slow pyrolysis	Ageing of bioliquids		[81]
Paige A. Case	2014		Pine sawdust pyrolysis oil		Stabilisation	In situ formate assisted/0 removal	[82]
Zuogang, Guo	2014		Pyrolysis oil/diesel		Stability	Emulsion fuels	[83]
Nathalia S. Tessarolo	2013		Bio-oils		Chemical detailing/2-D-GC		[84]
Ofei D. Mante	2012		Poultry litter	Fast pyrolysis	Storage stability bio-crudes		[85]

Continued

Table 7.6 Continued

First author	Year	Review	Feedstock	Liquefaction	Bio-oil focus	Upgrading	Reference
Kim	2012		Impregnated wood	Fast pyrolysis	Characterisation of liquid products		[86]
Tae-Seung Kim	2012		Bio-oil		Effect of storage time		[87]
R. N. Hilten/K.C. Das	2010		Bio-oil	Slow pyrolysis	Stability/accelerated ageing procedures		[88]
R. N. Hilten	2010		Pyrolysis vapour		Bio-oil quality/stability	In-line esterification of vapours	[89]
Ronghai He	2009		Switchgrass bio-oil		Stability and physicochemical properties	High-P homogenisation	[90]
Luc Moens	2009		Hardwood bio-oils		Stabilisation chemistry	Neutralisation by esterification	[91]
Garcia-Perez, M.	2007		Bio-oils		Analysis/lumping in chemical families		[92]
Zhang Qi	2007	Yes	Biomass		Bio-oil properties	General	[21]
Jacques Lede	2007		Biomass	Fast pyrolysis	Bio-oil properties/stability		[93]
McCormick, R.L.	2007		Biodiesel and blends		Stability/biodiesel and blends	Blending	[38]
Dinesh Mohan	2006	Yes	Wood/biomass	General	Physical and chemical properties		[22]
Boucher, M.E,	2000		Softwood bark	Vacuum pyrolysis	Stability and ageing	Blending with methanol	[94]
Diebold, J.P.	2000	Yes	Fast pyrolysis oils	Fast pyrolysis	Storage stability/physical and chemistry		[23]

and altering the oil physical properties. An extensive and detailed introduction to the chemistry of bio-oils is provided by Diebold [23].

Bio-oils are undergoing ageing processes once collected, resulting in increasing viscosity and average molecular weight. Typically, the carbonyl-containing species are decreasing, and the water-insoluble fraction is increasing [23, 53, 96, 97]. Reactions that normally would be expected to require catalysis is proceeding readily without. As the composition changes over time, the mutual solubility of the components changes, thus making phase separation more likely.

Oxygen-based chemistry is a key, and at least three principally different oxygen types must be addressed for understanding bio-oil degradation. One type is the already-present chemically bonded oxygen of the biomass, the other is oxygen in water, and the third is oxygen contact due to air exposure. These participate in different mechanisms, and their properties and effects should be understood independently. Stability can be improved by lowering the oxygen content, in particular the oxygen heteroatoms that contribute to the acidity and reactivity, and also by substantial removal of fines of solid materials like char. Reactions between oligomers are responsible for nearly all the viscosity increases observed during ageing.

Equilibrium-controlled chemical reactions constantly adjust to changes in temperature, water content, volatiles content, etc. In a real complex bio-oil, the possible chemical reactions than can proceed is very high, considering the large number of complex oligomers. Equilibrium constants of a number of liquid acetal formation reactions and corresponding normal boiling points or the resulting products are provided by Diebold [23]. Such data are also of relevance for treatments involving addition of solvents and alcohols.

7.3.2.2 *Specific reactions*

An extensive introduction to the detailed chemistry of bio-oils is reviewed nicely by Diebold [23]. Information relevant for equilibration and rates of selected reactions, to address potential correlations with time processes, has been extracted. Ageing is positively correlated to the temperature and is thus proceeding much faster at elevated such. The viscosity changes may be represented by Arrhenius exponential functions, which are strongly implying that chemical reactions are responsible, via increasing of the average molecular weight. In fact, the bio-oils are tending to change their chemical composition towards thermodynamic equilibrium. The originally single-phase oil can separate into tarry, sludgy, waxy and aqueous phase during ageing [23].

Reactions of bio-oils are categorised into principal types [23, 94]:

1. Ester formation by reactions between alcohols and acids or by transesterification of two esters.
2. Homopolymerisation of aldehydes to polyactals.
3. Aldehyde and alcohol addition to hemiacetals or acetals.
4. Polymerisation of phenol and aldehyde to resins.
5. Polymerisation of furan derivatives.
6. Alcohol reaction with unsaturated aldehyde to form alkoxy aldehyde/acetal.
7. Condensation of olefins.
8. Oxidation of alcohols and aldehydes to carboxylic acids.

Diebold [23] suggests approximately the same types of components and generic reactions, distinguishing between reactions involving chemically bonded internal oxygen, and oxidation reactions caused by air exposure. In addition, Diebold also suggests that organic sulphur components can form oligomers. It is distinguished between reactions that can form equilibrium compositions and those that are irreversible of nature. Stability relevant reactions and phenomenon are addressed in the following.

7.3.2.2.1 Off gases

Bio-oils are typically releasing light gases during storage. This can be free gases already dissolved in the oil, gradually releasing into the headspace. If methanol is used as solvent, it can adsorb large amounts of CO_2, that is, 5 vol of CO_2 per volume methanol at 283 K, a little above 1 wt%. Desorbing CO_2 can then tend to strip other absorbed gases from the bio-oil.

7.3.2.2.2 Esterification

This reaction is catalysed by acids, and the equilibrium constant is generally larger than 1, increasing to higher values in the range 4–5.2 for low-molecular-weight esters and also being nearly independent upon temperature. Reactions appear as second order with regard to acetic acid. Low-molecular-weight esters are volatile and difficult to condense by vapour cooling. Reaction rates can be slowed down by addition of basic components like ammonium acetate buffers and sodium acetate. The rate is gradually slower for larger components [23], formic acid > acetic acid > benzoic acid, etc.

$$ROH + \mathrm{RCOOH} = \mathrm{RCOOR} + \mathrm{H_2O}$$

7.3.2.2.3 Transesterification

This exchange reaction can proceed in a mixture with the presence of two or more esters:

$$\mathrm{R_1COOR_2} + \mathrm{R_3COOR_4} = \mathrm{R_1COOR_4} + \mathrm{R_3COOR_2}$$

7.3.2.2.4 Aldehyde homopolymerization

Aldehydes can undergo reactions with each other to form polyacetal oligomer and polymers. This is known as an aldol reaction, proceeding nicely under basic conditions and very slowly under acidic conditions:

$$\mathrm{nRCHO} + \mathrm{H_2O} = \mathrm{RHCHO}(OH)$$

The poly(oxymethylene) polymer has limited solubility in water. Methanol is used to stabilise formaldehyde commercially, with MeOH contents ranging from 6% to 15%.

7.3.2.2.5 Hydration

Another candidate reaction is hydration, that is, aldehydes or ketones react with water to form hydrates and also referred to as glycols:

$$\mathrm{RCOR} + \mathrm{H_2O} = RC(OH)(OH)\mathrm{R, R\,either\,H\,or\,alkyl\,groups}$$

It is claimed that this reaction is rapid [98] and exemplified by formaldehyde. Equilibrium constants for water reacting with formaldehyde is 41, 0.018 for acetaldehyde, and 0.000025 for acetone, respectively. This means that formaldehyde will prefer a hydrate form in the presence of water. Acetaldehyde will partly prefer to be in hydrated form.

7.3.2.2.6 Hemiacetal formation

In a mixture of aldehydes and alcohols, a hemiacetal can form via the following exothermal reaction:

$$ROH + \mathrm{RCHO} = RC(\mathrm{OR})(OH)\mathrm{H}$$

This reaction proceeds fast without catalyst. Examples of reactants are ethanol and acetaldehyde, isopropyl alcohol and acetaldehyde, ethanol and benzaldehyde, and ethanol and methoxy benzaldehyde.

Aldehydes and alcohols react to form acetals according to the following reaction:

$$2\,ROH + \mathrm{RCHO} = RC(\mathrm{OR})(\mathrm{OR})\mathrm{H} + \mathrm{H_2O}$$

In the presence of acidic catalysts, stable formals are formed from aldehydes and OH groups of sugars, starches, and cellulose. Cellulose reacted with CH_2O creates a cross-linked polymer. Catalysts for acetal formation at low temperatures include salts potentially present in biomass as ash, chloride salts of aluminium, iron, lithium, magnesium, manganese, and zinc, copper sulphate, sodium bisulphite, and monosodium phosphate. Trace amounts of strong mineral acids also catalyse the acetalisation reaction, for example, HCl, phosphoric acid, and nitric acid. Strong organic acids such as oxalic and tartaric acid also catalyse the reaction, but not acetic acid. Equilibrium constants for acetalisation are provided by Diebold [23].

7.3.2.2.7 Transacetalization

Reactions forming acetals are reversible. Alcohols can be reacted with acetals or different acetals with each other to generate new acetals. The average molecular weight of alcohols or acids can be lowered by adding a low-molecular-weight alcohol or aldehyde.

7.3.2.2.8 Phenol/aldehyde reactions and resins

Phenol reacts as an alcohol with formaldehyde to form hemiformal, which is an intermediate for formation of orto-methylol phenol, being a highly reactive partner with other phenols. In the presence of acidic catalysts, phenols and substituted phenols react with aldehyde hydrates to form novolak resins and water. Use of strong acids results in thermoplastic resins with molecular masses in the range of 500–5000, with methylene linkages constituting a mixture of ortho and para. The rate is inversely proportional to the water content and proportional to the concentrations of reactants and catalyst. Other aldehydes react similarly as formaldehyde, though at a slower rate. Because the reaction rate of formaldehyde with the polymer is much slower than with monomeric phenolics, novolaks have very little 3-D cross-linking between the chains.

The non-cross-linked novolaks maintain their solubility in solvents such as acetone [99]. At higher pH (4–7), phenol and formaldehyde react catalytically with metal salts forming 'high ortho' novolaks. Active sites for phenol reacting with formaldehyde are the 2, 4, and 6 positions. If phenol/formaldehyde reactions are important during storage, hardwood bio-oils should store better than softwood oils, due to the better chain-termination ability of syringols (2,6 dimethoxy phenols) compared with catechols (1,2 dihydroxy benzene).

Furfural has reactivity of its aldehyde functional group to form polymers with phenol.

7.3.2.2.9 Polymerisation of furan derivatives

Furfural alcohol undergoes condensation reactions with evolution of H_2O, analogue to condensation of ortho-methylol phenol, with linkages at the 2 and 5 positions adjacent to the oxygen of the furan ring. The reaction is catalysed by lactic acid, hydroxy acetic acid, formic acid, calcium chloride, and strong acids. Such a polymerisation takes place during 1 h at 373 K, producing an acetone-soluble resin.

7.3.2.2.10 Sulphur containing compounds

Organic sulphur can prevent formation of peroxides, but bio-oils typically contain very low levels of *S*. Especially effective are the nonyl sulphide and thiophenol (phenyl mercaptan). In reactions of mercaptans with peroxides, the H is abstracted, and the disulphide is formed, nearly doubling the molecular weight of the sulphur compound. Alternatively, the RS* free radical can attach to unsaturated bonds that build larger molecules. This terminates free radicals and prevents further chain reactions [23, 100].

7.3.2.2.11 Alcohol addition

Unsaturated aldehydes such as acrolein are very reactive because the vinyl group conjugates with the carbonyl group. In addition to forming acetals with alcohols, acrolein can react with alcohol in a reaction that is catalysed by acids or bases to form the alkoxy aldehyde, unsaturated acetal, or alkoxy acetal:

$$\mathrm{H_2C = CHCHO} + 3\,ROH = \mathrm{ROCH_2}CH_2\mathrm{C(OR)(OR)} + \mathrm{H_2O}$$

7.3.2.2.12 Condensation of olefinics

Unsaturated compounds can combine into polyolefins. Such components include acrolein, 2-butenone, propenyl-substituted phenolics, and 2-propene 1-ol. Acidic or basic medium and also free radicals can contribute to enhance these reactions.

Carboxylic acids catalyse olefinic condensation reactions, but hydroperoxides that form free radicals are the most deleterious oxygenated species in petroleum liquids. Sources of free radicals in the bio-oil include organic hydroperoxides, organic peroxides, and nitrogen compounds (*N* from eventual proteins in the original feedstock).

7.3.2.2.13 Air oxidation

Direct exposure to air can cause oxidation, particularly of the alcohols and aldehydes to carboxylic acids. A well-known example is auto-oxidation of ethanol in wine to vinegar (acetic acid), which is perceived as a sour taste, possibly already after a short overnight exposure.

Storage can be strongly affected by reactions forming hydroperoxides and alkylperoxides by auto-oxidation with air. These peroxides are not very stable and are spontaneously decomposing to form free radicals. In concentrated form, many organic peroxides are explosive. Organic or hydroperoxides can be formed by reacting air with olefins, ethers, ketones, nitrogen compounds, aldehydes, and organic acids. Thus, exposure to air would be expected to increase formation of polyolefins during bio-oil storage. Ease of forming peroxides is related to ease of activating the C–H bond, and the following order is suggested:

n-alkanes < branched alkanes < aralkanes = alkenes < alkynes.

7.3.2.2.14 Gas forming reactions

Most carboxylic acids are stable regarding tendency for decarboxylation, but some dicarboxylic acids are relatively unstable and convert easily into the monoacid with release of gaseous CO_2 at moderate temperatures. This is believed to occur only when the same carbon atom is hosting two carboxyl groups. Similarly, beta-keto acids also decompose a low temperature to form ketone and CO_2.

Ferulic acid, which is claimed to be a primary pyrolysis product from lignin, starts to decompose noticeably at 473 K to 4-vinyl guaiacol and CO_2. The decomposition rate was faster in the presence of oxygen, indicating the presence of free radicals, according to Fiddler [101]. This is also indicating that the reactions can proceed at lower temperatures in the presence of oxygen.

7.3.2.2.15 Hydrogen

The presence of metals such as Fe, Zn, and Al in the bio-oil can presumably trigger reactions forming organic acid salts and free hydrogen. Hydrogen then might escape as a gas to the headspace of the container.

7.3.2.3 Stability issues

According to Huhnke [7], the last 5–10 years have contributed to the development of promising measuring methods of direct relevance for bio-oil thermal or oxidative stability so that improvements in the storage and transportation stability of bio-oil can be better quantified [51, 102]. However, no formally standardised methods for quantitative measurement of bio-oil stability are yet available. Existing procedures for testing of storage stability of hydrocarbon fuels are available as ASTM standards [103, 104]. The tests are related to thermal and oxidative stability. Stability tests with accelerated procedures are available in ASTM standards [105, 106].

Thermal stability is defined as the ability to withstand temperature stress [52]. Accelerated thermal ageing processes have been broadly adapted [51, 107, 108]. Properties such as composition, carbonyl and water-insoluble components before and after exposure to high temperature are measured, compared, and evaluated. In a round-robin test, bio-oils were subject to heating in sealed bottles at 353 K for 24 h, thereafter cooling to room temperature. Correlations can be established between composition and physical parameters. Composition can be correlated to molecular weight and carbonyl content with viscosity.

7.3.2.3.1 Oxidative bio-oil stability

This issue is concerning oxidative degradation processes. Zahira et al. [37] provide a nice outlook into fuel parameters affecting and indicating the extent of fuel stability and methods for stability prediction. Potential products are carboxylic acids, esters, and also peroxides that can facilitate polymerisation of unsaturated species. Limited work has focussed on oxidative stability of bio-oils [88, 92, 109], the reason being obvious is that it as a crude is already in an extensive oxidised state. A parameter of interest is the oxidation onset temperature (OOT), measured by an ASTM procedure for hydrocarbons [110]. The sample is heated from 298 to 523 K in oxygen-containing atmosphere at a rate of 10 K/min. The OOT is determined as the level where oxidation contributes to the increase. High OOT is favourable for stability. Formation of solid materials is suggested as an indicator of oxidative stability.

A general mechanism can be learned from studies of biodiesel [38], that is, instability is due to chemical structures with an ability to cause initiation of free-radical reactions via peroxide-type intermediates. Thereafter, the mechanism follows the classical, propagation, chain branching, and termination scheme. The chemistry of fat/lipids is fairly well understood. The multiunsaturated chains are the most reactive in this regard. Oxidation rates can correlate to when the double bonds are in allylic position.

A few studies [111, 112] were recently confirming that free radicals are responsible for bio-oil instability during storage. Sealing to avoid exposure to oxygen and also well-controlled temperatures are required.

7.3.2.3.2 Accelerated ageing studies

There are a number of reports on accelerated ageing studies of thermal stability of bio-oils [7]. This is typically done by using a temperature level well above ambient, such as 353 K. Selected properties are measured before and after the exposure to the thermal stress. Correlations can then be made between the compositional changes and governing parameters.

The influence of ageing time on the content and composition of the low-boiling volatiles of pyrolysis oil was investigated by Li et al. [76]. These relationships are illustrated in Figs. 7.2 and 7.3. Accelerated ageing of liquid samples at 353 K was performed followed by atmospheric distillation. A total yield of 52% was obtained from the fresh pyrolysis oil. The fraction containing the low-boiling volatiles decreased with ageing time (fraction 1 in Fig. 7.3). GC–MS measurements confirmed the disappearance of carbonyl compounds like acetaldehyde, 2-butenal, and benzaldehyde (Fig. 7.3). An accompanying increase in viscosity and water-insoluble solids was attributed to polymerisation and condensation reactions.

It is obvious from Fig. 7.3 that aldehydes and ketones are steadily declining, while, for example, the esters are very much increasing. It is a little surprising that the carboxylic acids are not much affected though.

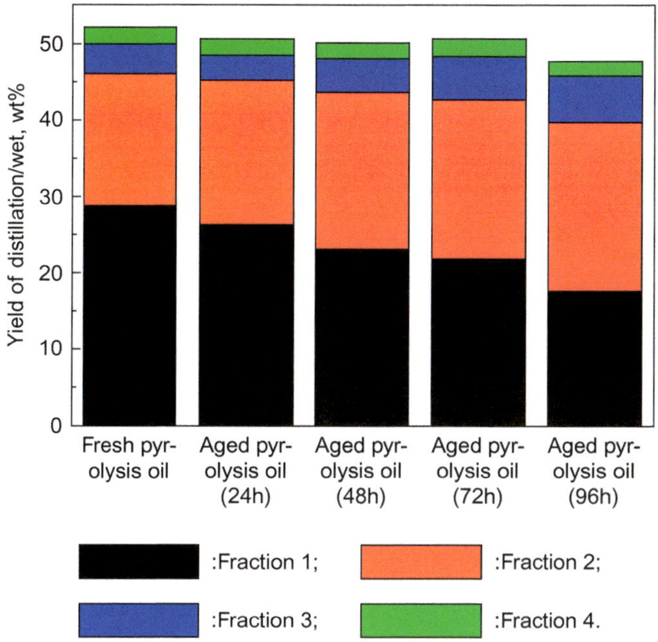

Fig. 7.2 The distribution of different fractions by atmospheric distillation for fresh pyrolysis oil and aged pyrolysis oil kept at 353 K for up to 96 h [76].

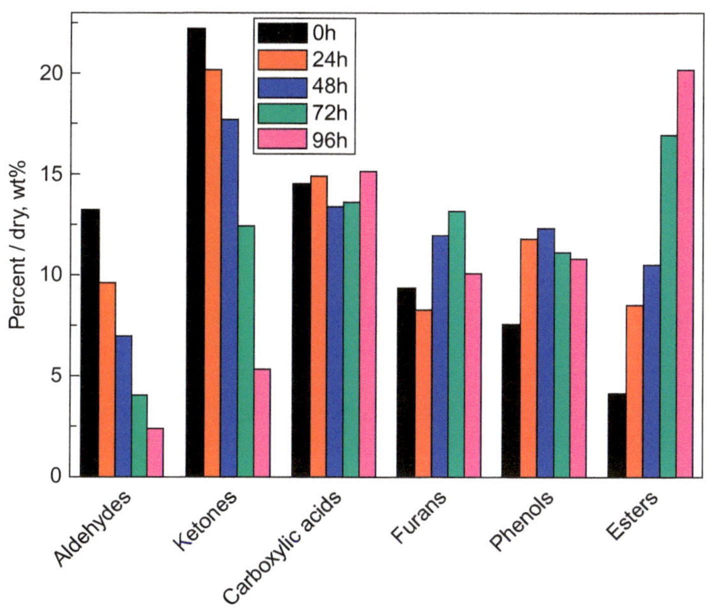

Fig. 7.3 The distribution of organic compounds in fraction 1 over ageing time [76].

7.4 Stabilisation methods

7.4.1 General

It is a highly desired goal to develop processes and methods that can be used to improve the bio-oil quality and time dependent behaviour. This is critical for subsequent handling and providing predictability of end-use application. Methods for improving stability are mainly targeted on reduction of viscosity and removal of highly reactive groups or components, thus modifying the bio-oil towards lower reactivity and acidity. Removal of solids like the char residue is another issue of importance for stability.

Reactive and non-reactive approaches, that is, chemical and physical, are targeted. In addition, thermal approaches can be considered of relevance. Chemical methods encompass, for example, esterification, mild hydrotreatment, and catalytic cracking. These methods can either be performed in situ in the pyrolysis process or as a dedicated post-treatment step. Physical methods encompass addition of a wide range of polar solvents, emulsification, removal of solid materials, and high-pressure homogenisation [113].

This chapter is reviewing and highlighting selected results and developments in techniques and methods for improving stability and quality. A selection of relevant publications is given in Table 7.6.

7.4.2 Chemical (Reactive) stabilisation

Bio-oil instability is strongly linked to the oxygen content, but is highly composition-dependent. Different classes of oxygenates and their respective reactive properties are determining. Chemical methods for bio-oil stabilisation is therefore targeting on achieving transformation and removal of specific functional groups, such as those present in carboxylic acids, alcohols, aldehydes, ketones, olefins, and phenols.

7.4.2.1 Esterification

Hilten et al. [89] used simple reactive condensation for decreasing the concentration of reactive components in the oily phase, thereby increasing storage stability, heating value, and overall quality. The pyrolysis vapours were esterified in situ with ethanol during condensation, that is, resulting in the formation of water and esters. Adding 23wt.% ethanol decreased water content and viscosity in the oily phase by 16% and 56%, respectively. The pH of the oil phase was then also increasing, from ca. 2.5 to 3.1. Results were interpreted to be consistent with formation of esters. A maximum conversion level of ca. 40% of the acetic acid was indicated.

The high acidity of crudes is causing limitations for further processing. It is therefore important to neutralise or remove the carboxylic acid components. A minireview on the field of carboxylic removal by reacting with alcohols is given by Xun Hu [6], focussing on the chemical reaction pathways. Fig. 7.4 [6] provides a simple illustration of a model esterification reaction.

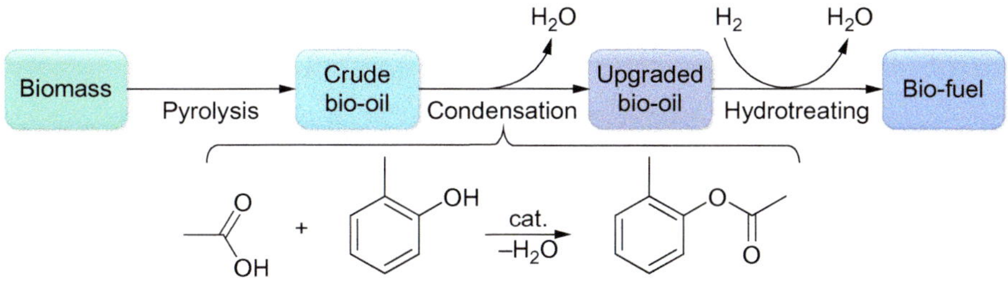

Fig. 7.4 Esterification of acetic acid with o-cresol [6].

Options are to treat the oil with inexpensive alcohols, for example, methanol, ethanol, or butanol, thus enabling conversion of carboxyl and carbonyl groups to esters and acetals or ketals, respectively, either by noncatalytic or catalytic means. These reactions are equilibrium controlled but can be assisted by removing water using azeotropic distillation or reactive distillation for removing the esters and acetals, leaving the water in the bio-oil [18]. An advantage of catalytic esterification is a higher conversion of the undesirable components, resulting in improved stability and other properties. Highly polar organics are converted into components of lower polarity, which is shifting the tendency to phase separation of the mixture. Catalytic esterification uses milder operating conditions than, for example, deep hydrotreatment and catalytic cracking, but with a lower potential for oxygen removal. The reaction can proceed readily at mild conditions, even at room temperature. One challenge of esterification is the instability of the produced acetals and their tendency to revert back to carbonyls. Aldehydes can inhibit esterification in the low-temperature range.

Esterification by reacting the pyrolysis oil with alcohols can be performed as a dedicated pretreatment step before further processing or alternatively already in situ in the primary conversion step. Esterification as a dedicated step (ex situ) has been targeted by Ehsan Reyhanitash [114], comparing addition of alcohols at high and low loadings. The main desire was to prevent the self-oligomerisation of reaction intermediates during the upgrading process. Effects of methanol, ethanol, and glycerol at a 5 wt% level relative to the oil were investigated in the presence of Amberlyst-35 DRY as a solid acid catalyst. It was concluded that significantly positive effects were obtained, that is, a lowered molecular weight because of a lower degree of self-polymerisation.

Equilibration is proceeding when adding either methanol or removing/adding water. Diebold [23] studied the equilibration behaviour of a pseudo bio-oil consisting of 25% water, 3% acetaldehyde, 2% methanol, and 8% acetic acid as function of added methanol in the range up to 20 wt%. Increasing methanol shifted acetic acid and acetaldehyde to methyl acetate and dimethoxyethane.

Esterification of bio-oils with typical water content was also investigated for by using methanol [91]. It was concluded that it was inefficient, and one reason pointed at was the high levels of water causing equilibration limitations.

Esterification of acids and acetalisation/ketalisation of carbonyl compounds are reversible reactions causing formation of water, being illustrated in Fig. 7.5. Conversion improvements can be achieved by simultaneous removal of water in multifunctional reactors combining chemical reaction and separation in the same unit. Reactive

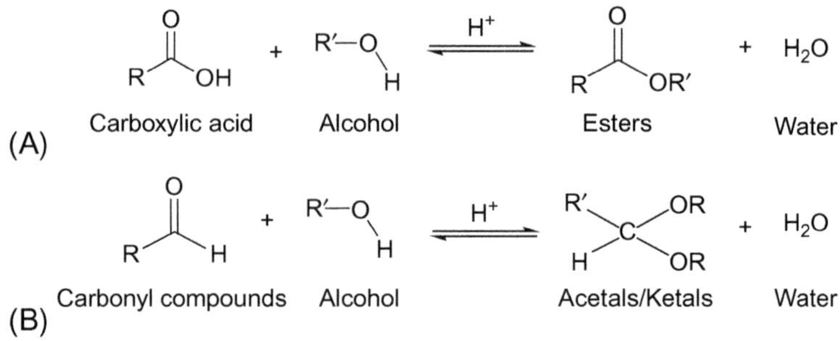

Fig. 7.5 (A) Esterification of carboxylic acids. (B) Acetalisation/ketalisation of carbonyl compounds [6].

distillation (RD) is an example of such a concept. The distillation column packing material is then embedded with a catalyst capable of facilitating the desired chemical reaction. Reactive chromatography (RC) requires a packing material acting as both a catalyst and adsorbent capable of selective adsorption of one of the reaction products.

Esterification can lead towards reduced acidity, improved volatility and heating value, and also a better miscibility with diesel type fuels. It has been performed over heterogeneous acid/base type catalysts, homogenous ionic liquid catalysts, and also aluminium silicate systems. Application of solid catalysts has been limited due to extensive catalyst coking and inefficient regeneration. New catalytic approaches are desired.

According to Xun et al. [6], some major future challenges of esterification include further detailing and understanding of the complex chemistry, for example, of the heavy sugars and the pyrolytic lignin. Polymerisation reactions typically take place during esterification. It is a requirement to develop better solid, that is, heterogeneous catalysts, being active and hydrothermally stable. Important in this respect is also regeneration feasibility. Separation issues to be improved include alcohols recycling. Most importantly, esterification alone is not sufficient to obtain a bio-oil with properties for direct use as a fuel. Therefore, it is important to establish more completing upgrading concepts that lead to a final high-quality bio-oil.

7.4.2.2 Catalytic cracking

Catalytic cracking can be performed at typical temperatures of 623–923 K at relatively high pressures. Oxygen is rejected as CO_2. Zeolite-based catalysts such as the protonated HZSM-5 is a typical choice. Cracking is causing decarboxylation and decarbonylation reactions, which is challenging for maintaining high carbon efficiency. Cracking leads typically to a product mixture with lower H/C and higher O/C ratio than hydrodeoxygenation (HDO).

Catalytic fast pyrolysis applies catalysts in situ to intensify parallel cracking reactions. An overview of work on catalytic pyrolysis is shown in Table 7.7. A review on the field of in situ pyrolysis with zeolite catalysts is provided by Galabdina et al. [116].

Two principally different approaches are cracking and hydrocracking. Hydrocracking is performed in the presence of H_2 at pressure over an acidic catalyst that typically also contains a metal function to activate the molecular hydrogen.

Table 7.7 A summary of some laboratory-scale catalytic pyrolysis experiments [18]

Reactor	Catalyst	Comments	References
BFB (2 kg/h) with catalyst bed. Fresh catalyst charged to bed	HZSM-5 at 450–500 °C	Yield structure, liquid (33 wt%), gas (53 wt%), char and coke (12.5 wt%): oils are stable and have low viscosity; contain mainly phenolics, creosols, methyl-substituted phenols; distillable (100% up to 640 °C) without char or solid formation; ageing not significant over 10 months. TAN number is 41 (compared with 90 for raw bio-oil)	[161, 162]
BFB with catalyst bed	Beta, Y.ZSM-5 Mordenite	Catalyst structure influenced composition of oils but not the yield structure; ketones were higher, and acids and alcohols are lower over ZSM-5; mordenite yielded low amounts of polyaromatic hydrocarbons, most active deoxygenation, zeolite > Y > ferrite	[163]
BFB with separate upgrading zone within reactor	Zeolite	Bio-oil yield ranged between 43.5 and 52.7 wt%; more selectivity for organics directly related to more water and CO; most active deoxygenation, B-zeolite > Y > ferrierite	[164]
CFB pilot with catalyst regeneration	FCC catalyst	Bio-oil yield decreased (73–49 wt%) with introduction of catalyst; gas, coke, and char production increased rapidly due to secondary reactions; catalytic bio-oils contain more hydrocarbons and less oxygenated compounds; FCC catalyst may be too active since it yielded a lot of water and coke	[165]
BFB and packed bed vapour upgrading	FCC catalyst	Lower amounts of acid (14.5–3.021 wt%) and increased aromatic hydrocarbons (0.488–16.795 wt%) compared with pyrolysis	[166]

Continued

Table 7.7 **Continued**

Reactor	Catalyst	Comments	References
BFB and fixed bed vapour upgrading	HZSM-5. HY, Ga/HZSM-5. Ga/HY	HZSM-5 is more effective at upgrading the bio-oil than HY; oxygen is mainly being converted to H_2O, CO, and CO_2	[167]
BFB with catalyst bed	CoMo-S/Al_2O_3	Max yield of BTXN was 6.3 wt% at 590 °C	[168]
BFB with catalyst bed	HZSM-5	Max liquid yield was 56.8 wt%; oxygen content of oil was 14.69 wt%; H/C and O/C of oil were 1.51 and 0.15; HHV was 34.6 MJ/kg, reduced concentrations of ketones and phenols, dramatic increase in aromatics in oil (7.62–74.22 wt%)	[169]
BFB with catalyst bed	FCC	Optimal catalyst to biomass ratio was 1:10 for fresh FCC and 1:5 for spent FCC; max. Oil yields for fresh FCC (11.8 wt%) catalyst were lower than for spent FCC (18 wt%); hydrocarbons in the oil increased with increasing catalyst mass; less C and 0 were transferred to oil in the presence of catalyst	[170]

Hydrocracking leads to a larger degree of saturation reactions than the base case cracking, and as such, it should also lead to a more stabilised product mixture.

The produced bio-oil can also be subject to upgrading by dedicated cracking. A recent publication on the use of Cu-modified zeolite on upgrading of pyrolysis oils is given by Wahyu et al. [117].

It has been claimed that pyrolysis over basic catalysts can drastically reduce the carboxylic acid and oxygen content. Catalysts such as MgO, CaO, and CsX or hydrotalcites have been employed in the pyrolysis step [118]. Such treatment can provide formation of more stable oils that are suitable for further treatment.

7.4.2.3 Hydrotreatment (HT)

A general introduction to catalytic hydrotreatment of bio-oils is given by Lødeng et al. [119]. Oxygen removal and saturation can be obtained by mild hydrodeoxygenation, abbreviated HDO. In principle, it can be done in situ by cofeeding hydrogen in the pyrolysis step or as a dedicated subsequent step at mild conditions. The temperature applied is determining the severity, and mild HDO will typically be performed well below 523 K but at high pressures of H_2 (<250 bars) over a heterogeneous catalyst. Mild HDO is intended to improve stability and storability by the removal of the most reactive gum precursors. Catalytic stabilisation of low-quality bio-oils is mainly done via hydrogenation and mild deoxygenation causing saturation and/or elimination of water and/or CO_2. Mild deoxygenation can be performed with relatively low consumption of H_2, avoiding saturation of aromatics. The resulting bio-oil is an intermediate product that can be stored, transported, and further processed. A requirement before further processing is, for example, to reduce the potential for carbonaceous formation in the subsequent processing steps. Dedicated studies of stabilisation are provided by Yakovlev et al. [120], Ardiyanti et al. [121], and Venderbosch et al. [122].

A typical strategy is to treat the fast pyrolysis bio-oil in stages of stepped-up severity, to some degree adapted by need to the reactivity levels of critical components or fractions. First, raw bio-oil stabilisation at low temperature, <473 K, and then subsequently at higher severity according to the target quality, proceeding with co-formation of considerable amounts of water. Mild hydrogenation converts carbonyl groups into more stable alcohols. Preferably, also carboxylic acids should be converted. However, a multiple number of steps are challenging for maintaining a high overall processing efficiency.

Oh et al. [65] studied storage performance before and after HDO, using noble metal catalysts. Liquid samples were stored in sealed glass bottles at 296 K for a total period of 12 weeks. Water content, viscosity, acidity, and chemical composition were measured during this period. HDO resulted in a substantial reduction in these physical properties, and they remained relatively stable under storage. The acidity measured as total acid number (TAN) actually decreased and was attributed to a reduction in phenolic hydroxyl groups caused by dimerisation and trimerisation of phenols.

The prospect of performing low-temperature, low-pressure hydrogenation has been explored by Rover et al. [70]. The phenolic fractions of pyrolysis oil from corn stover and red oak were exposed to hydrogen at 294 K and 1 bar pressure over a Pd/C catalyst. This low-severity treatment resulted in carbon yields in excess of 98% and a dramatical drop in viscosity. Changes in functional groups were evaluated with FTIR and gave evidence of conversion of carbonyl groups into alcohols. HNMR showed increased aliphatic protons supporting that substantial hydrogenation of the samples took place even at these mild conditions. The authors conclude that their findings support their hypothesis that an initial upgrading step must be performed at much milder conditions than conventionally employed.

Routray et al. [41] used a bifunctional Ru/C supported catalyst in a low-temperature catalytic step at 403 K. Carbonyl-containing components were converted, such as hydroxyacetone to propylene glycol, hydroxyacetaldehyde to ethylene glycol, and levoglucosan to sorbitol. Detailed gas-phase and liquid-phase selectivities are provided. The majority of the carbonyl groups were unaccounted for. It was considered possible that these groups were consumed by aldol condensation reactions, that is, in pathways leading towards coke formation. It was observed that the water content in the liquid increased from 21 to 30 wt% after reaction. The reasons for water increase were suggested to be caused either by HDO, via carbon losses due to coking, or by dehydrating or aldol condensation reactions. Coke formation was considered the most likely to dominate. Hydrolysis products were not observed, due to low hydrogenolysis activity of the Ru/C catalyst. Coking was identified as a dominant problem, and a suggested solution was to produce a more stable pyrolysis oil upstream in the first place. A requirement is to remove the destabilising carboxylic acids and other carbonyl-containing components.

7.4.2.4 Hybrid methods

The US company IH2 is aiming at integrating fast pyrolysis of biomass with catalytic hydrotreatment for direct production of products [123]. Hydrogen in then cofed with the biomass in the fluidised-bed hydropyrolysis reactor at pressure in the presence of a hydrotreatment catalyst. Typical temperatures are given as 653–723 K at pressures of 20–35 bars. Liquid yields of 86% are claimed [123]. However, no specific information about oxygen contents of the product oil was found. Liquid viscosity comparable with conventional diesel is claimed.

Esterification can also be coupled to other methods such as preoxidation, hydrogenation, reactive distillation, and also going from the typical subcritical to supercritical conditions [6]. The potential of both reactive distillation and reactive chromatography in esterification of acidic acid with different alcohols in a synthetic bio-oil was explored by Praveen et al. [69]. Amberlyst-15, a strong-acid ion-exchange resin, was used as catalyst. Catalyst stability was a main challenge in reactive distillation. Polymerisation of furfural present in the feed was identified as the cause of catalyst deactivation. The vapour-liquid equilibrium dictates the operating temperature, and in the case of *n*-butanol, the column temperature reached 388 K resulting in a relative high deactivation rate. Switching to lower-boiling alcohols improved the catalyst stability, but declining activity was still observed.

Reactive chromatography does not suffer from the temperature limitation as mentioned above, and the same catalyst showed sufficient activity with ethanol and methanol at lower temperature rendering a much more stable operation. Separation of produced ethyl acetate and water was also established. Development of continuous processes based on these concepts still requires much work.

Simultaneous conversion of acids, aldehydes, and phenols by combining hydrogenation-esterification reactions, by employing bifunctional catalysts, offers a simplified stabilisation step. This was addressed by, for example, Wen Chen et al. [78]. Physically mixed Ru/C and sulphated zirconium catalysts were able to convert a model feed consisting of phenol, *n*-butanal, and acidic acid into alcohols and esters in 95% yields.

Using cofeed with hydrogen-donating components or sacrificing internal bio-oil components are two modes of hydrostabilisation/upgrading. Of relevance for this is the concept of co-pyrolysis of plastics with a high hydrogen content together with lignocelluloses. This provides an opportunity for hydrogen transfer from the molecules/ polymer chains of the plastic to the unsaturated lignocellulose derived.

More specific solvents dedicated for the purpose can be applied. However, solvent availability must be evaluated in relevance to the processing scale and applicability potential. According to Khairuddin et al. [63], tetralin is a good model of an effective hydrogen donor. In case of biomass liquefaction, the hydrogen transfer from the H-donor 'solvents' stabilises the free radicals of the fragmented biomass. H-donor solvents can also have dual functions, both stabilising fragments and also assisting thermal cleavage. During this process, the H-donor is dehydrogenated to an unsaturated compound. Tetralin can give rise to increased conversions during pyrolysis processes. A study at supercritical water addition above 673 K gave high overall biomass conversion above 90%, with 28% recovered as heavy oil, and 14% estimated to be the light oil yield. Oxygen content of the heavy oil was ca. 16 wt%. Results indicated that similar conversion could be obtained in the presence of supercritical water alone, that is, without added H-donor.

Reddy et al. [75] studied transfer hydrogenation reactions over supported mono- and bimetallic catalyst containing Ni and Cu with *p*-cresol and furfural as model compounds. Isopropanol was utilised as the hydrogen-donor solvent. Yields of >95% of hydrogenated and deoxygenated products were obtained with the Ni-Cu/Al_2O_3 catalyst. This catalyst was also active in the presence of water, although the activity decreased when concentration increased. Treating a complex pyrolysis oil over Ni-Cu/C catalyst gave a product with higher H/C ratio and lower O/C ratio, due to contributions to hydrogenation and deoxygenation of the pyrolysis oil. Polymerisation reactions leading to solid formation were also observed.

Utilising in situ generated hydrogen from compounds acting as hydrogen donors and thus avoiding the need for externally produced molecular hydrogen is an attractive concept, with the promise of a simpler and thus more cost-effective process. Organic oxygenates, like carboxylic acids and alcohols, present in the aqueous phase can be a potential hydrogen source. The resulting reduction in concentration of the corrosive acids will be a synergetic benefit.

Generation of hydrogen by aqueous-phase reforming (APR) of methanol coupled with hydrogenation of the light oxygenates has been reported by Xu Ying [71]. They performed experiments with both model compounds and complex bio-oil in an autoclave setup over a Ni-Raney catalyst. Operation conditions favouring simultaneous APR and hydrogenation of acetone and phenol in the model feed were identified. Processing of the complex bio-oil at the same conditions resulted in compositional changes with decreased content of aldehydes, ketones, and organic acids. An increasing content of alcohols and esters was attributed to hydrogenation and esterification reactions taking place.

Peng et al. [124] studied the catalytic upgrading of bio-oil by HZSM-5 in sub- and supercritical ethanol, that is, by multihybrid concepts. Tang et al. [125] studied one-step bio-oil upgrading through hydrotreatment, esterification, and cracking, as an example of another multihybrid system.

7.4.3 Thermal stabilisation

7.4.3.1 Pyrolysis optimization schemes (Noncatalytic)

Bio-oil properties are dependent upon many parameters and aspects. This includes the pyrolysis process conditions, as well as the feedstock types and their quality. Optimisation by cofeeding of lignocellulose with other resources is one option being investigated.

7.4.3.2 Supercritical (H_2O, CO_2, alcohols)

A high pressure above conventional levels is needed to establish supercritical conditions. However, when approaching the critical point, for example, water has properties such as low viscosity and high solubility for organic substances. It can then serve as an excellent medium for establishing homogeneous conditions, thereby improving efficiency of reactions. By moving from subcritical to supercritical temperatures (647 K) at high pressures (>22 Mpa), both the rate of hydrolysis and phase partitioning and solubility of components can be controlled. Potentially, more favourable pathways to liquid and gaseous products can then be realised [63]. According the Kruse and Dahmen [126], water is a magic solvent for biomass conversion.

Remon et al. [127,128] studied the effect of operating conditions of sub- and supercritical water on the upgrading of lignocellulosic bio-oil from pinewood, the oil being produced by the Biomass Technology Group (BTG). This includes the temperature (583–723 K), pressure (200–260 bars), and catalyst/bio-oil mass ratio (0–0.25), as well as reaction time (0–60 min). It was claimed that the change from sub- to supercritical exerted a significant influence on the process. An oxygen removal degree of 70% was achieved. Carboxylic acids were lowered significantly, by the formation of light gases. A Ni–Co catalyst was also of importance for control of gas versus liquid and oil properties [127].

Hermawan et al. [129] and also Qing Zhang et al. [130] studied upgrading of fast pyrolysis oil in supercritical ethanol, in the absence of catalysts and hydrogen. At 673 K and after 30 min, almost all organic species in the pyrolysis oil were converted, resulting in a high yield of upgraded oil (83 wt%) with an enhanced HHV of 34.1 MJ/kg (compared with 24.3 MJ/kg of the original oil). Major chemical components were alcohols, esters, phenols, hydrocarbons, and aromatics. Beforehand, an effective removal of the carboxylic acids was expected. The oil stability was also targeted experimentally, and after ageing at 353 K for 1 week, a marginal increase in the viscosity of the upgraded bio-oil was observed, which demonstrated a significant improvement in the stability compared with the original oil. Fig. 7.6 shows an overview of results of Hermawan et al. [129]. Fig. 7.7 shows the results of the stability tests as the viscosity changes. Results with the use of ethanol were claimed to be encouraging.

A comparison test was also made by using supercritical methanol instead of ethanol, since the first is potentially a much less expensive medium. However, oil yields obtained were far lower, ca. 57 wt%, compared with >80 wt%, the difference believed to be due to the lack of carbon chain of methanol. This indicates that methanol, as a C_1 component, is being far less powerful than ethanol for carbon-carbon chain building.

Cui et al. [131] was addressing esterification in CO_2 as the supercritical medium. Esterification with ethanol of individual components acetic acid (AC), propanoic acid (PA), and both of these two mixed with acrylic acid, as well as real oil, under supercritical CO_2 conditions was investigated. The carboxylic acids converted at a much higher degree at supercritical conditions, compared with atmospheric. Total acid conversions up to 87% were obtained, from a starting point of 27% carboxylics in real oil. The pH value increased from 3.78 to 5.1. It is suggested that the positive effect is due to transfer of the formed esters into the CO_2 phase. Results are shown in Figs. 7.8– 7.10.

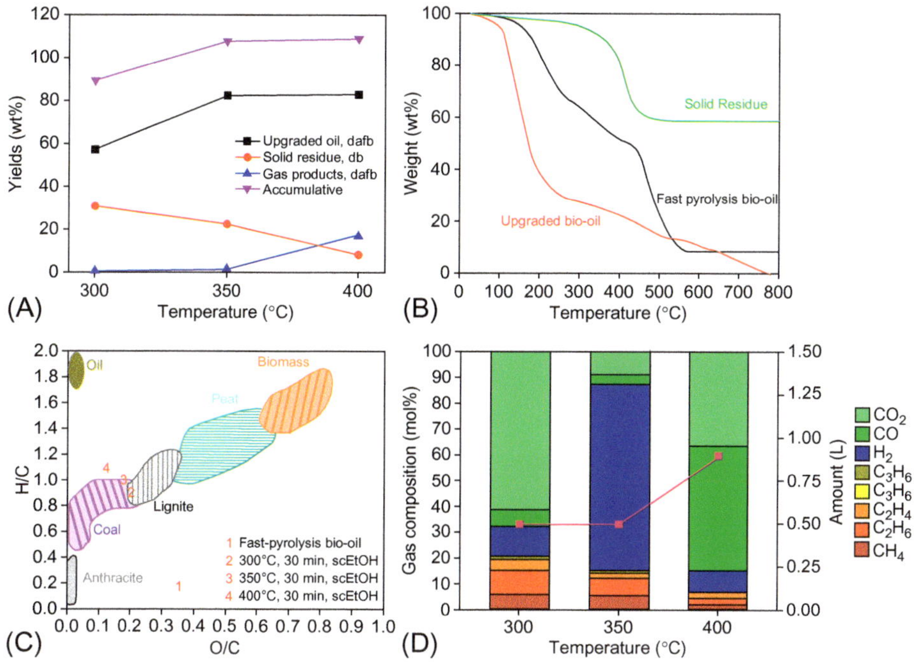

Fig. 7.6 Effect of the reaction temperature on (A) product yields, (B) TGA profiles of fast pyrolysis oils, upgraded oil and solid residue produced at 673 K and 30 min, (C) H/C and O/C ratios on the van Krevelen diagram, and (D) gas composition [129].

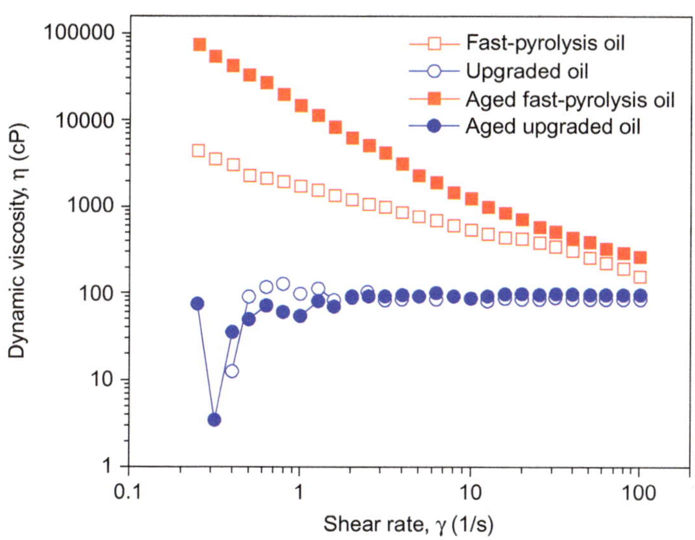

Fig. 7.7 Dynamic viscosity of the fast pyrolysis bio-oil and the upgraded oil before and after ageing. The viscosity was measured at 323 K [129].

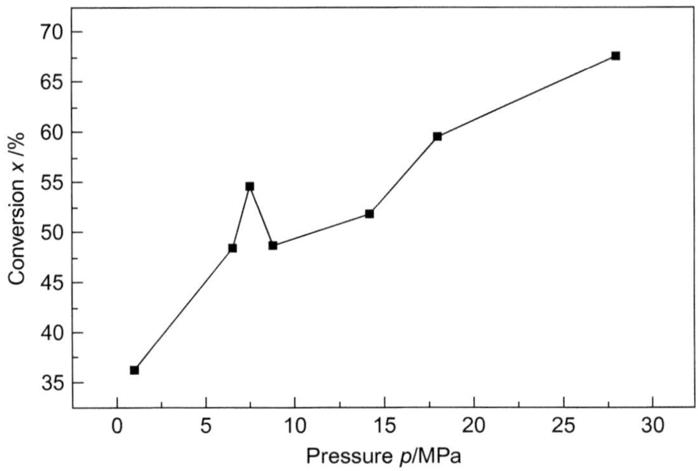

Fig. 7.8 Esterification conversion of acetic acid as a function of scCO2 pressure. Experimental condition: 75.8 g acid-ethanol-water solution in which acid/water mass ratio was 1:0.95; ethanol/acid molar ratio was 4:1, 0.6 g PTS, 333 K, and 3.0 h [131].

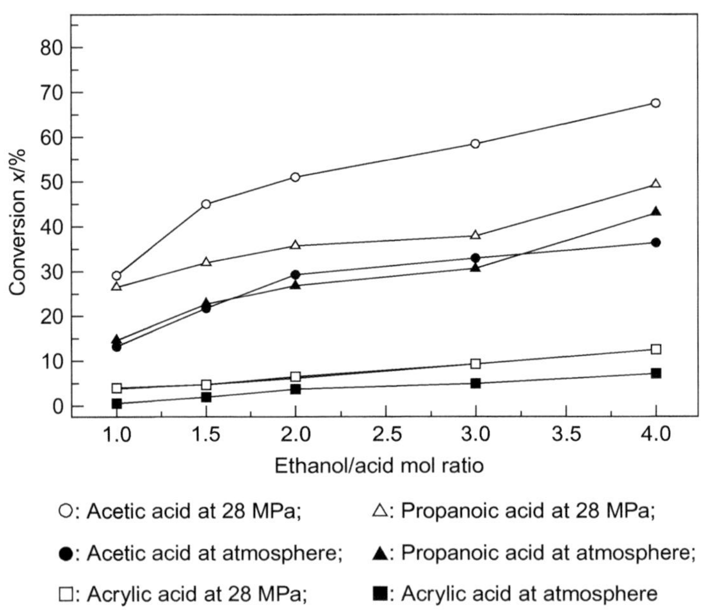

Fig. 7.9 Acid conversion as a function of the molar ratio of ethanol to acid both under supercritical CO2 condition and at atmospheric pressure [131].

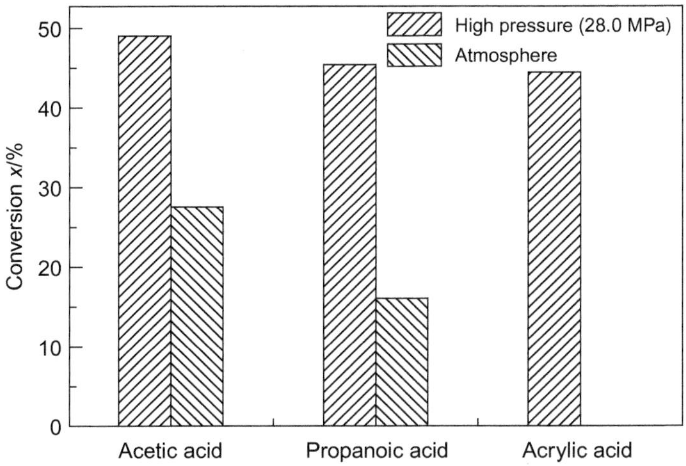

Fig. 7.10 Conversion comparison of organic acids at 28.0 MPa and atmospheric pressure. Experimental conditions: molar ratio of ethanol/total acid = 4:1, and mass ratio of water/total acid = 0.95:1 at 333 K, 3.0 h, and 0.6 g PTS [131].

The best conditions were determined to be 353 K and 28 MPa for 3 h, causing 87% of the total acids to be removed.

7.4.3.3 Biomass co-pyrolysis

Pyrolysis of lignocellulose with other resources like municipal and industrial waste, agricultural residues, plastics, and coal is aiming at improving bio-oil quality. Both thermal and catalytic concepts are targeted [12,132]. According to Hassan [132], plastic waste can be attractive as a cofeedstock from a chemical perspective because it contains very low oxygen amounts and a rather high H/C ratio; thus, it can potentially provide hydrogen to the other biomass resource and finally result in a beneficial C/H/O ratio and an improved bio-oil quality.

Co-pyrolysis employs two or more feedstock types, and the pyrolysis is performed in a similar way as normal pyrolysis, at moderate temperature pressure- and oxygen-free conditions. For a synergic case, the liquid yield is typically higher than from normal pyrolysis, up to 22 wt% [12]. In addition, a higher heating value of the bio-oil can be obtained, exemplified by 26–35 MJ/kg [133]. The synergetic effect between plastics and biomass can be attributed to free-radical chemistry interacting during co-pyrolysis.

Co-pyrolysis can be further developed by implementing a catalyst, microwave-assisted heating, hydrogen addition, solvent additions, etc. Several improvements have been reported by using a catalyst for the quality of the co-pyrolysis compared with ordinary pyrolysis, with improvements in relative contents for aromatics, whereas the contents of coke and oxygenated PAHs were reduced. A catalyst can strongly influence yields and selectivity, but conditions that prevent catalyst deactivation is still to be improved. Optimisation of the catalyst acidity and also pore structure, such as advanced hierarchical systems, are considered as viable and promising directions of development. A significant near completing lowering in oxygen content has been reported by Solak [134] in co-pyrolysis of cellulose and PE over different clay catalysts. FTIR confirmed that groups such as –OH, C=O, and C-O-C were efficiently removed. It is of interest to look closer into the effect of a catalytic metal function as well, thus intensifying hydrogen handling reactions.

7.4.3.4 High-pressure homogenization (HPH)

HPH is typically performed by forcing a liquid through a narrow nozzle at high pressure and by such establishing high shear stress. Typical pressures are moderate and between 15 and 40 bars. This can be used to stabilise bio-oil as emulsions, and the droplet size can be adjusted by the levels of pressure and energy input. Weak bonds in macromolecules can be broken by applying sufficiently high pressures, that is, both physical and chemical changes can be controlled. This means that the technique has a certain potential as a stabilisation technique. Ronghai et al. [90] studied the homogenisation of bio-oil from switchgrass. Viscosity, the ethanol-insoluble fraction, and average molecular weight decreased. The chemical composition is altered, by decreasing acetic acid and 1,2-ethanediol and increasing amounts of furfural, levoglucosan, and diethoxymethyl acetate. Test of stability indicated that HPH had a significantly positive effect. The viscosity of bio-oil homogenised at 100 MPa increased only by ca. 14% after storage at 313 K for 60 days, while that of unhomogenised oil increased 56% after a similar period.

It was concluded that further investigations are necessary for optimising HPH conditions and also for assessing the possibilities of applying catalysts to selectively improve reactions of high relevance for the bio-oil stability.

7.4.4 Physical (Nonreactive) stabilisation

7.4.4.1 Introduction

While reactive chemical methods are appearing as more effective for achieving long-term stability, physical methods can be used for obtaining stability in a simpler way and in a shorter time perspective. Such methods include, for example, ways of viscosity lowering by solvents or blending and also solid removal of, for example, the char particles.

In-depth knowledge on solubility is a key to understand bio-oil physical stability. This closely links to the chemical processes that are gradually changing the physical properties. The mutual solubility of a mixture of components relates to chemical structures and relative amounts. Low-molecular-weight components are generally more soluble than the higher. Similar polarities or similar molecular structures tend to be favourable for mutual solubility:

- High polarity: Water, alcohols, and organic acids
- Intermediate: Esters, ethers, and phenolics
- Nonpolar: Hydrocarbons

Some compounds favour mutual solubility between different categories of polarities due to long hydrocarbon chain with an oxygen-containing functional group, such as OH.

Bio-oils with high oxygen content on a dry basis can keep more water in solution than oils with lower content. Some water appears to be present in the form of aldehyde hydrates. Acetaldehyde can react with water forming the *gem* diol, that is, formula $CH_3C(OH)_2$. Solubility is a combination of contributions from non-polar and dispersion interactions, polar interactions, and hydrogen bonding interactions. Sum of contributions is the total solubility parameter, that is, the sum of the so-called Hansen parameters, and can assist in predicting the solubility of solutes (j) and solvents (i). If the Hansen parameters fall within the 'spherical volume' of the solute, solubility is to be expected [23]. Water and *n*-hexane have extreme and diametrical properties. An overview of Hansen parameters for typical bio-oil components and others is provided by Diebold [23].

Chemical reactions are changing the polarity of the bio-oil during ageing. Esterification converts highly polar organic acids and alcohols into components with relatively low polarity and extremely polar water. The polarity of the organic material is decreasing when the water content is increasing, which increases the tendency for phase separations. Formation of larger molecules during ageing causes the mutual solubility to decrease. This also increases the tendency for phase separation, of a light highly polar fraction and a less polar heavier fraction. Light waxy phases and heavy sludge phases can also form during storage [135].

7.4.4.2 Addition of solvents and chemicals

Important properties to consider during blending are stability, ageing, phase separation, and thermal shock. Three phases can in principle be included, that is, the bio-oil, the aqueous phase, and the solvent. The pyrolytic aqueous fraction can contribute to lower combustion temperature and thus reduce NOx, SO_3, and PMs. The solvent, such as methanol, improves homogeneity, facilitates combustion, and enhances stability [13]. Majhi et al. [136] have studied blending optimisation of Hempel distilled bio-oil and commercial diesel.

Solvent addition is an efficient way of obtaining a homogenous lower viscosity mixture. Polar solvents are typically decreasing viscosity and slowing down the chemical changes. Many solvents have been tested including ethyl acetate, methyl isobutyl ketone, acetone, methanol, ethanol, and isopropanol [23]. Methanol was found the most promising, slowing down the ageing rate at 363 K to about 5% of the reference level [23]. Ageing was then related to changes in the viscosity. Fluidity also improved by increasing concentrations of methanol. To some degree, the solvent itself can participate in reactions that outcompete the less desired ones. One challenge that arises from the addition of a solvent is an eventual following decrease in the flash point, which is an indicator of the fire hazard.

Oasmaa et al. [137] have studied addition of different alcohols like methanol, ethanol, and isopropanol. For added levels up to 10%, a linear decrease was observed for viscosity, density, flash point, and pour point. It also increased the heating value and homogeneity of the blend. Higher levels of ethanol, that is, 25% increased heating value with 16%, decreased kinematic viscosity with 40%. Improved combustion stability and reduced emissions were also obtained.

It was reported that a bio-oil content of 37.5% as a blend with diglyme (ether) can be used for reliable operation of engines without modifications. The main purpose of the diglyme was to improve the self-ignition characteristics. Similar experience that was obtained for 30% bio-oil blends from ultra-fast and fast pyrolysis with diesel fuel [13].

The composite effect of adding ethanol acetonitrile and methyl acetate to address bio-oil stability was studied by Renzhan et al. [66]. A systematic study on the effect of pH, water content, and viscosity was carried out. An optimised composite addition was ca. 6.6 wt% ethanol, 1 wt% acetonitrile, and ca. 2.4 wt% methyl acetate, found to be superior to the sum of individual effects. It was claimed that components such as 4-hydroxy-butanoic acid, p-cresol, and vanillin were virtually eliminated by the composite additive. Studies on single chemicals like isopropanol [97], acetone [23], methanol [87], ethanol [85], and acetone-methanol mixture [135] provided insight on effects on viscosity and stability.

7.4.4.3 Antioxidants

Oxidation stability is the resistance to chemical changes induced by oxygen. Due to their chemical composition and specific structural features, bioliquids like biodiesel (FAME type) are easily subject to oxidation during storage upon exposure to air. This is mainly due to oxidation of the polyunsaturated carbon chains, thus causing deterioration of fuel quality properties. Addition of antioxidants to the fuel is applicable for slowing down or preventing this mechanism, thus better maintaining the quality over time. Sterically hindered phenolics and secondary aromatic amines have properties that make them acting as free-radical scavengers. An example of such a structure is tert-butyl catechol (TBC), used by the olefin polymer industry. Other examples are tert-butyl hydroxyquinone (TBHQ), butylated hydroxytoluene (BHT), and pyrrogallol (PY) [115, 138–140]. Free radicals are then attacking easily available H from HO and NH groups of the scavengers instead of the fuel carbon chain. Typical levels for obtaining a desired extent of the effects are in the range of 200–1000 ppm. Garcia et al. [62] have studied the extracted phenolic fraction from pyrolysis oils (produced by BTG, nL) for utilising it as an antioxidant in biodiesel. Different extracting solvents, including acetate esters, were applied to also ensure a proper miscibility with the biodiesel. It was shown that <4 wt% bio-oil in biodiesel led to an improvement in the oxidation stability of 475%, which in that case helped meeting the European standard requirement (EN 14214). Some main conclusions were that the antioxidants were not deteriorating over 5 months of cold storage. The induction time of the biodiesel was remaining close to constant for 2 months. A key to good performance of the antioxidant is the initial oxidation state of the biofuel.

7.4.4.4 Char and inorganics removal

Char and ash particles are contributing to oil instability [18]. The char itself is a complex material. Also of importance is that different metal ions typically are embedded in the char particles. These can act as catalysts for various condensation and polymerisation reactions. Char can also act as a catalyst for cracking volatiles and such reduce the bio-oil yield. Alkali also causes corrosion and deposits on internal surfaces and can even cause plugging of pipelines and equipment. Cyclone separators are one common method for char removal, but are not able to remove the micro- and submicroparticles.

7.4.4.5 Emulsification

Challenging properties of bio-oils such as high viscosity, high acidity, and high water content can be handled by emulsifying the bio-oil with fossil diesel fuel or with biodiesel (bioemulsions). Another fuel, such as diesel, can be used to emulsify the bio-oil and in this way decreasing its viscosity and increasing the energy content. A study of emulsification of biodiesel under various combinations of emulsifiers has been performed by Bo-Jhih Lin et al. [141]. Another option in this respect is using biodiesel consisting of esters and fatty acids, of which both are renewable, environmentally friendly, and compatible with existing fossil fuel.

Two immiscible liquids are involved in an emulsification process. Droplets of the dispersed phase are encapsulated by the continuous phase. Mechanical agitation and an emulsification agent, a surfactant, are required. The quality of the emulsified mixture depends upon a number of parameters including the blending ratio, surfactant type and concentration, agitation intensity, temperature, and time. In the ageing process, the bio-oil becomes saturated with large low-polarity molecules, and there is a segregation of polar molecules from less polars, with multifunctional molecules to form micelles, analogous to petroleum fuel oils [142]. The surface of char particles can act as condensation nuclei by attracting low-polarity molecules.

Micelles form as very small droplets that gradually aggregate and coalesce into larger ones. This proceeds until the droplet's total surface area is low enough to obtain a stable coating by multipolar molecules, thus stabilising the emulsion. This point of stabilisation of the two-phase mixture can be reached either before or after the droplets are large enough to plug filters and phase separate, indicating that a critical practical droplet size exists.

Both ionic and nonionic components can be used as emulsifiers. Ionic ones are metal soaps or sulphur-containing detergents. Nonionic can be linear block copolymers, where each individual block has different solubility characteristics. Only a very small fraction of the bio-oil is soluble in typical diesel fuels. Some mixtures of bio-oil in diesel oil have shown to be stable emulsions for >90 days. Levels of bio-oil up to 40% with 1% emulsifiers were studied by Ikura et al. [143]. Extending the useful lifetime of bio-oils as an apparent single phase appears feasible by using emulsifying agents. Challenges relate to the costs of surfactants and also the energy input of the mechanical work required for agitation [13].

A more extensive introduction to emulsification can be found in literature [144–146]. Butler is pointing to stabilisation of emulsions by catalysis of reactions at the liquid/liquid interface with recoverable nanoscale catalysts [147, 148].

7.4.4.6 Fractionation

More or less stable fractions can be established by separations. The pyrolysis oil can, for example, be separated into an oil-rich and an aqueous rich fraction prior to further processing. The oxygen-rich components that are key contributors to instability, such as carboxylic acids, are then enriched in the water phase. Such separations were performed by Vispute and Huber [149], and the aqueous phase is made up of ca. 50% of the total carbon present in the original pyrolysis oil. The remaining carbon was in the heavy phase. The aqueous fraction was treated with a Ru/C catalyst, which mainly yielded polyols, like propylene glycol, ethylene glycol, and sorbitol. On the other hand, use of the same catalyst on the lignin phase yielded mainly naphthenic components [42, 150]. The oil phase itself can also be separated into different fractions. A concept for achieving this is temperature swing extraction, well presented by Shushil et al. [151].

7.4.4.7 Filtration

Separation by filtration is feasible on different scales ranging from macro-, via ultra- and micro- [18], to nanofiltration [152] and reverse osmosis. Both solid materials and molecular range components can be separated.

Char particles of different size is forming during pyrolysis. The presence of such particles, as retaining in the bio-oil, is undesired, since they create problems for both storage, further processing, and end use. Examples of problems are related to metal corrosion, impact on engine ignition, blockage of valves and restricted openings, and blocking of subsequent upgrading reactors/units (e.g. hydrotreatment, HDO). Deep removal of particles is therefore highly required, but it is still particularly challenging to obtain efficient removal of particles in the size range below 10 μm. It is considered a necessity for different reasons, including stability considerations, to reduce the char content of the bio-oil. Particles including chars >10 μm can be removed by hot gas-vapour filtration, that is, already before quenching and liquid formation. Alternatively, cyclone separation can be applied for the larger particles. The major challenge is the smaller particles. Particles below 10 μm can still cause the presence of 0.2 wt% ash. Both micro- and nanofiltration concepts have been proposed.

It is demonstrated that a major quantity of particles in pyrolysis bio-oils can be removed by microfiltration [153, 154]. This has been widely used for the removal of fines (0.02–10 μm particles) in fossil oils and water industry. A membrane is used as a barrier to separate two bulk phases and to control transport of components. The driving force can be pressure, temperature, concentration, or electric field. Javaid et al. [154] studied the liquid-phase microfiltration in tubular ceramic membranes and were able to remove a significant fraction of the char particles and of the ash components. Membrane fouling by caking proceeds, which required development of cleaning protocols. A pyrolysis oil used by Routray et al. [41] was subject to microfiltration with a 0.8 μm membrane resulting in the removal of char particles, amounting to ca. 3 wt%. Biomass pretreatment by washing with water or dilute acid is effective for removing ash from the raw biomass, according to Scahill [155].

The importance of establishing efficient filtration can by exemplified with Mortensen et al. [156], who observed irreversible deactivation from 0.05 wt% S and reversible deactivation by K and Cl.

According to Teella [153], nanofiltration with polymeric membranes has been applied for the last two decades and for different applications such as the removal of contaminants from water, deacidification of vegetable oils, and sugar fractionation. There is a lack of systematic investigation on using nanofiltration directly applied for pyrolysis oils; however, Teella [153] performed an investigation by using commercial NF membranes for the removal of carboxylic acids from the aqueous sugars containing fraction of bio-oils, that is, with an indirect relevance. The study included chemical compatibility of the membranes to different organic compounds. Retention factors of glucose were above 90% and even negative for acetic acid at moderate pressure conditions (40 bars). Fluxes were linearly dependent upon transmembrane pressure and decreased with increasing solute concentration. It was identified that guaiacols caused irreversible damage to the membranes. Thus, seemingly, the phenolic components should preferably not be present or alternatively be removed prior to the acetic acid and sugar separation.

Alternative separation methods dedicated to the removal of carboxylic acids are involving concepts such as ion-exchange resins, extraction, neutralisation, vacuum evaporation, charcoal adsorption, and overliming. Such separation by distillation is difficult because the typical temperatures >353 K required lead to oil decomposition and coke/light gas formation.

7.4.4.8 Microwave and Plasma

Concepts with alternative ways of providing the energy are also investigated. Microwave pyrolysis and plasma pyrolysis are two concepts that have recently been reported. Yin et al., Tuan et al., and Oost et al. can be addressed [157–159] for further information. Hybrid microwave catalytic pyrolysis is also reported [160].

7.5 Conclusions and outlook

Bio-oils produced from lignocellulosic biomass by state-of-the-art pyrolysis liquefaction processes are by thermochemical nature chemically instable, with a tendency to undergo nondesired self-reactions at ambient conditions, immediately upon production. This general property constitutes a serious bottleneck for widespread implementation. Efficient solutions to solve or circumvent this issue are needed for future development for enabling utilisation of lignocellulosic biomass as a source for renewable transport fuels and chemicals.

The challenge of bio-oil instability has been targeted by the research community for already >30 years. Considerable knowledge and information is available, including the chemical and physical mechanisms, methods to quantitatively measure stability, the difference between stabilisation and upgrading of bio-oil, and recent developments in techniques and methods to improve the storage and transportability. Even though the problem is well understood, an emerging quick fix is not apparent.

Stabilisation approaches in the academic literature can be classified into chemical, physical, and thermal, of both in situ and ex situ nature. Parameters of high relevance for stability encompass the different types and levels oxygen, water, viscosity, unsaturated groups, hydrogen deficit, and solid inorganic content. Some concepts or methods are merged or combined into hybrid approaches, targeted at harvesting synergies. Strategies both dedicated to stabilisation or as a problem sought handled as part of a conversion step towards final products are followed. Stabilisation itself is attractive since this would create a situation of a high degree of compatibility with regard to existing infrastructure.

Major challenges are the wide variability in feedstock and obtained oil qualities, as well as the general low hydrogen content in lignocellulose. It is seemingly difficult to establish standardised solutions. Differences in regional feedstock and bio-oil variability are highly important and make general applicability of methods difficult.

7.5.1 Hybrid solutions

The state of the art is pointing to the need of addressing multifunctional approaches. One-pot processing has been forecasted as a challenging vision. Simple solutions are particularly attractive, but such are still in want. Both chemical and separation concepts should be addressed and possibly also integrated in multifunctional steps. For overall efficiency, it is however highly important to keep the number of processing steps at a minimum.

7.5.2 Hydrogen

A hydrogen strategy is appearing as a key for stabilisation and obtaining carbon efficiency in a chemical perspective but can also be a potential major cost. As such, the availability of hydrogen is a bottleneck. Different H_2 strategies are proposed, not only with molecular hydrogen but also by in situ transhydrogenation from other components by using either solvents or co-pyrolysis with hydrogen-rich feedstock. It should be possible to add H_2 at different places along the 'chain', already from the primary pyrolysis steps. Conditions of very high H_2 pressures should preferably be relieved (<100 bars). As an already refined component or fuel itself, H_2 is also a key factor regarding cost and energy efficiency if used in large amount in the process.

7.5.3 Integration strategies

Smart ways of efficient integration with other plants in (bio)refineries and clusters is seemingly a way forward, probably an absolute requirement. This could establish required conditions with availability of steam, hydrogen-rich streams, and oxygenates for esterification. Preferable side or waste streams of otherwise low value should be identified and applied. Biomass waste could find a place in such a picture.

As a widely available resource with high hydrogen content, methane could be used to assist opening up efficient and sustainable opportunities. Biorefineries integrated with natural gas or biomethane utilisation could improve conditions for continuous operation of liquid value chains. It could facilitate establishing plants of flexible scale, by solving the issue of hydrogen availability.

7.5.4 Catalytic solutions

Catalytic approaches are principally promising but need further development on directions of flexibility, robustness/deactivation, and selectivity at mild process conditions, that is, moderate hydrogen pressures and temperatures. Conventional catalysts are not optimised for the conditions encountered in the liquid chain and particularly less so when considering the primary steps. New materials are obviously required to handle the levels of heteroatoms and other challenging components in the liquid fractions.

References

[1] Briens C, Piskorz J., Berruti F., Biomass valorization for fuel and chemicals production—a review, Int J React Eng 2008:8 (Review R2).

[2] Energy and Climate Policies beyond 2020 in Europe—overall and selected countries, Danish Energy Agency (2015) https://ens.dk/sites/ens.dk/files/Globalcooperation/eu_energy_and_climate_policy_overview.pdf.

[3] Anjani RK, Gollakota MR, Subramanyam MD, Kishore N. A review on the upgradation techniques of pyrolysis oil. Renew Sust Energ Rev 2016;58:1543–68.

[4] Kan T, Strezov V, Evans TJ. Lignocellulosic biomass pyrolysis: a review of product properties and effects of pyrolysis parameters. Renew Sust Energ Rev 2016;57:1126–40.

[5] Krutof A, Hawboldt K. Blends of pyrolysis oils, petroleum, and other bio-based fuels: a review. Renew Sust Energ Rev 2016;59:406–19.

[6] Hu X, Gunawan R, Mourant D, Mahmudul Hasan MD, Wu L, Song Y, et al. Upgrading of bio-oil via acid catalyzed reaction in alcohols—a mini review. Fuel Process Technol 2017;155:2–19.

[7] Yang Z, Kumar A, Huhnke RL. Review of recent developments to improve storage and transportation stability of bio-oil. Renew Sust Energ Rev 2015;50:859–70.

[8] Kim J-S. Production, separation and applications of phenolic rich bio-oil—a review. Biosource Technol 2015;178:90–8.

[9] Yang H, Yao J, Chen G, Ma W, Yan B, Qi Y. Overview of biomass pyrolysis oil of biomass. Energy Procedia 2014;61:1306–9.

[10] Kanaujia PK, Sharma YK, Garg MO, Tripathi D, Singh R. Review of analytical strategies in the production and upgrading of bio-oils derived from lignocellulosic biomass. J Anal Appl Pyrolysis 2014;105:55–74.

[11] Lehto J, Oasmaa A, Solantausta Y, Kytö M, Chiaramonti D. Review of fuel oil quality and combustion of fast pyrolysis bio-oils from lignocellulosic biomass. Appl Energy 2014;116:178–90.

[12] Abnisa F, Wan Daud WM. A review on co-pyrolysis of biomass: an optional technique to obtain a high-grade pyrolysis oil. Energy Convers Manag 2014;87:71–85.

[13] No S-Y. Applications of bio-oils from lignocellulosic biomass to transportation, heat and power generation—a review. Renew Sust Energ Rev 2014;40:1108–25.

[14] Kanaujia PK, Sharma YK, Agrawal UC, Gard MO. Analytical approach to characterizing pyrolysis oil from biomass. Trends Anal Chem 2013;42:125–36.

[15] Xiu S, Shahbazi A. Bio-oil production and upgrading research: a review. Renew Sust Energ Rev 2012;16:4406–14.

[16] Bridgwater AV. Review of fast pyrolysis of biomass and product upgrading. Biomass Bioenergy 2012;38:68–94.

[17] Akhtar J, Amin NAS. A review on process conditions for optimum bio-oil yield in hydrothermal liquefaction of biomass. Renew Sust Energ Rev 2011;15:1615–24.
[18] Butler E, Devlin G, Meier D, McDonnell K. A review of recent laboratory research and commercial developments in fast pyrolysis and upgrading. Renew Sust Energ Rev 2011;15:4171–86.
[19] Zakzeski J, Bruijnincx PCA, Jongerius AL, Weckhuisen BM. The catalytic valorization of lignin for the production of renewable chemicals. Chem Rev 2010;110:3552–99.
[20] Elliott DC. Historical developments in hydroprocessing bio-oils. Energy Fuel 2007;21:1792–815.
[21] Qi Z, Jie C, Wang T, Xu Y. Review of biomass pyrolysis oil properties and upgrading research. Energy Convers Manag 2007;48:87–92.
[22] Mohan D, Pittman Jr. CU, Steele PH. Pyrolysis of Wood/Biomass for bio-oil: a critical review. Energy Fuel 2006;20:848–89.
[23] Diebold JP. A review of the chemical and physical mechanisms of the storage stability of fast pyrolysis bio-oils, NREL, report; 2000.
[24] Bridgwater, A.V.,Fast pyrolysis processes for biomass, Renew Sust Energ Rev, 4 (2000) 1–73.
[25] Elliott DC, Beckman D, Bridgwater AV, Diebold JP, Gevert SB, Solantausta Y. Developments in direct thermochemical liquefaction of biomass: 1983–1990. Energy Fuel 1991;5:399–410.
[26] Elliott DC, Hart TR. Catalytic hydroprocessing of chemical models for bio-oil. Energy Fuel 2009;23:631–7.
[27] Huber GW, Iborra S, Corma A. Synthesis of transportation fuels from biomass: chemistry, catalyst and engineering. Chem Rev 2006;106:4044–98.
[28] Huber GW, Corma A. Synergies between bio- and oil-refineries for the production of fuels from biomass. Angew Chem Int Ed 2007;46:7184–201.
[29] Alonso DM, Bond JQ, Dumesic JA. Catalytic conversion of biomass to biofuels. Green Chem 2012;12:1493–513.
[30] Mortensen PM, Grunwaldt JD, Jensen PA, Knudsen KG, Jensen AD. A review of catalytic upgrading of bio-oil to engine fuels. Appl Catal A: General 2011;407:1–19.
[31] Toor SS, Rosendahl L, Rudolf A. Review—hydrothermal liquefaction of biomass: a review of subcritical water technologies. Energy 2011;36:2328–42.
[32] Venderbosch RH, Prins W. Fast pyrolysis technology development. Biofuels Bioprod Biorefin 2010;4:178–208.
[33] Cai W, Liu R. Performance of a commercial scale biomass fast pyrolysis plant for bio-oil production. Fuel 2016;182:677–86.
[34] IEA web; https://www.iea.org/topics/renewables/subtopics/bioenergy/.
[35] BP - http://www.bp.com/content/dam/bp-country/en_au/media/fuel-news/long-term-storage-diesel.pdf.
[36] Friberg SE, Jones S. Emulsions. In: Kirk and Othmer's Encyclopedia of Chemical Technology. 4th ed. Vol. 9. New York: John Wiley and Sons; 1994. p. 393–413.
[37] Yaakob Z, Narayanan BN, Padikkaparambil S, Surya Unni K, Mohammed Akbar P. A review on the oxidation stability of biodiesel. Renew Sust Energ Rev 2014;35:136–53.
[38] McCormick RL, Westbrook SR. Empirical study of the stability of biodiesel and biodiesel blends, NREL, milestone report NREL/TP-540-41619; May 2007.
[39] Arturi KR, Toft KR, Nielsen RP, Rosendahl LA. Characterization of liquid products from hydrothermal liquefaction (HTL) of biomass via solid-phase microextraction (SPME). Biomass Bioenergy 2016;88:116–25.
[40] Mullen CA, Boateng AA. Energy Fuel 2008;22:2104–9.

[41] Routray K, Barnett KJ, Huber GW. Hydrodeoxygenation of pyrolysis oils. Energ Technol 2016;4:1–15.
[42] Chen W, McClelland DJ, Azarpira A, Ralph J, Luo Z, Huber GW. Green Chem 2016;18:271–8.
[43] Scholze B, Hanser C, Meier D. J Anal Appl Pyrolysis 2001;58–59:387–400.
[44] Oasmaa A, Kuoppala E, Solantausta Y. Energy Fuel 2003;17:433–43.
[45] Oasmaa A, Peacocke C, Gust S, Meier D, McLellan R. Norms and standards for pyrolysis liquids. End-user requirements and specifications. Energy Fuel 2005;19:2155–63.
[46] Charon N, Ponthus J, Espinat D, Broust F, Volle G, Valette J, et al. Multi-technique characterization of fast pyrolysis oils. J Anal Appl Pyrolysis 2015;116:18–26.
[47] Lazzari E, Schena T, Primaz CT, da Silva Maciel GP, Machado ME, Cardoso CAL, et al. Production and chromatographic characterization of bio-oil from the pyrolysis of mango seed waste. Ind Crop Prod 2016;83:259–536.
[48] Undri A, Abou-Zaid M, Briens C, Berruti F, Rosi L, Bartoli M, et al. A simple procedure for chromatographic analysis of bio-oils from pyrolysis. J Anal Appl Pyrolysis 2015;114:208–21.
[49] Haarlemmer G, Guizani C, Anouti S, Deniel M, Roubaud A, Valin S. Analysis and comparison of bio-oils obtained by hydrothermal liquefaction and fast pyrolysis of beech wood. Fuel 2016;174:180–8.
[50] Stedile T, Ender L, Meier HF, Simionatto EL, Wiggers VR. Comparison between physical properties and chemical composition of bio-oils derived from lignocellulose and triglyceride sources. Renew Sust Energ Rev 2015;50:92–108.
[51] Elliott DC, Oasmaa A, Preto F, Meier D, Bridgwater AV. Results of the IEA round robin on viscosity and stability of fast pyrolysis bio-oils. Energy Fuel 2012;26:3769–76.
[52] Czernik A., Bridgwater A., Overview of applications of biomass fast pyrolysis oil, Energy Fuel 18 2004 590–598.
[53] Czernik A, Johnson DK, Black S. Stability of wood fast pyrolysis oil. Biomass Bioenergy 1994;7:187–92.
[54] Demirbas A. Chemical valorization of wood for bio-fuels and bio-chemicals. Energy Sources A: recovery Util Environ Effects 2009;32:1–9.
[55] Oasmaa A, Korhonen J, Kuoppala E. An approach for stability measurement of wood based pyrolysis fast pyrolysis bio-oils. Energy Fuel 2011;25:3307–13.
[56] Vallinayagam R, Vedharaj WMY, Roberts WL, Dibble RW. Feasibility of using less viscous and lower cetane (LVLC) fuels in a diesel engine: a review. Renew Sust Energ Rev 2015;51:1166–90.
[57] Abdullah N, Gerhauser H, Sulaiman F. Fast pyrolysis of empty fruit bunches. Fuel 2008;87:2606–13.
[58] Wigley T, Yip ACK, Pang S. Pretreating biomass via demineralization to improve the quality of crude pyrolysis oil. Energy 2016;109:481–94.
[59] Song H, Jiang L, Yi W, Sheng S, Sun L, Boyang X, et al. Effects of inherent alkali and alkaline earth metallic species on biomass pyrolysis at different temperatures. Bioresour Technol 2015;192:23–30.
[60] Michailof M, Lappas AA. Optimization of bio-oil yields by demineralization of low quality biomass. Biomass Bioenergy 2015;83:105–15.
[61] Stefanidis SD, Heracleous E, Patiaka DT, Kalogiannis KG, Chrysoula Saddawi A, Jones JM, et al. Influence of alkali metals on the kinetics of the thermal decomposition of biomass. Fuel Process Technol 2012;104:189–97.
[62] Garcia M, Botella L, Gil-Lalaguna N, Arauzo J, Gonzalo A, Sanchez JL. Antioxidants for biodiesel. Additives prepared from extracted fractions of bio-oil. Fuel Process Technol 2017;156:407–14.

[63] Isa KM. Pyrolysis oil upgrading using supercritical water, with tetralin and 1 methylnaphthalene as a baseline study. Energy Convers Manag 2016;117:558–66.

[64] Le Zhang, Renzhan Yin, Yuanfei Mei, Ronghou Liu, Wenjuan Yu, Characterization of crude and ethanol-stabilized bio-oils before and after accelerated ageing treatment by comprehensive two-dimensional gas-chromatography with time-of-flight mass spectrometry, J Energy Institute xxx (in correction) (2016) 1–14.

[65] Shinyoung O, Choi HS, Kim U-J, Choi I-G, Choi JW. Storage performance of bio-oil after hydrodeoxygenative upgrading with noble metal catalysts. Fuel 2016;182:154–60.

[66] Yin R, Le Z, Liu R, Mei Y, Wenjuan Y. Optimization of composite additives for improving stability of bio-oils. Fuel 2016;170:1–8.

[67] Gao W, Zhang M, Hongwei W. Fuel properties and ageing of bioslurry prepared from glycerol/methanol/bio-oil blend and biochar. Fuel 2016;176:72–7.

[68] Nazari L, Yuan Z, Souzanchi S, Ray MB, Xu C. Hydrothermal liquefaction of woody biomass in hot-compressed water: catalyst screening and comprehensive characterization of bio-crude oils. Fuel 2015;162:74–83.

[69] Praveen G, Vijaya Lakshmi C, Ganesh A, Mahajani S. Stabilization of pyrolysis oil: comparison of reactive distillation and reactive chromatography. Chem Eng And Processing 2015;95:327–38.

[70] Rover MR, Hall PH, Johnston PA, Smith RG, Brown RC. Stabilization of bio-oils using low temperature, low pressure hydrogenation. Fuel 2015;153:224–30.

[71] Ying X, Jinxing L, Qiying L, Yanbin L, Chenguang W, Qi Z, et al. In-situ hydrogenation of model compounds and raw bio-oil over Raney Ni catalyst. Energy Convers Manag 2015;89:188–96.

[72] Le Roux E, Chaouch M, Diouf PN, Stevanovic T. Impact of pressurized hot water treatment on the quality of bio-oil produced from aspen. Biomass Bioenergy 2015;81:202–9.

[73] Zhang M, Hongwei W. Effect of major impurities in crude glycerol on solubility and properties of glycerol/methanol/bio-oil blends. Fuel 2015;159:118–27.

[74] Boscagli C, Raffelt K, Zevaco TA, Olbrich W, Otto TN, Sauer J, et al. Mild hydrotreatment of the light fraction of fast pyrolysis oil produced from straw over nickel-based catalysts. Biomass Bioenergy 2015;83:525–38.

[75] Hari P, Kannapu R, Mullen CA, Elkasabi Y, Boateng AA. Catalytic transfer hydrogenation for stabilization of bio-oil oxygenates: reduction of p-cresol and furfural over bimetallic Ni-Cu catalysts using isopropanol. Fuel Process Technol 2015;137:220–8.

[76] Li H, Xia S, Li Y, Ma P, Zhao C. Stability evaluation of fast pyrolysis oil from rice straw. Chem Eng Sci 2015;135:258–65.

[77] Cao X, Zhong L, Peng X, Sun S, Li S, Liu S, et al. Comparative study of the pyrolysis of lignocellulose and its major components: characterization and over distribution of their biochars and volatiles. Bioresour Technol 2014;155:21–7.

[78] Chen W, Luo Z, Chunjiang Y, Li G, Zhang J, Kongyu L. Catalytic transformations of acids, aldehydes, and phenols in bio-oil to alcohols and esters. Fuel 2014;135:55–62.

[79] Liu R, Fei W, Shen C. Influence of acetone addition on the physicochemical properties of bio-oils. J Energy Inst 2014;87:127–33.

[80] Chen D, Zhou J, Zhang Q, Zhu X. Evaluation of methods and research progresses in bio-oil storage stability. Renew Sust Energ Rev 2014;40:69–79.

[81] Cordella M, Berrueco C, Santarelli F, Paterson N, Kandiyoti R, Millan M. Yields and ageing of the liquids obtained by slow pyrolysis of sorghum, switchgrass and corn stalks. J Anal Appl Pyrolysis 2013;104:316–24.

[82] Paige A, Case M, Wheeler C, DeSisto WJ. Formate assisted pyrolysis of pine sawdust for in-site oxygen removal and stabilization of bio-oil. Bioresour Technol 2014;173:117–24.

[83] Guo Z, Wang S, Wang X. Stability mechanism investigation of emulsion fuels from biomass pyrolysis oil and diesel. Energy 2014;66:250–5.

[84] Tessarolo NS, Santos LRMD, Silva RSF, Azevedo DA. Chemical characterization of bio-oils using comprehensive two-dimensional gas chromatography with time-of-flight mass spectrometry. J Chromatography A 2013;1279:68–75.

[85] Mante OD, Agblevor FA. Storage stability of biocrude oils from fast pyrolysis of poultry litter. Waste Manag 2012;32:67–76.

[86] Kim J-Y, Kim T-S, Eom I-Y, Kang SM, Cho T-S, Choi IG, et al. Characterization of pyrolytic products obtained from fast pyrolysis of chromated copper arsenate (CCA) and alkaline copper quaternary compounds (ACQ) treated wood biomass. J Hazard Mater 2012;227–229:445–52.

[87] Kim T-S, Kim J-Y, Kim K-H, Lee S, Choi D. The effect of storage duration on bio-oil properties. J Anal Appl Pyrolysis 2012;95:118–25.

[88] Hilten RN, Das KC. Comparison of three accelerated ageing procedures to assess bio-oil stability. Fuel 2010;89:2741–9.

[89] Hilten RN, Bibens Brian P, Kastner James R, Das KC. In-line esterification of pyrolysis vapor with ethanol improves bio-oil quality. Energy Fuel 2010;24:673–82.

[90] He R, Ye P, Harte F, English B. Effects of high-pressure homogenization on physicochemical properties and storage stability of switchgrass bio-oil. Fuel Process Technol 2009;90:415–21.

[91] Moens L, Black SK, Myers MD, Czernik S. Study of the neutralization and stabilization of a mixed hardwood bio-oil. Energy Fuel 2009;23:2695–9.

[92] Garcia-Perez M, Chaala A, Pakdel H, Kretschmer D, Roy C. Characterization of bio-oils in chemical families. Biomass Bioenergy 2007;31:222–42.

[93] Lede J, Broust F, Ndiaye F-T, Ferrer M. Properties of bio-oils produced by biomass fast pyrolysis in a cyclone reactor. Fuel 2007;86:1800–10.

[94] Boucher ME, Chaala A, Pakdel H, Roy C. Bio-oils obtained by vacuum pyrolysis of softwood bark as a liquid fuel for gas turbines. Part II: stability and ageing of bio-oil and its blends with methanol and pyrolytic aqueous phase. Biomass Bioenergy 2000;19:351–61.

[95] Batts B, Fathoni AZ. A literature review on fuel stability studies with particular emphasis on diesel oil. Energy Fuel 1991;5:2–21.

[96] Oasmaa A. Fuel oil quality properties of wood-based pyrolysis liquids. Jyväsky: Finland, University of Jyväskylä; 2003.

[97] Oasmaa A, Peacocke C. A guide to physical property characterization of biomass-derived fast pyrolysis liquids. Technical research centre of Finland: Espoo; 2001.

[98] Adkins H, Broderick AE. Hemiacetal formation and the refractive indices and densities of mixtures of certain alcohols and aldehydes. J Am Chem Soc 1928;50:499–503.

[99] Granger FS. Condensation of phenols with formaldehyde. Resinification of phenol alcohols. Ind Eng Chem 1937;29:860–6.

[100] Mushrush GW, Speight JG. Petroleum products: instability and incompatibility. Washington, DC: Taylor and Francis; 1995390.

[101] Fiddler W, Parker WE, Wasserman AE, Doerr RC. Thermal decomposition of ferulic acid. J Agric Food Chem 1967;15(5):757–61.

[102] Meier D., van de Beld B., Bridgwater A.V., Elliott D.C., Oasmaa A., Preto F., State of the art of fast pyrolysis in IEA bioenergy member countries. Renew Sust Energ Rev 2013 20 619–641.

[103] ASTM Standard D64686468–08. Standard test method for high temperature stability of middle distillate fuels. ASTM International: West Conshohocken, PA; 2013.

[104] ASTM D873–12. Standard test method for oxidation stability of aviation fuels (potential residue method). ASTM International: West Conshohocken, PA; 2012.
[105] D3374-14 ASTM. Standard test method for oxidation stability of distillate fuel oil (accelerated method). West Conshohocken, PA: ASTM International; 2015.
[106] ASTM D525-12a. Standard test method for oxidation stability of gasoline (induction period method). West Conshohocken, PA: ASTM International; 2012.
[107] Alsbou E, Helleur B. Accelerated ageing of bio-oil from fast pyrolysis of hardwood. Energy Fuel 2014;28:3224–35.
[108] Meng J, Moore A, Tilotta D, Kelly S, Park S. Toward understanding of bio-oil ageing. Accelerated ageing of biomass fractions. ACS Sustain Chem Eng 2014; 2:2011–8.
[109] Jena U, Das K. Comparative evaluation of thermochemical liquefaction and pyrolysis for bio-oil production from microalgae. Energy Fuel 2011;25:5472–82.
[110] 2009-02 ASTME. Standard test method for oxidation onset temperature of hydrocarbons by differential scanning calorimetry. Easton, MD: American Society for Testing and Materials; 2002.
[111] He W, Liu Q, Shi L, Liu Z, Ci D, Lievens C. Understanding the stability of pyrolysis tars from biomass in a view point of free radicals. Bioresour Technol 2014;156:372–5.
[112] Meng J, Smirnova TI, Song X, Moore A, Ren X, Kelley S. Identification of free radicals in pyrolysis oil and their impact on bio-oil stability. RSC Adv 2014;4:29840–6.
[113] Paquin P. Technological properties of high pressure homogenizers: the effect of fat globules, milk proteins, and polysaccharides. Int Dairy J 1999;9:329–35.
[114] Reyhanitash E, Tymchyshyn M, Yuan Z, Albion K, van Rossum G, Xu C. Hydrotreatment of fast pyrolysis oils: effects of esterification pre-treatment of the oil using alcohol at a small loading. Fuel 2016;179:45–51.
[115] Maia ECR, Borsato D, Moreira I, Spacino KR, Rodriques RPR, Lazarin-Gallina A. Study of the biodiesel B100 oxidative stability in mixture with antioxidants. Fuel Process Technol 2011;92:1750–5.
[116] Galabdina A, Muraza O. In-situ fast pyrolysis of biomass with zeolite catalysts for bio-aromatics/gasoline production: a review. Energy Convers Manag 2015; 105:338–54.
[117] Widayatno WB, Guan G, Rizkiana J, Yang J, Hao X, Tsutsumi A, et al. Upgrading of bio-oil from biomass pyrolysis over cu-modified beta-zeolite catalysts with high selectivity and stability. Appl Catal B Environ 2016;186:166–72.
[118] Kocal JA. Brandvold T.A. In: (UOP) WO2013019328; 2013.
[119] Rune Lødeng, Lenka Hannevold, Håkon Bergem and Michael Stöcker, Catalytic hydrotreatment of bio-oils for high quality fuels production by, Book Chapter 11 in 'The Role of Catalysis for the Sustainable Production of Biofuels and Biochemicals (Elsevier)', edited by K. Triantafyllidis, A. Lappas and M. Stöcker, 2013, 352–395.
[120] Yakovlev VA, Khromova SA, Sherstyuk OV, Dundich VO, Ermakov DY, Novopashina VM, et al. Development for new catalytic systems for upgraded bio-fuels production from bio-crude-oil and bio-diesel. Catal Today 2009;144:362–6.
[121] Ardiyanti AR, Khromova SA, Venderbosch RH, Yakovlev VA, Melian-Cabrera IV, Heeres HJ. Catalytic hydrotreatment of fast pyrolysis oil using bimetallic Ni-cu catalysts on various supports. Appl Catal A: General 2012;449:121–30.
[122] Venderbosch RH, Ardiyanti AR, Wildschut J, Oasmaa A, Heeres HJ. Stabilization of biomass-derived pyrolysis oils. J Chem Technol Biotechnol 2010;85:674–86.

[123] Marker Terry, Roberts, M., Wangerow J., Ortiz-Toral P., Linck M., Swanson D. (GTI), biomass to gasoline and diesel using IH2® key pilot data tests on the road to commercialization, Presentation at the International Conference on Thermochemical Conversion Science, Westin Chicago River North, 2–5 November, 2015, 2015.
[124] Peng J, Chen P, Lou H, Zheng X. Catalytic upgrading of bio-oil by HZSM-5 in sub- and super-critical ethanol. Bioresour Technol 2009;100:3415–8.
[125] Tang Z, Lu Q, Zhang Y, Zhu X, Guo Q. One step bio-oil upgrading through hydrotreatment, esterification, and cracking. Ind Eng Chem Res 2009;48:6923–9.
[126] Kruse A, Dahmen N. Water—a magic solvent for biomass conversion. J Supercrit Fluids 2015;96:36–45.
[127] Remon J, Arcelus-Arrillaga P, Garcia L, Arauzo J. Production of gaseous and liquid bio-fuels from the upgrading of lignocellulose bio-oil in sub- and supercritical water: effect of operation conditions on the process. Energy Convers Manag 2016;119:14–36.
[128] Remon J, Arauzo J, Garcia L, Arcelus-Arrillaga P, Millan M, Suelves I, et al. Bio-oil upgrading in supercritical water using Ni-co catalysts supported on carbon nanofibres. Fuel Process Technol 2016;154:178–87.
[129] Prajitno H, Insyani R, Park J, Ryu C, Kim J. Non-catalytic upgrading of fast pyrolysis oil in supercritical ethanol and combustion behaviour of the upgraded oil. Appl Energy 2016;172:12–22.
[130] Zhang Q, Xu Y, Li Y, Wang T, Zhang Q, Ma L, He M, Li K. Investigation on the esterification by using supercritical ethanol for bio-oil upgrading. Appl Energy 2015;160:633–40.
[131] Hong-you C, Jing-hua W, shu-qin W, Shu-ping Z, Li Z-h, Li-hong W, et al. Upgrading bio-oil by esterification under supercritical CO_2 conditions. J Fuel Chem Technol 2010;38(6):673–8.
[132] Hassan H, Lim JK, Hameed BH. Review. Recent progress on biomass co-pyrolysis conversion into high-quality bio-oil. Bioresour Technol 2016;221:645–55.
[133] Chen W, Shi S, Zhang J, Chen M, Zhou X. Co-pyrolysis of waste newspaper with high-density polyethylene: synergistic effect and oil characterization. Energy Convers Manag 2016;112:41–8.
[134] Solak A, Rutkowski P. The effect of clay catalysts on the chemical composition of bio-oil obtained by co-pyrolysis of cellulose and polyethylene. Waste Manag 2014;34(2):504–12.
[135] Diebold J.P., Czernik S., Additives to Lower and Stabilize the viscosity of pyrolysis oils during storage, Energy Fuel, 11 2007 1081–1091.
[136] Majhi A, Sharman YK, Naik DV. Blending optimization of Hempel distilled bio-oil with commercial diesel. Fuel 2012;96:264–9.
[137] Oasmaa A, Kuoppala E, Selin J-F, Gust S, Solantausta Y. Fast pyrolysis of forestry residue and pine. Improvements of the product quality by solvent addition. Energy Fuel 2004;18:1578–83.
[138] Rizwanul-Fattah IM, Masjuki HH, Kalam MA, Hazrat MA, Masum BM, Imtenan S. Effect of antioxidants on oxidation stability of biodiesel derived from vegetable and animal based feedstocks. Renew Sust Energ Rev 2014;30:356–70.
[139] Dinkov R, Hristov G, Stratiev D, Boynova-Aldayri V. Effect of commercially available antioxidants over biodiesel/diesel blends stability. Fuel 2009;88:732–7.
[140] Focke WW, van der Westhuizen I, Lofte-Grobler AB, Kshoane KT, Reddy JK, Luyt AS. The effect of synthetic antioxidants on the oxidative stability of biodiesel. Fuel 2012;93:227–33.
[141] Lin B-J, Chen W-H, Budzianowski WM, Hsieh C-T, Lin P-H. Emulsification analysis of bio-oil and diesel under various combinations of emulsifiers. Appl Energy 2016;178:746–57.

[142] Radlein D. The production of chemicals from fast pyrolysis bio-oils. In: Bridgwater A, Szernik S, Diebold J, Meier D, Oasmaa A, Peakcocke C, Piskorz J, Radlein D, editors. The fast pyrolysis of biomass: a handbook. Newbury: CPL Press; 1999. p. 164–88.

[143] Ikura M., Mirmiran S., Stanciulescu, M., Sawatzky, H., (1998) 'Pyrolysis liquid-in-diesel oil microemulsions', U.S. Patent 5,820,640,.

[144] Chiaramonti D, Oasmaa A, Soantausta Y. Power generation using fast pyrolysis liquids from biomass. Renew Sust Energ Rev 2007;11:1056 1086.

[145] Chiaramonti D, Bonini M, Fratini E, Tondi G, Gartner K, Bridgwater AV, et al. Development of emulsions from biomass pyrolysis liquids and diesel and their use in engines—part 1: emulsion production. Biomass Bioenergy 2003;25:85–99.

[146] Chiaramonti D, Bonini M, Fratini E, Tondi G, Gartner K, Bridgwater AV, et al. Development of emulsions from biomass pyrolysis liquids and diesel and their use in engines—part 2: test in diesel engines. Biomass Bioenergy 2003;25:101–11.

[147] Faria J., Ruiz M.P., Zapata P., Shen M., Resasco D.E., Bio-oil refining on recoverable catalysts that stabilize emulsions in bi-phasic liquid systems. In: TCS 2010, symposium on thermal and catalytic sciences for biofuels and biobased products.

[148] Crossley S, Faria J, Shen J, Resasco DE. Solid nanoparticles that catalyse biofuel upgrade reactions at the water/oil interface. Science 2010;327:68–72.

[149] Vispute TP, Huber GW. Production of hydrogen, alkanes and polyols by aqueous phase processing of wood-derived pyrolysis oils. Green Chem 2009;11:1433–45.

[150] Yan N, Yuan YA, Dykeman R, Kou YA, Dyson PJ. Hydrodeoxygenation of lignin-derived phenols into alkanes by using nanoparticle catalysts combined with brønsted acidic ionic liquids. Angew Chem Int Ed 2010;49:5549–53.

[151] Kumar S, Lange J-P, van Rossum G, Kersten SRA. Bio-oil fractionation by temperature swing extraction:principle and application. Biomass Bioenergy 2015;83:96–104.

[152] Teella A, Huber GW, Ford DM. Separation of acetic acid from the aqueous fraction of fast pyrolysis bio-oils using nano-filtration and reverse osmosis membranes. J Membr Sci 2011;378:495–502.

[153] Ford DM, Javaid A, Berg G, Ryan T, Teella A, Huber GW. Membrane processes for the stabilization of fast pyrolysis bio-oil. In: TCS 2010, symposium on thermal and catalytic science for biofuels and biobased products. 2010.

[154] Javaid A, Ryan T, Berg G, Pan X, Vispute T, Bhatia SR, et al. Removal of char particles from fast pyrolysis bio-oil by microfiltration. J Membr Sci 2010;363:120–7.

[155] Scahill J, Diebold J, Feik C. Removal of residual char fines pyrolysis vapors by hot gas filtration. Dev Thermochem Biomass Conversion 1997;253–66.

[156] Mortensen PM, Gardini D, Carvalho HWP, Damsgaard CD, Grunwaldt J-D, Jensen PA, et al. Cat Sci Technol 2014;4:3672–86.

[157] Yin C. Microwave-assisted pyrolysis of biomass for liquid biofuels production. Bioresour Technol 2012;120:273–84.

[158] Tian Y, Zuo W, Ren Z, Chen D. Estimation of novel method to product bio-oil from sewage sludge by microwave pyrolysis with the consideration of efficiency and safety. Bioresour Technol 2011;102:2053–61.

[159] Oost GV, Habovsky M, Kopecky V, Konrad M, Hlina M, Kavka T. Pyrolysis of waste using a hybrid agron-water stabilized torch. Vacuum 2006;80:1132–7.

[160] Mohamed BA, Kim CS, Ellis N, Bi X. Microwave-assisted catalytic pyrolysis of switch-grass for improving bio-oil and biochar properties. Bioresour Technol 2016;201:121–32.

[161] Agblevor FA, Mante O, Abdoulmoumine N, McClung R. Production of stable biomass pyrolysis oils using fractional catalytic process. Energy Fuels 2010;24:4087–9.

[162] Agblevor FA, Beis S, Mante O, Abdoulmoumine N. Fractional catalytic pyrolysis of hybrid poplar wood. Ind Eng Chem Res 2010;49:3533–8.
[163] Aho A, Kumar N, Eränen K, Salmi T, Hupa M, Murzin DY. Catalytic pyrolysis of woody biomass in a fluidized bed reactor: influence of the zeolite structure. Fuel 2008;87:2493–501.
[164] Aho A, Kumar N, Lashkul AV, Eränen K, Ziolek M, Decyk P. Catalytic upgrading of woody biomass derived pyrolysis vapours over iron modified zeolites in a dual-fluidized bed reactor. Fuel 2010;89:1992–2000.
[165] Lappas AA, Bezergianni S, Vasalos IA. Production of biofuels via co-processing in conventional refining processes. Catal Today 2009;145:55–62.
[166] Li H-y, Yan Y-j, Ren Z-w, et al. Online upgrading of organic vapors from the fast pyrolysis of biomass. J Fuel Chem Technol 2008;36:666–71.
[167] Park HJ, Dong J-I, Jeon J-K, Yoo K-S, Yim J-H, Sohn JM, et al. Conversion of the pyrolytic vapor of radiata pine over zeolites. J Ind Eng Chem 2007;13:182–9.
[168] Wang C, Hao Q, Lu D, Jia Q, Li G, Xu B. Production of light aromatic hydrocarbons from biomass by catalytic pyrolysis. Chin J Catal 2008;29:907–12.
[169] Zhang H, Xiao R, Huang H, Xiao G. Comparison of non-catalytic and catalytic fast pyrolysis of corncob in a fluidized bed reactor. Bioresour Technol 2009;100:1428–34.
[170] Zhang H, Xiao R, Wang D, Zhong Z, Song M, Pan Q, et al. Catalytic fast pyrolysis of biomass in a fluidized bed with fresh and spent fluidized catalytic cracking (FCC) catalysts. Energy Fuels 2009;23:6199–206.

Further reading

[171] https://en.wikipedia.org/wiki/Energy_density.
[172] Bulushev DA, Ross JRH. Catalysis for conversion of biomass to fuels via pyrolysis and gasification: a review. Catal Today 2011;171:1–13.

Recent developments in the catalytic hydrotreatment of pyrolysis liquids

8

W. Yin, R.H. Venderbosch[†], H.J. Heeres**
*University of Groningen, Groningen, The Netherlands, [†]Biomass Technology Group BV, Enschede, The Netherlands

8.1 Introduction

Due to the depletion of fossil resources and environmental issues, for example, global warming as a result of greenhouse gases emissions, there is an urgent need to identify, develop, and implement alternative resources for energy, transportation fuel, and chemicals [1]. Solar, wind, hydroelectric, and geothermal activity and biomass are considered renewable and sustainable resources and are expected to (partly) replace fossil resources in the near future [2]. In this respect, lignocellulosic biomass, for example, wood, agricultural waste, and forest waste, (see Table 8.1) is of particular interest as it is the only sustainable resource containing renewable organic carbon [3].

Lignocellulosic biomass is a polymeric material composed of three main constituents: cellulose, hemicellulose, and lignin. The composition of some typical biomass resources are shown in Table 8.1[3]. Cellulose is a linear polysaccharide with β-1,4 linkages of D-glucopyranose monomers. Inter- and intramolecular hydrogen bonding render it highly crystalline and as such has a high reluctance to hydrolysis reactions. Hemicellulose is a biopolymer consisting of a wide range of sugars, namely, five-carbon sugars (usually xylose and arabinose) and six-carbon sugars (galactose, glucose, and mannose). Compared with cellulose, hemicellulose is amorphous and highly branched, which makes it relatively easy to be hydrolysed to monomeric sugars by acids. Lignin is a three-dimensional polymer consisting of propyl-phenol units with various types of linkages and is found in the cell walls of particularly woody biomass [3].

However, the use of biomass also has some drawbacks. For instance, collection is costly; various types of biomass show major differences in structure and composition; it has a low energy density compared with fossil resources and often contains significant amounts of water and ash, as shown in Table 8.1. To overcome some of these drawbacks, it is crucial to develop commercially attractive ways to convert lignocellulosic biomass into a liquid product with a higher volumetric energy density and a lower level of contaminants [4].

Fast pyrolysis is considered as a very attractive technology for this purpose. The liquid product, known as pyrolysis liquids (PLs), bio-oil or bio crude, contains negligible amounts of ash and has a volumetric energy density 5–20 times higher than the original biomass [5]. In the following parts of this chapter, a short introduction on fast pyrolysis technology and the composition and properties of PLs will be reported.

Direct Thermochemical Liquefaction for Energy Applications. https://doi.org/10.1016/B978-0-08-101029-7.00007-2

Table 8.1 **Chemical composition, energy content, and yield of various terrestrial biomass species**

Biomass component	Corn grain	Corn stover	Switchgrass	Sugarcane	Sweet sorghum	Eucalyptus	Pine
Productivity (dry metric tons/ha-year)	7	13–24	8–20	73–87	43.8	40.0	11.6
Lower heating value (MJ/dry kg)	17.0	17.5	≈17	16.8	17.3	18.1	18.6
Energy inputs (MJ/dry kg)	1.35	1.20		0.346	2.82	5.57	7.43
Energy content (GJ/ha-year)	120	228–420	136–340	1230–1460	760	720	210
Energy content (boe/ha-year)	20	40–70	23–58	210–250	128	123	37
Representative components (dry wt%)							
Celluloses	3	36	40–45	22	35	48	46–50
Hemicelluloses	6	23	31–35	15	17	14	19–22
Extractives (starches and terpenes)	72	6	0	43	23	2	3
Lignins	2	17	6–12	11	17	29	21–29
Uronic acid			0	0	1	4	3
Proteins	10		5–11				
Ash		10	5–6	9	5	1	0.3

Adapted with permission from Huber GW, Iborra S, Corma A. Chem Rev 2006;106:4044–98. Copyright 2006 American Chemical Society.

8.2 Fast pyrolysis technology

Fast pyrolysis is a very versatile biomass conversion technology, which is carried out in the absence of oxygen at a temperature ranging from 450°C to 600°C under atmospheric pressure. A short vapour residence time (< 2 s) in combination with high biomass heating rates ensures a high yield of liquid product, which can be up to 70 wt% on dry biomass feed [4]. The process parameters of various pyrolysis technologies and the product distributions are shown in Table 8.2. Clearly, the liquid yields are highest when using fast pyrolysis technology.

Process development activities in the field of fast pyrolysis culminated in the construction of successful (semi-) commercial-scale plants in the United States and Canada (Pyrovac, Dynamotive, and Ensyn), Finland (Valmet), Germany (KIT), and the Netherlands (BTG, EMPYRO).

A successful example of PLs production by fast pyrolysis technology in the Netherlands is reported by the Biomass Technology Group (BTG). The company has developed proprietary pyrolysis technology with distinct advantages compared with conventional processes. The reactor technology (rotating cone) excludes the use of inert gases as required for fluidised-bed reactors. Instead the biomass particles are fed and mechanically mixed near the bottom of the rotating cone with an excess sand. Pyrolysis vapours pass through a cyclone and are quenched by cold PLs. The sand together with the char formed during the pyrolysis process are transported to a separate fluidised-bed combustor where the char is combusted and the sand, reheated to reactor temperature, is recycled to the pyrolysis reactor. A process flow diagram of BTG's fast pyrolysis technology is shown in Fig. 8.1.

In 2015, BTG-BTL (BTG bioliquids) designed and built a 5 t/h pyrolysis unit in Hengelo (the Netherlands); see Fig. 8.2. The unit produces 22,000 t/year of PLs from woody biomass waste, which will be used to replace natural gas in an industrial boiler of the dairy firm FrieslandCampina.

Fast pyrolysis technology is moving to maturity. The technology is on the verge of commercial exploitation, and large quantities of PLs will thus become available in the near future.

Table 8.2 Comparison of thermochemical techniques

Mode	Conditions	Liquid (wt%)	Solid (wt%)	Gas (wt%)
Fast	~500°C, short hot vapour residence time ~1 s	75	12	13
Intermediate	~500°C, hot vapour residence time ~10–30 s	50 in two phases	25	25
Carbonisation (slow)	~400°C, long vapour residence time hours to days	30	35	35
Gasification	~750–900°C	5	10	85
Torrefaction (slow)	~290°C, solids residence time ~10–60 min	0 unless condensed, then up to 5	80	20

Adapted with permission from Bridgwater AV. Biomass Bioenergy 2012;38:68–94. Copyright 2012 Elsevier.

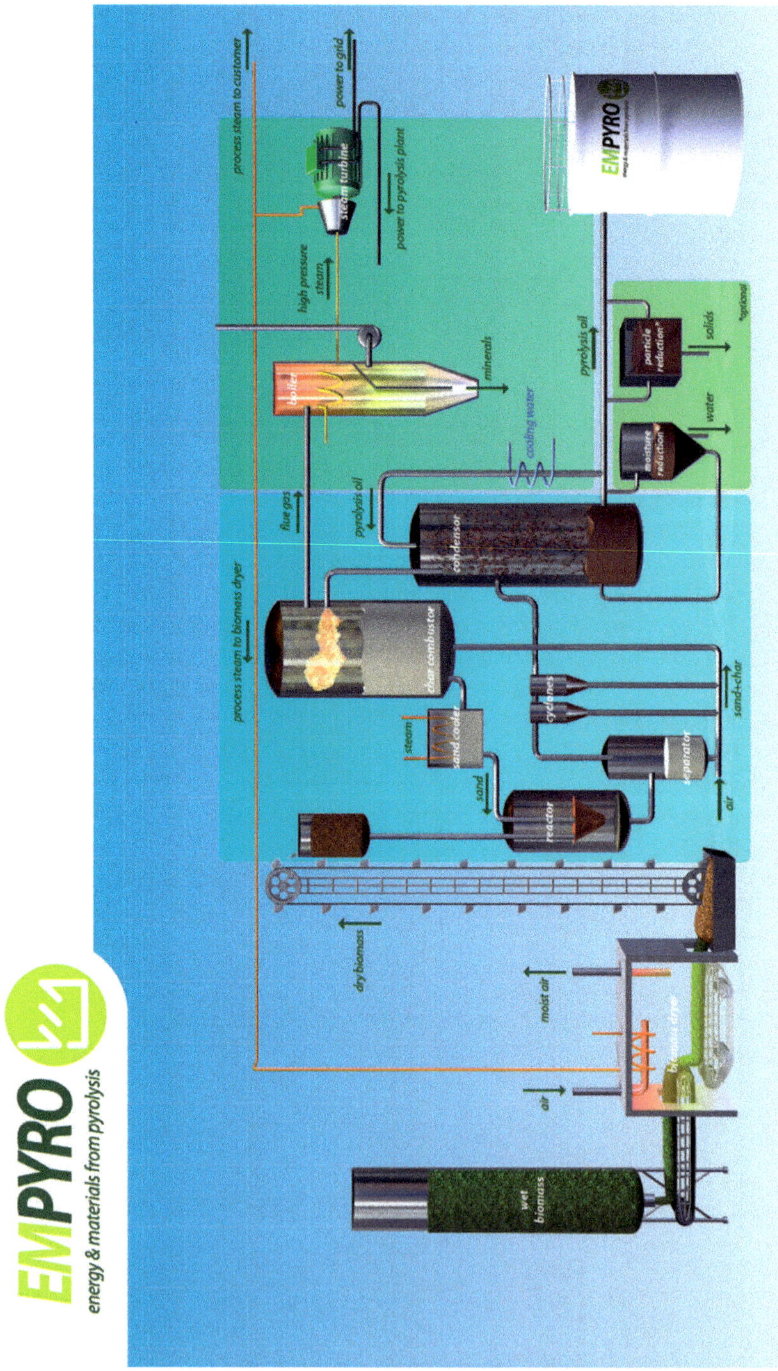

Fig. 8.1 Process flow diagram of BTG's fast pyrolysis process [6].

Fig. 8.2 BTG-Bioliquids 5 t/h wood pyrolysis plant (BTG, EMPYRO, the Netherlands).

8.3 The composition of PLs

Lignocellulosic biomass is composed of three main biopolymers: cellulose, hemicellulose, and lignin. Decomposition of these primary building blocks during fast pyrolysis leads to the formation of significant amounts of mono- and oligomeric, oxygenated functional molecules. Examples are organic acids, aldehydes, ketones, esters, alcohols, furans, sugars, and phenolics; see Fig. 8.3 for details. The total amount of low-molecular-weight GC detectables in PLs is around 45–65 wt%; the amount of oligomeric compounds, including sugar and lignin oligomers, is estimated to be around 15–25 wt%. The remainder is water, either feedstock moisture or reaction water [7].

For cellulose pyrolysis, the most abundant volatiles are levoglucosan, furans, and light oxygenates (such as glycolaldehyde and formic acid). The mechanism of levoglucosan formation from cellulose involves the cleavage of a glycosidic linkage between pyranose rings to form a glucosyl cation, which is stabilised by the formation of a 1,6-anhydride. The subsequent cleavage of the glycosidic bond linking the anhydride to the end of the polymer chain yields levoglucosan; see Scheme 8.1 for details.

The formation pathways to furanics and other light oxygenates is less straightforward [8]. Proposed reaction pathways for the formation of C1–C6 organic compounds from glucose are shown in Scheme 8.2.

The pyrolysis products from hemicellulose can be classified into three main categories: (i) low-molecular-weight compounds, for example, formic acid, acetic acid, acetaldehyde, and acetol; (ii) furan/pyran ring derivatives; and (iii) anhydrosugars [9]; see Scheme 8.3 for details. Patwardhan et al. [9] reported that a double dehydration product of xylose, the most abundant building block of hemicellulose, was the major product [3].

For lignin pyrolysis, the main products are volatile compounds (e.g. methanol, acetone, and acetaldehyde), monophenols (e.g. phenol, guaiacol, syringol, and catechol), and polysubstituted phenols; see Scheme 8.4 for details [10].

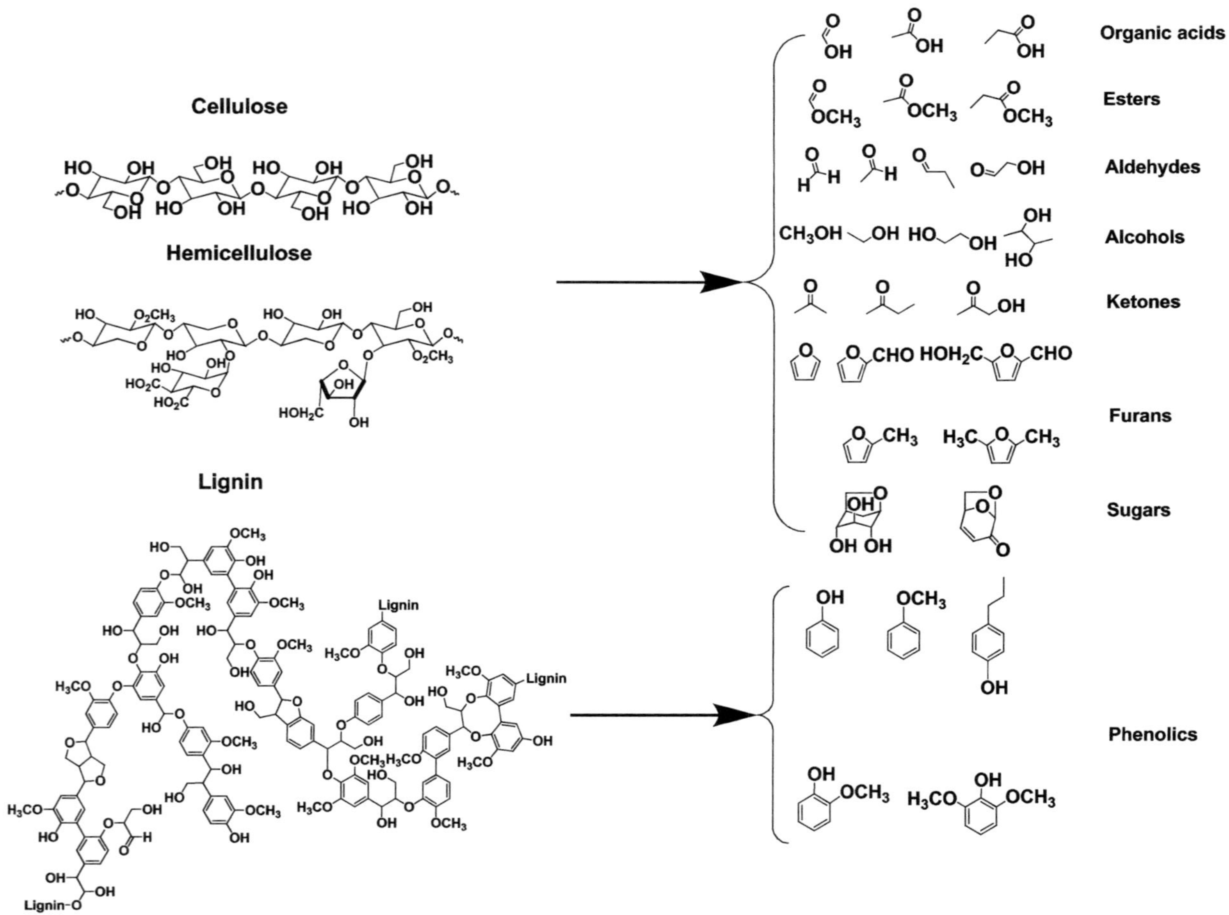

Fig. 8.3 Typical products formed from the pyrolysis of the three main building blocks of lignocellulosic biomass. Adapted with permission from Talmadge MS, Baldwin RM, Biddy MJ, McCormick RL, Beckham GT, Ferguson GA, Czernik S, Magrini-Bair KA, Foust TD, Metelski PD, Hetrick C, Nimlos MR. Green Chem 2014;16:407–53. Copyright 2014 Royal Society of Chemistry.

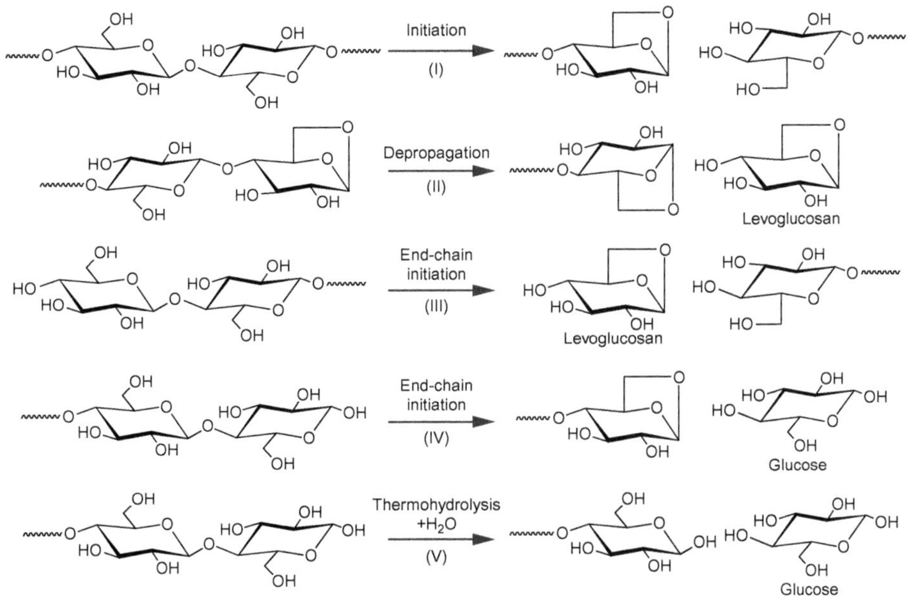

Scheme 8.1 Mechanisms of formation of levoglucosan and glucose during fast pyrolysis. Adapted with permission from Vinu R, Broadbelt LJ. Energy Environ Sci 2012;5:9808–26. Copyright 2012 Royal Society of Chemistry.

Breaking of the β-*O*-4 linkage by a radical mechanism is considered to be the primary step of the depolymerisation of the lignin structure. Free-radical reactions will lead to the formation of lignin monomers such as vanillin and 2-methoxy-4-methylphenol, while the recombination of free radicals leads to the formation of oligomeric products and/or char. Patwardhan et al. [11] reported that volatile monomeric compounds are the primary products of lignin pyrolysis, whereas secondary reactions of such monomers lead to the formation of oligomers.

8.4 Properties of PLs

The presence of various oxygenated organics has a major impact on product properties of PLs. A comparison of typical properties of PLs and crude oil is shown in Table 8.3.

8.4.1 Acidity of PLs

PLs are acidic in nature ($pH < 3$) because of the presence of around 3–6 wt% of volatile organic acids mainly from the degradation of the hemicellulose fraction. Examples are formic acid, acetic acid, and glycolic acids. Other compounds like phenolics and fatty and resin acids also have a certain acidity. The total acid number (TAN) is the preferred indicator for acidity in PLs. The determination of TAN is generally carried out by a titration method using a basic solution, for example, KOH or NaOH [12].

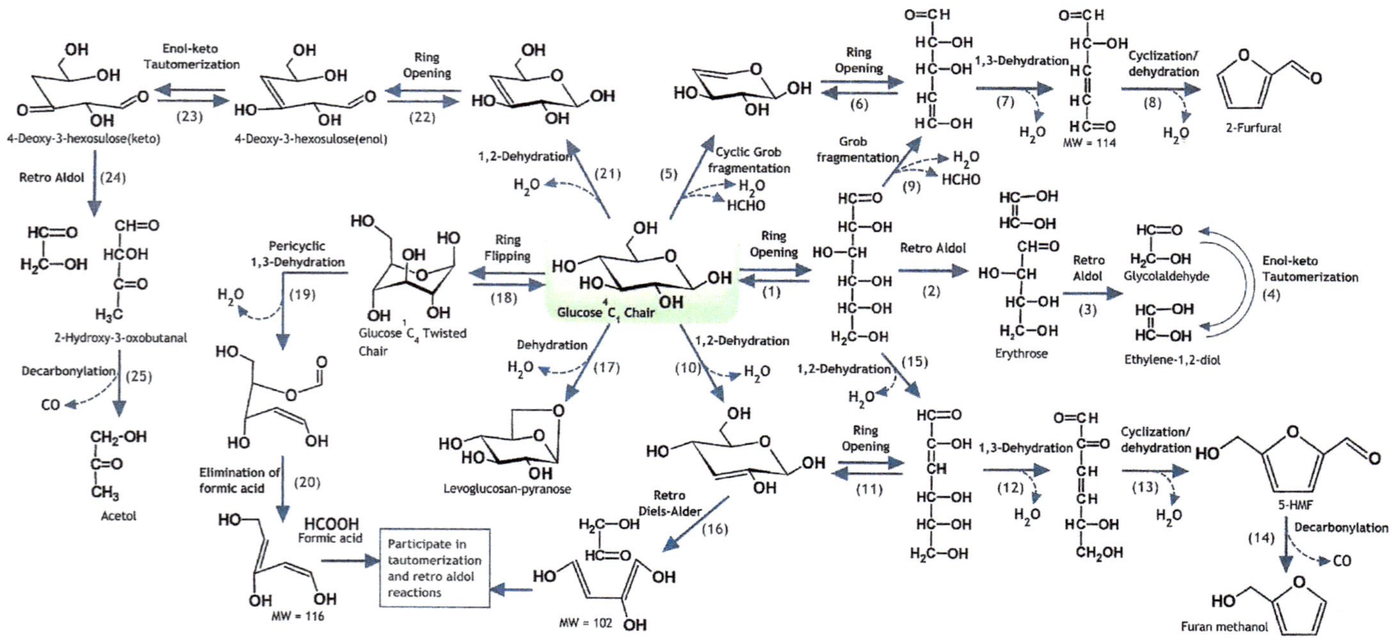

Scheme 8.2 Mechanism of formation of various C1–C6 organic compounds starting from glucose (pyranose form).
Adapted with permission from Vinu R, Broadbelt LJ. Energy Environ Sci 2012;5:9808–26. Copyright 2012 Royal Society of Chemistry.

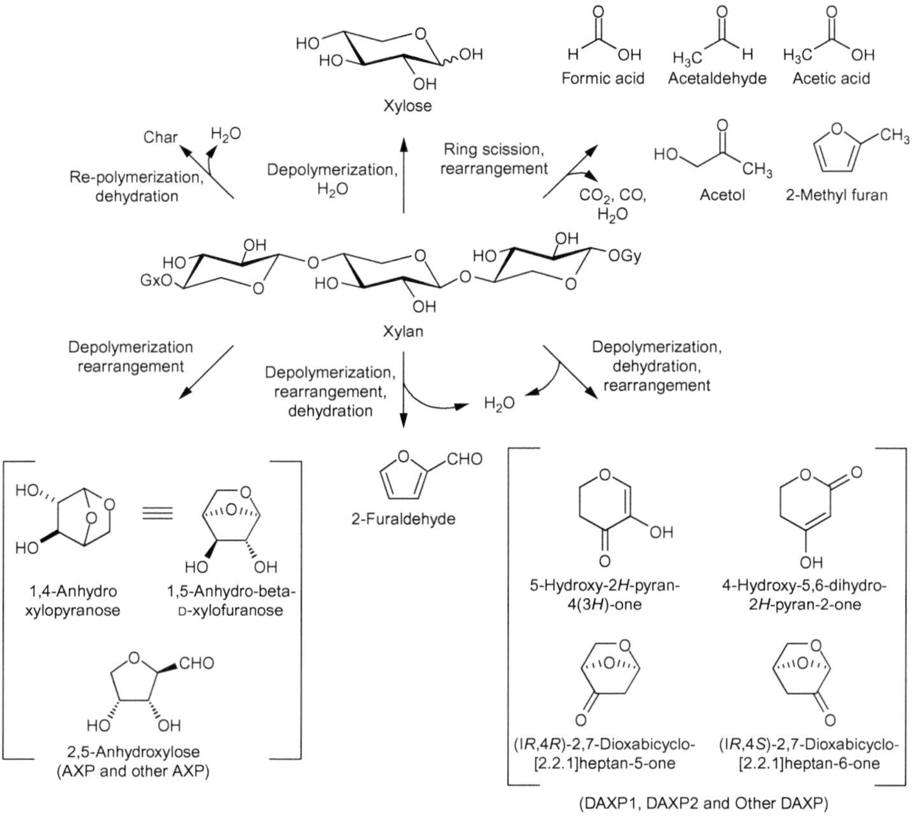

Scheme 8.3 Proposed reaction scheme, showing primary pyrolysis reactions of switchgrass hemicellulose.
Adapted with permission from Patwardhan PR, Brown RC, Shanks BH. ChemSusChem 2011;4:636–43. Copyright 2011 Wiley.

The acidic properties of PLs are a major drawback as it potentially leads to corrosion in pipelines and storage vessels.

8.4.2 Heating value of PLs

The typical lower heating value of PLs is between 16 and 19 MJ/kg, which is around half of that of crude oils (44 MJ/kg). The low heating value of PLs results from the presence of oxygenated organic molecules (28–40 wt% on dry basis) and water (15–30 wt%); see Table 8.3 for details. Due to the higher density (1.05–1.25 kg/L compared with 0.86–0.94 kg/L for crude oil), the heating value of PLs on a volumetric basis is about 60% of that for crude oil [13].

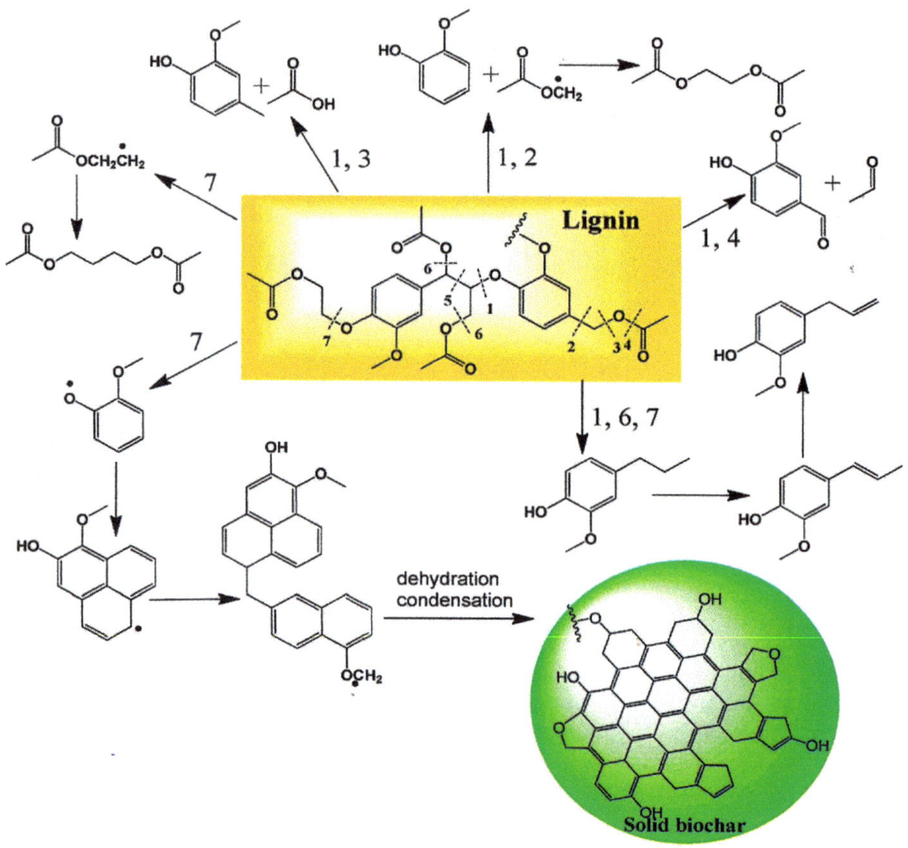

Scheme 8.4 Possible pathways for the decomposition of lignin during the pyrolysis process. Adapted with permission from Liu W-J, Jiang H, Yu H-Q. Green Chem 2015;17:4888–907. Copyright 2015 Royal Society of Chemistry.

Table 8.3 Comparison of typical properties of PLs and crude oil

Composition	PLs	Crude oil
Water (wt%)	15–30	0.1
pH	2.8–3.8	–
Density (kg L^{-1})	1.05–1.25	0.86–0.94
Viscosity 50°C (cP)	40–100	180
HHV (MJ kg^{-1})	16–19	44
C (wt%)	55–65	83.86
O (wt%)	28–40	<1
H (wt%)	5–7	11–14
S (wt%)	<0.05	<4
N (wt%)	<0.4	<1
Ash (wt%)	<0.2	0.1
H/C	0.9–1.5	1.5–2.0
O/C	0.3–0.5	~0

Adapted with permission from Liu C, Wang H, Karim AM, Sun J, Wang Y. Chem Soc Rev 2014;43:7594–623. Copyright 2014 Royal Society of Chemistry.

8.4.3 Stability of PLs

PLs are known to age during storage [14], and this ageing process is accelerated at elevated temperature [5,15]. Unlike crude oils and derivatives, PLs contain a significant amount of reactive oxygenated functional molecules, which can react to form larger molecules during handling or storage and at elevated temperature (polymerisation). Other transformations upon storage involve etherification and esterification reactions. Water is formed as a byproduct for all these reactions. Ageing has a negative effect on product properties and, for instance, leads to an increase in the viscosity and water content.

Fig. 8.4 shows the main chemical compositional changes of PLs during ageing [16]. A clear increase in water insolubles is observed after 12 months of storage. Reactive carbonyl-containing molecules, for example, aldehydes and ketones, are expected to play a major role in the ageing of PLs. This is supported by the ageing tests given in Fig. 8.4, which shows that the amounts of carbonyl compounds decrease monotonously in time.

Fig. 8.5 gives a detailed comparison of the composition of PLs derived from pinewood and the product oil after a high-pressure thermal treatment (HPTT) using a fractionation method [5,15]. Upon heating, sugars (indicated as ether insolubles) are partly converted to water insolubles and water. This observations is rationalised by assuming that the anhydrosugars, for example, levoglucosan, are initially hydrolysed

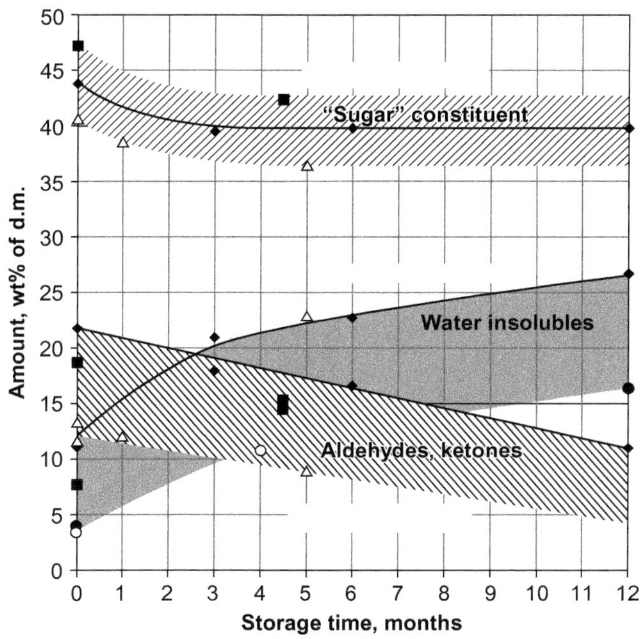

Fig. 8.4 Main changes in forestry residue PLs during storage.
Adapted with permission from Oasmaa A, Korhonen J, Kuoppala E. Energy Fuels 2011;25:3307–13. Copyright 2011 American Chemical Society.

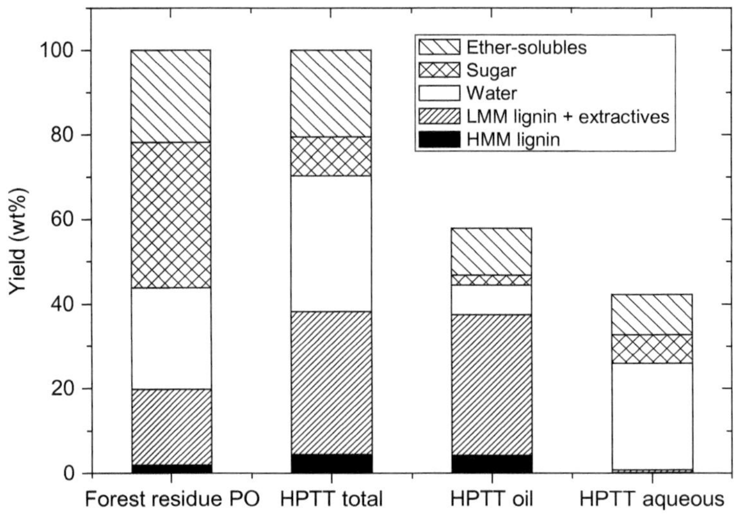

Fig. 8.5 Composition of a typical pine PLs and high-pressure thermal treatment (HPTT) products from PLs according to a fractionation method [5].

to sugar monomers like glucose, which in acidic media and elevated temperatures, further dehydrate to 5-HMF and repolymerisation products thereof like humin type of oligomers and polymers [17].

8.5 Characterisation of PLs by solvent fractionation

As depicted in Section 8.4 of this chapter, PLs are complex mixtures, composed of water, various volatiles and nonvolatile, oligomeric components from both sugar and lignin fractions. For the determination of the moisture content of PLs, a Karl Fischer titration is typically used. Quantification of monomeric components (Fig. 8.3) is best done using gas chromatography (GC) or high-performance liquid chromatography (HPLC) techniques. However, oligomeric compounds (15–25 wt% of PL) are nonvolatile, so analysis is not possible by conventional GC techniques.

An alternative procedure involves solvent fractionation prior to detailed analyses. An example is a method developed by Oasmaa et al. [7]; see Scheme 8.5 for details, which was successfully used for the characterisation of PLs and their hydrogenated product oils [5,18]. Solvent fractionation combined with conventional analysis techniques, for example, GC, NMR, GPC, and HPLC, gives a very good understanding on the molecular composition of PLs.

Seven distinct fractions were obtained using this solvent fractionation scheme and include (a) water; (b) extractives; (c) acids and alcohols; (d) aldehydes, ketones, and lignin monomers; (e) sugars; (f) low-molecular-weight lignin; and (g) high-molecular-weight lignin. The detailed composition of each fraction is provided in Table 8.4.

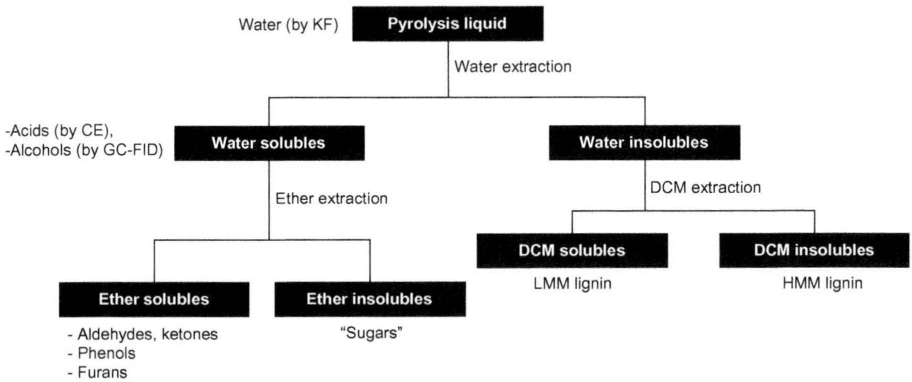

Scheme 8.5 Example of a solvent fractionation protocol for PLs [18].

8.6 Biorefinery concepts for PLs

PLs can be used as the input for a biorefinery, similar to the use of crude oil in an oil refinery. A PL-based biorefinery concept is shown in Scheme 8.6. In this concept, PLs are valorised in an integrated process for the production of heat, power, biofuels, and bio-based chemicals.

PLs obtained from fast pyrolysis of biomass can easily be combusted in boilers, diesel engines, gas turbines, and stirling engines for heat and power generation. Research activities have been carried out by VTT Energy (Finland), MIT (the United States), Orenda Aerospace Corporation (Canada), ZSW (Germany), etc. and have been reviewed by Czernik and Bridgwater [14,20].

Gasification of PLs and further processing of the product synthesis gas in Fischer-Tropsch processes has been investigated and claimed to be technically and economically feasible [21–25]. Catalytic steam reforming of PLs or fractions thereof to syngas using a conversion strategy analogous to catalytic steam reforming of alcohols and oxygenated aromatics has been studied. The syngas is rich in CO_2 and contained hydrogen and minor amounts of CO and CH_4 [25]. By adjusting the H_2/CO ratio by including a water-gas shift reaction, production of Fischer-Tropsch fuels, DME, and alcohols was claimed to be feasible.

PLs contain substantial amounts of valuable oxygenated compounds; some of them are actually existing bulk chemicals. Examples are organic acids (formic acid, acetic acid, and glycolic acid), glycolaldehyde, anhydrosugars, and phenolics. Solvent extraction is widely used for the separation and isolation of individual chemicals from PLs. Rasrendra et al. [26] reported that reactive extraction of organic acids, particularly acetic acid, from the aqueous phase of PLs, is technically feasible by using a long-chain aliphatic tertiary amine. The best results were obtained with tri-*n*-octylamine (TOA) in 2-ethyl-hexanol (40 wt%) giving 84% acetic acid recovery at equilibrium conditions at room temperature. Other organic acids, for example, formic acid and glycolic acid, were also coextracted with 92% and 69% efficiency, respectively. The same reactive extraction system was used by Vitasari et al. [27–29] to extract glycolaldehyde from the aqueous phase of PLs. It was shown that an amine is not essential

Table 8.4 **Chemical composition of reference pine PLs and its fractions [5].**

	wt% (wet basis)	Compound types	C	H	N	O
			wt% (dry basis)			
Water solubles	75–85					
Acids, alcohols	5–10	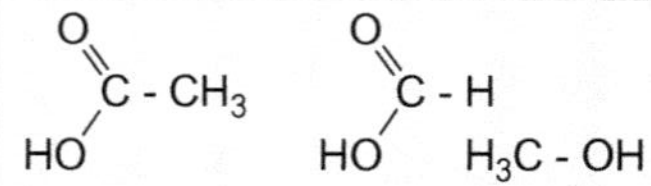	36.0	6.0	0	58.0
Ether solubles	5–15	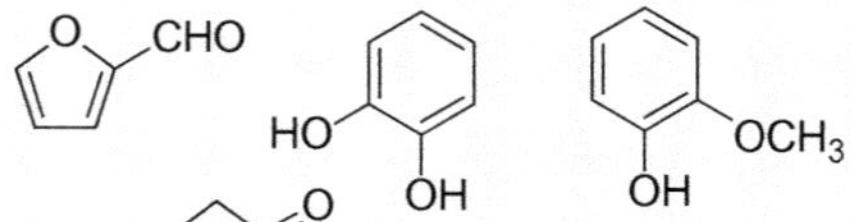	60.0	6.0	0.1	33.9
Ether insolubles	30–40	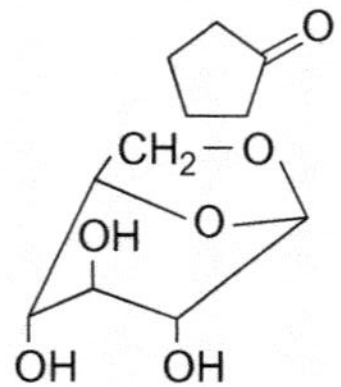	46.0	6.3	0.3	47.4
Water	20–30		0	11.1	0	88.9
Water insolubles	15–25		66.2	6.6	0.3	26.9
n-Hexane solubles	2–6	CH_3-$(CH_2)_n$-CO_2H $n = 10–30$ $HOCH_2$-$(CH_2)_n$-CO_2H $n = 10–28$ 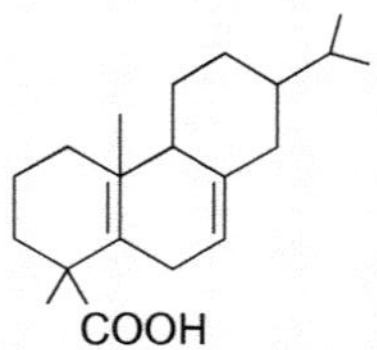	77.4	10.4	0	12.2
DCM solubles	5–10	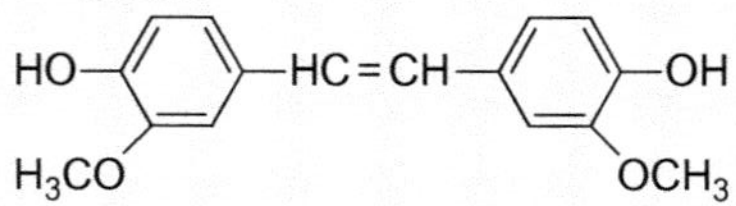	68.1	6.7	0.4	24.7
DCM insolubles	2–10	Degraded lignin	64.1	5.9	1.5	28.4

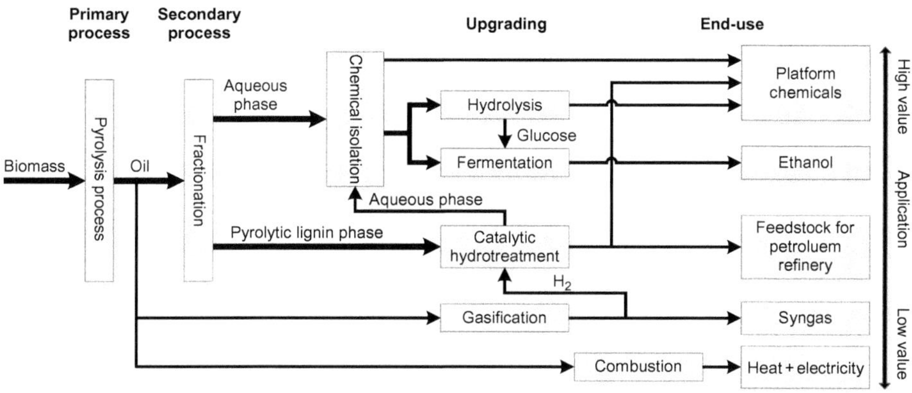

Scheme 8.6 Possible biorefinery concept for PLs [19].

and the highest yields of glycolaldehyde was obtained using 2-ethyl-1-hexanol as the solvent. Bennett et al. [30–33] reported that a simple water extraction is a convenient way to obtain levoglucosan from PLs. Optimisation of the water-to-oil ratio, temperature and contact time yielded an aqueous phase enriched in levoglucosan. Wang et al. [34] reported the extraction of monophenolics from the water-insoluble fraction of PLs by a multiple-step extraction using aqueous acids and bases combined with dichloromethane.

The above-mentioned products obtained by various extraction procedures can be used directly or further processed to high-value chemicals. For example, glycolaldehyde is a precursor for ethylene glycol by hydrogenation [35], and the hydrolysis of levoglucosan gives glucose, which can be fermented to fuel ethanol [36].

PLs upgrading to either a stabilised oil or a transportation fuel by catalytic hydrotreatment has been studied extensively. For instance, in the European sixth framework project BIOCOUP [37], the main objective was to develop processes/catalysts for the catalytic hydrotreatment of PLs to stabilised oils with a reduced oxygen content range (~10 wt%), to be used as a cofeed with, for example, vacuum gas oil (VGO) in existing fluidised catalytic cracking (FCC) units. The stabilised product oil can also be processed further by a deep hydrotreatment to hydrocarbons that can be used as a transportation fuel.

8.7 Catalytic hydrotreatment

The major objectives of the catalytic hydrotreatment process are to reduce the oxygen content, increase the H/C ratio, reduce the molecular weight, and improve the thermal stability of PLs. Other thermal processes have been developed for PLs upgrading; examples include a HPTT and zeolite cracking (see Table 8.5 for details). Among all these upgrading technologies, catalytic hydrotreatment is considered the most promising because it yields a product with a higher H/C ratio compared with HPTT, and char formation is much lower when compared with zeolite cracking [38].

Table 8.5 **High-temperature processes for PLs upgrading [38].**

	HPTT	Zeolite cracking	Hydrotreatment/HDO
Temperature (K)	473–623	573–873	523–673
Pressure (MPa)	15–30	No	10–30
Catalyst	No	Yes	Yes
Coke formation	High	High, 26–39 wt%	Low, <5 wt%

8.8.1 Reactor configurations

Catalytic hydrotreatment of PLs is carried out in the presence of a hydrogenation catalyst at elevated temperatures (523–673 K) under high-pressure hydrogen (10–30 MPa). A simplified representation is given in Scheme 8.7.

High pressure has a positive effect on the overall reaction rates and is also believed to be necessary to keep the water in the liquid state, which is expected to decrease char formation [39].

The overall reaction stoichiometry for the catalytic hydrotreatment of PLs may be generalised by Eq. (8.1):

$$CH_{1.47}O_{0.56} + 0.44H_2O + 0.319H_2 \rightarrow 0.647CH_{1.48}O_{0.11} + 0.183CH_{3.02}O_{1.09} + 1.164H_2O \tag{8.1}$$

The catalytic hydrotreatment of PLs has been reported in batch autoclaves and continuous reactor configurations, particularly in packed-bed reactors. In batch reactors, operating conditions and concentrations of all species are time-dependent. The experiments are often carried out in a closed system, and hydrogen consumption is not compensated for. As such, it is difficult to assess catalyst activity (often taken as an average over the batch time) and catalyst stability. Moreover, heating and cooling down of the batch reactor content using conventional heating often has a timescale in the order of >15–30 min, particularly when reaction temperatures up to 400°C are used. Isothermal conditions throughout an experiment are thus not met, and this may affect quantification of catalyst performance. In contrast, continuous experiments are performed in a steady-state mode with by far time dependencies of concentrations and temperatures.

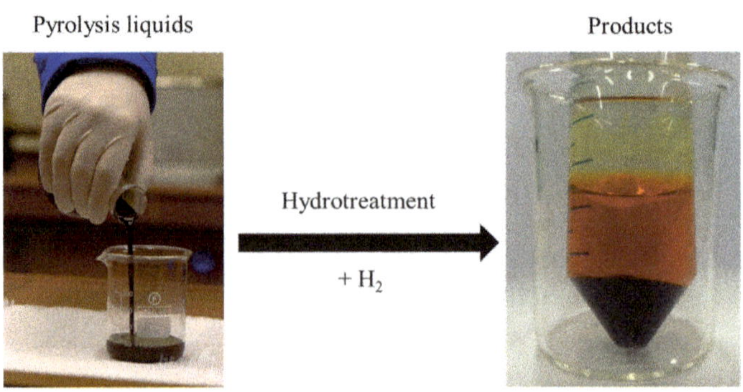

Scheme 8.7 A simplified process of the catalytic hydrotreament of PLs.

A major difference in catalytic experiments using heterogeneous catalysts between batch and continuous set-ups also includes the size of the catalyst particles. It is often in the micrometer range for batch slurry systems and up to several millimeters in a continuous packed-bed reactor to reduce pressure drops. This will have a major impact on catalyst performance, particularly for the faster hydrotreatment reactions (e.g. aldehyde reduction) using larger particle as intraparticle mass-transfer limitations may affect the overall reaction rates to a large extent. Notwithstanding these differences, the continuous reactor experiments provide valuable insight in catalyst stability and the suitability of a particular continuous reactor set-up for subsequent scale-up studies.

Due to the thermal instability of PLs as depicted in Section 8.3 of this chapter, the direct hydrotreatment of PLs at elevated temperatures (>300°C) is not the best choice as polymerisation of thermally labile components in PLs results in the formation of considerable amounts of char/humins. To avoid this, a two-step hydrotreating process for the upgrading of PLs was developed by both Elliott and Venderbosch and is characterised by a low- and high-temperature step [5,40,41]. A schematic representation of a packed-bed set-up for the catalytic hydrotreatment of PLs using a staged concept is shown in Fig. 8.6.

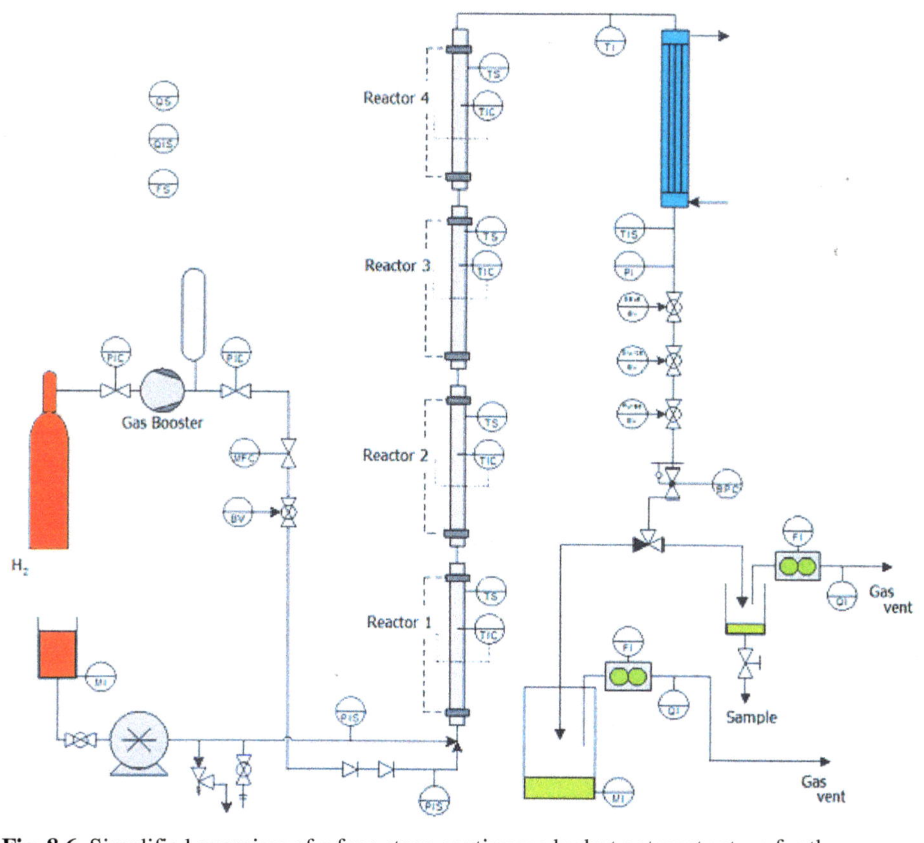

Fig. 8.6 Simplified overview of a four-stage continuous hydrotreatment set-up for the catalytic hydrotreatment of PLs.

In the low-temperature stage (around 200°C), thermally labile components are hydrogenated to compounds with improved stability. In the high-temperature stage (up to 400°C), further hydrodeoxygenation of stabilised products and hydrocracking occur, leading to products with reduced oxygen contents and lower molecular weights. In this staged concept, char formation is considerably decreased, which has a positive effect on both operability and catalyst stability.

8.7.2 Chemical reactions during catalytic hydrotreatment

Catalytic hydrotreatment involves a number of reactions (Fig. 8.7), which can be classified as hydrogenations, hydrodeoxygenations, decarbonylations, decarboxylations, cracking, hydrocracking, and polymerisation reactions.

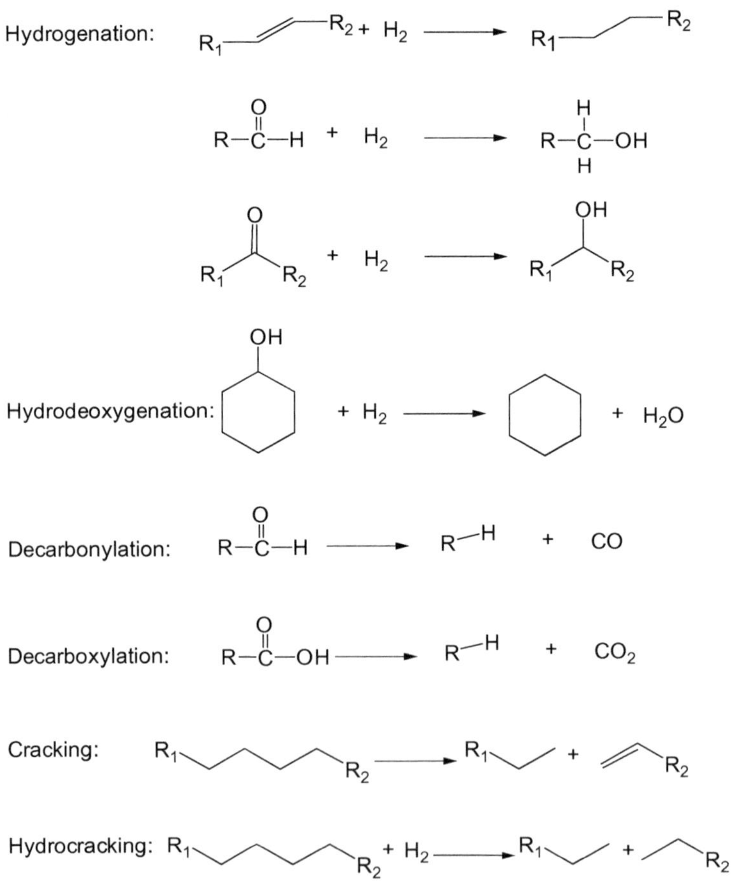

Fig. 8.7 Major reactions occurring during catalytic hydrotreatment of PLs.
Adapted with permission from Zacher AH, Olarte MV, Santosa DM, Elliott DC, Jones SB. Green Chem 2014;16:491–515. Copyright 2014 Royal Society of Chemistry.

Venderbosch et al. [5] investigated the catalytic hydrotreatment of PLs using a Ru/C catalyst in a four-stage packed-bed reactor, with different process severities, especially temperature, in the four stages. The experiments revealed that catalytic hydrogenation and thermal, noncatalytic polymerisation occur in a parallel mode in the initial phase of the hydrotreatment process; see Scheme 8.8 for details.

Polymerisation leads to the formation of higher-molecular-weight fragments that upon further condensation leads to char. This route is as such not preferred, and the rate of polymerisation should be reduced as much as possible. The preferred pathway involves the hydrogenation of the thermally labile components in the PL feed to stable molecules that are not prone to polymerisation. Subsequent reactions (hydrogenations and hydrocracking) on a timescale of hours lead to products with reduced oxygen contents and ultimately to higher H/C ratios.

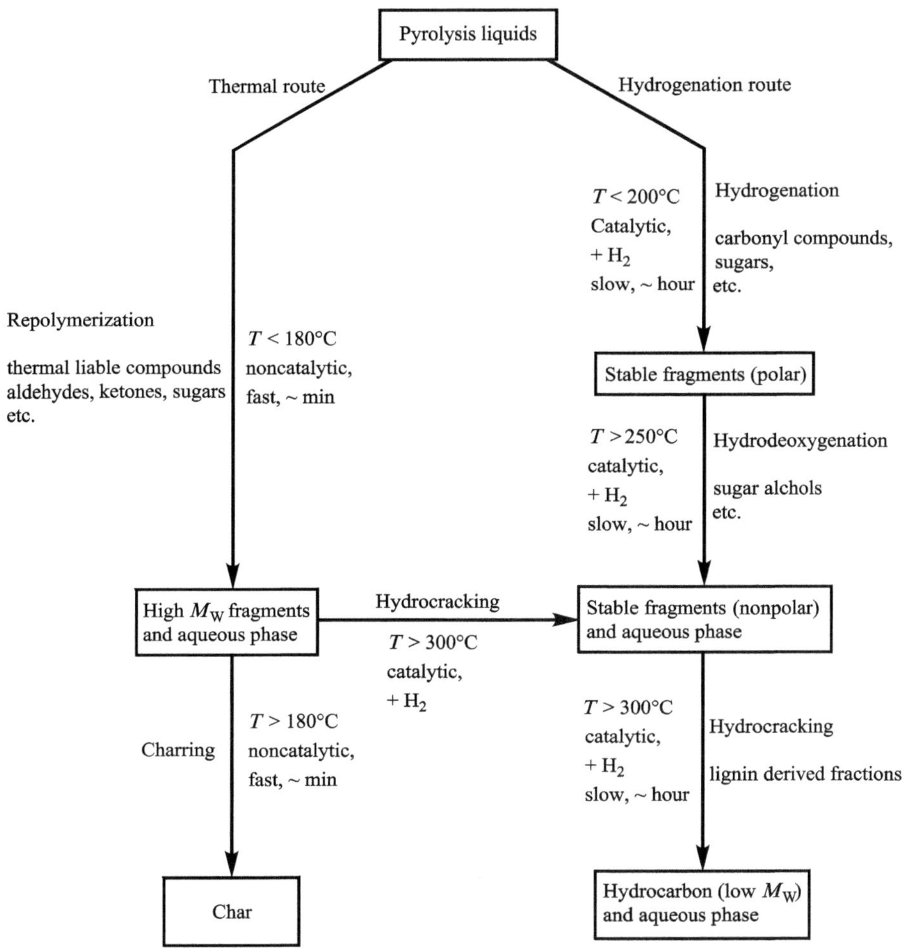

Scheme 8.8 Proposed reaction network for the catalytic hydrotreatment of PLs [42].

Among all the components in PLs, carbonyl-containing molecules and sugars (the so-called water-soluble fraction or pyrolytic sugars) play a major role in polymerisation reactions. Examples are reactions of monomeric sugars, which are known to form hydrochars upon heating. These reactions involve sugar dehydrations (e.g. glucose to HMF) accompanied by the formation of humin type of products [17].

8.7.3 Reactivity of various component classes in PLs

As mentioned earlier, PLs can be fractionated using various methods, for example, by solvent extraction. A simple method involves the addition of water to PLs, which will result in the formation of a brown precipitate (pyrolytic lignin) and a water fraction enriched in oxygenates. Catalytic hydrotreatment of the two fractions, supported by model component studies, has given valuable insights into the reactivity of various component groups. The water-soluble fraction (pyrolytic sugars) is known to be by far more reactive at low temperature (< 200°C) than the water-insoluble fraction (pyrolytic lignin). Carbonyl-containing molecules like small aldehydes and ketones in the water-soluble fraction are readily converted, most likely hydrogenated, to the corresponding alcohols at temperatures below 180°C. Examples are the hydrogenation of glycolaldehyde to ethylene glycol [35] and hydroxyacetone to 1,2-propanediol [43]. Sugar monomers like glucose are known to be hydrogenated at temperatures as low as 80°C to, for example, sorbitol [44–46], but higher temperatures (around 250°C) are required for the full conversion of anhydrosugars, for example, levoglucosan [42,47] and sugar oligomers [48,49]. The pyrolytic lignin fraction is less reactive, and hydrocracking reactions occur at temperatures above 300°C, resulting in monophenolics followed by hydrocarbon formation by hydrodeoxygenation reactions [42]. Among all component classes in PLs, organic acids (except formic acid) are the most persistent, and temperatures higher than 400°C are required to observe reactivity [42]. An overview of the reactivity of various organic component classes is given in Fig. 8.8 [42].

The high reactivity of the water-soluble fractions was further confirmed by Wildschut et al. [50]. Carbonyl-containing molecules and sugars are readily converted using Ru/C as the catalyst at 350°C in 4 h, while no significant changes were observed for the water-insoluble, pyrolytic lignin fraction, indicated as the DCM insolubles in Fig. 8.9 [50]. The total amounts of hydrocarbons increased at prolonged reaction times at the expense of carbonyl-containing molecules and sugars.

Thus, we can conclude that in order to obtain stabilised oils with good thermal stability, the catalyst should possess a high activity for carbonyl-containing and sugar components of PLs.

8.7.4 Application of product oils from catalytic hydrotreatment reactions

The typical products from a catalytic hydrotreatment reaction include a gas phase, an oil phase, an aqueous phase, and char. For hydrotreatments below 180°C, the organic and aqueous phases are miscible and form one liquid phase. For the oil phase, a high energy density, a high H/C ratio, a good thermal stability, a limited amount of organic

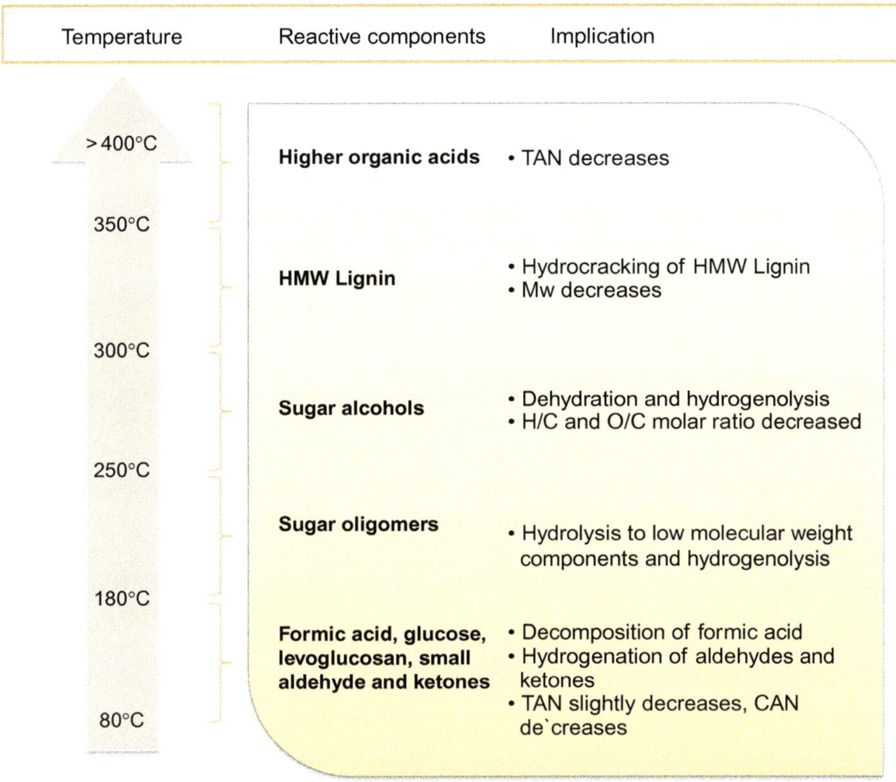

Fig. 8.8 Reactivity of the various organic component classes in PLs versus reaction temperature [42].

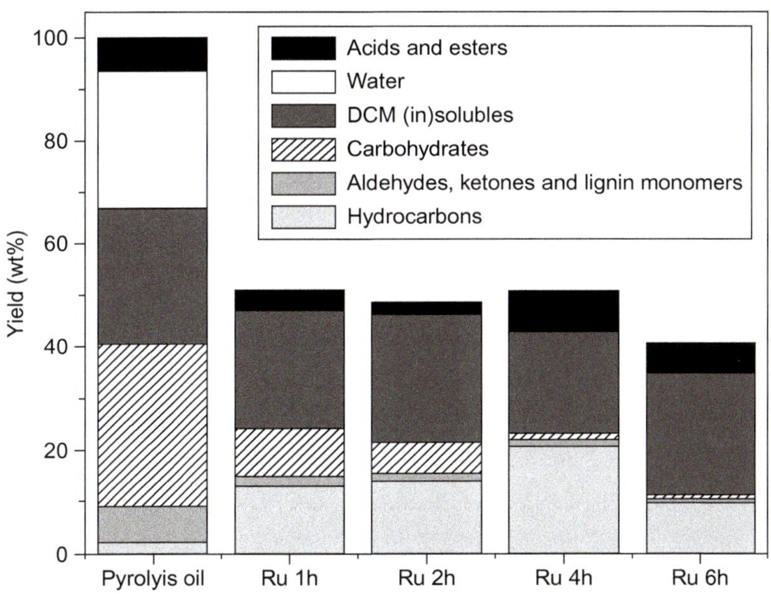

Fig. 8.9 Composition of PLs and hydrotreated product oils (Ru/C, 350°C, 200 bar) at various reaction times using solvent-solvent extraction [50].

acids, and a high solubility in a polar solvents are preferred. The actual oxygen content of the product oil can be tuned, for instance, to the 10 wt% range (for cofeeding purposes in existing FCC units) and to <1 wt% for direct use as a transportation fuel, by adjustment of the process severity.

8.8 Catalysts development for the catalytic hydrotreatment of PLs

PLs are complex mixtures of monomeric and oligomeric molecules, with a high diversity in molecular structures. Examples are carboxylic acids, aldehydes, ketones, sugars, and lignin monomers and oligomers. In addition, the water content of PLs is reported as high as 30 wt%. These features put serious constraints on the design of efficient catalysts for the hydrotreatment reactions. Catalysts developed for the hydrotreatment of various fossil (crude) oil fractions are not necessary suitable for use with PLs. A good catalyst should have a good hydrogenation activity for a broad range of organic compounds with different reactivities in the low-temperature range, be stable in acidic water at elevated temperatures, and removal of coke deposits should be possible, preferably by simple oxidative treatments [51].

Catalyst stability is thus of prime importance. Main deactivation mechanisms mentioned in the literature for catalysts used in catalytic hydrotreatment processes are (a) sintering of active metals, resulting in a decrease in surface area of active metals; (b) leaching of active metals due to the presence of carboxylic acids in the PL feed; (c) limited stability of the support material in acidic media under hydrothermal conditions, leading to a collapse of the structure of the catalysts; (d) accumulation of inorganic components from the PL feed on the catalysts surface that block the active sites; and (e) coke/char deposition on the catalysts leading to pore blockage [52–54].

A range of catalysts have been reported in the literature for the catalytic hydrotreatment [55]. In the following part of this chapter, the results on the catalytic hydrotreatment of PLs derived from lignocellulosic biomass using noble metal catalysts (e.g. Ru, Pt, Pd, and Rh) and nonnoble-metal, mainly Ni-based catalyst, are summarised. The catalytic hydrotreatment of PLs diluted by solvents like alcohols [56–59], formic acid [60], and decalin [61] are also included, while studies on the hydrotreatment of PL fractions like the water-soluble [62] and water-insoluble fraction [63,64] are beyond the scope of this chapter. Model component studies are also not reported; details can be found in recent review papers from Wang [65] and Robinson [66].

8.8.1 *Noble metal based catalysts*

Supported noble metals (Pt, Pd, Ru, and Rh) are widely used for industrial hydrogenation processes because of their high catalytic activity, allowing operation at low temperature and pressure and good catalyst stability [67]. Besides, some noble metals are resistant to poisoning with O-, S-, and N-containing compounds [55]. As such, noble-metal-based catalysts have been studied in detail in the last two decades.

An overview of catalytic hydrotreatment studies of PLs using supported noble metal catalysts is given in Table 8.6. A distinction is made between batch and continuous operation. Noble metals include Pt, Pd, Rh, and Ru supported on various supports like activated carbon, TiO_2, Al_2O_3, and ZrO_2.

8.8.1.1 Studies in batch set-ups

Wildschut et al. [68] reported the catalytic hydrotreatment of PLs using Ru/C (commercial and homemade catalysts) in a batch reactor set-up (Table 8.6, entry 1). The catalytic hydrotreatment of PLs using a commercial Ru/C catalyst at 350°C and 200 bar hydrogen yielded about 55 wt% of an oil phase with a H/C ratio of 1.24. Solid/char formation was in the order of 3 wt%. After two catalyst recycle runs, the oil yield decreased to around 30 wt% with a H/C ratio of 1.08, whereas the amounts of char increased to 20 wt%, indicative for substantial catalyst deactivation. Furthermore, nine Ru/C catalysts were synthesised using different Ru precursors ($RuCl_3$, $Ru(NO)(NO_3)_3$ and $Ru(acac)_3$) and Ru loadings (1, 3, and 5 wt%) to determine the effect of catalyst precursor on catalyst performance for the hydrotreatment of PLs (350°C, 200 bar for 4.3 h). The best results with respect to hydrogenation activity, as expressed by the H/C ratio of the product (1.32), were obtained for the catalyst prepared with $RuCl_3$ (5 wt%).

Elkasabi et al. [69] studied the catalytic hydrotreatment of PLs from switchgrass (320°C, 145 bar of H_2 for 4 h) using Pt, Ru, or Pd on carbon catalysts in a batch autoclave (Table 8.6, entry 2). Organic product yields between 36.2 and 46.6 wt% were obtained. Pt/C performed the best in terms of hydrogen consumption and deoxygenation level (O/C ratio of 0.15).

Mercader et al. [70] reported the catalytic hydrotreatment of PLs and fractions thereof at different temperatures (220°C, 270°C, and 310°C) under a constant total pressure of 190 bar using a 5 wt% Ru/C catalyst in a batch autoclave (Table 8.6, entry 3). Around 53–58 wt% of product oil with oxygen contents between 26.7 and 15.6 wt% was obtained. The oils from the PL fractions were evaluated by coprocessing with an excess of fossil feeds in catalytic cracking and hydrodesulphurisation (HDS) lab-scale units. All coprocessing experiments were conducted successfully without operational problems.

Mullen et al. [71] reported the mild batch hydrogenation of the PLs produced from white oak at 138 bar H_2 and 320°C over Ru/C, Pd/C, and Pt/C catalysts (Table 8.6, entry 4). The results revealed that Ru and Pd catalysts were more effective than Pt for deoxygenation reactions. The highest H/C ratio of 1.87 and lowest O/C ratio of 0.32 were obtained using the Ru/C catalyst.

Capunitan et al. [72] studied the hydrotreatment of corn stover PLs and distilled fractions thereof using noble metal catalysts such as Ru and Pd on carbon support at 125 bar, 4 h reaction time, and two different temperatures (200°C and 300°C) in a batch reactor (Table 8.6, entry 5). Results showed that Ru performed better than Pd at higher temperature (300°C) giving a deoxygenation level of about 25%–26%. In addition, higher yields (54–67 wt%), higher H/C molar ratios (1.46–1.48), and lower O/C molar ratios (0.065–0.08) were obtained using Ru. Analysis of the chemical composition of the products indicated that the major reactions involve hydrogenation of

Table 8.6 Overview of catalytic hydrotreatment of PLs using supported noble metal catalysts

Entry	PLs	Reactor type	Catalyst	Condition	References
1	Beech wood-derived PLs	Batch	Ru/C	350°C, 200 bar, 4 h	[68]
2	Switchgrass-derived PLs	Batch	Ru/C, Pt/C, Pd/C	320°C, 145 bar, 4 h	[69]
3	Forest residue-derived PLs	Batch	Ru/C	220–310°C, initial H_2 pressure of 100 bar followed by 190 bar for 4 h	[70]
4	White oak-derived PLs	Batch	Ru/C, Pt/C, Pd/C	320°C, 128 bar, 15 h	[71]
5	Corn stover-derived PLs	Batch	Ru/C, Pd/C	200°C or 300°C, 125 bar, 4 h	[72]
6	Beech wood-derived PLs	Batch	Ru/C, Ru/TiO_2, Ru/Al_2O_3, Pt/C and Pd/C	250°C, 100 bar, 4 h and 350°C, 200 bar, 4 h	[73]
7	Pinewood-derived PLs	Batch	Pt/ZrO_2,Pd/ZrO_2, Rh/ZrO_2,RhPt/ZrO_2, RhPd/ZrO_2, and PdPt/ZrO_2	350°C, 200 bar total pressure, 4 h	[74]
8	Beech wood-derived PLs	Batch	Ru/C	350°C, 200 bar total pressure, 1–6 h	[50]
9	Forest residue-derived PLs	Batch	Ru/C	230–340°C, 4 h, 290 bar total pressure	[75]
10	Oak-derived PLs	Semibatch	Pd/C, Pd/char, Pt/char, and Ru/char	150–280°C for stabilisation, 340–400°C for hydrodeoxygenation, 69–169 bar	[76]
11	Mixed wood-derived PLs, Second corn stover-derived PLs Poplar-derived PLs (hot-filtered)	Continuous	Pd/C	One stage packed-bed reactor First stabilisation, 0.25 LHSV, 340°C, 138 bar Then, HDO, 0.2 LHSV, 405°C, 103 bar	[77]

12	Forestry residue-derived PLs	Continuous	Ru/C	Two fixed-bed reactors Stabilisation, WHSV of 3.3 h^{-1}, 100°C, 200 bar HDO, WHSV of 3.3 h^{-1}, 150–360°C, 200 bar	[78]
13	Pinewood-derived PLs	Continuous	Ru/C	Four stages packed-bed reactor in series Temperature, 175/225/375/400; WHSV 5.4 kg oil kg^{-1} cat h^{-1}, 241 bar	[5]
14	Pinewood-derived PLs	Continuous	Ru/C + (NiMoS/γ-Al_2O_3, CoMoS/C)	Two stages packed-bed reactors in series Temperature, 170°C in the first stage and 250–400°C in the second stage LHSV, 0.19 $L_{PL}/(L_{cat\,h})$ 75–150 bar	[53]
15	Wood-derived PLs	Continuous	Ru/TiO_2	Two stages packed-bed reactors in series 120–160°C, 10.3 MPa, liquid hourly space velocity of 0.40 h^{-1}	[79]
16	Pine-derived PLs	Continuous	Ru/C	81–83°C, 84 bar, LHSV, 0.48–0.95 L/L/h	[80]

ketones to alcohols. The hydrotreated product from the distillate fraction with Ru as the catalyst at 300°C gave the lowest oxygen content (7.0 wt%). Product properties were determined, and for instance, the water content was 2.6%, the acidity was 5.3 mg KOH/g sample, and the heating value was 40.2 MJ/kg, which makes the product a potentially interesting feed for coprocessing with crude oils in existing refineries.

Wildschut et al. [73] investigated the hydrotreatment of PLs using a series of heterogeneous noble metal catalysts including Ru/C, Ru/TiO_2, Ru/Al_2O_3, Pt/C, and Pd/C. The reactions were carried out at two temperatures (250°C and 350°C) and pressures (100 and 200 bar) in a batch autoclave. Among all the noble metal catalysts studied, the Ru/C catalyst was found to be superior to other catalysts with respect to oil yield (up to 60 wt%) and deoxygenation level (up to 90 wt%) (Table 8.6, entry 6) at 350°C.

Ardiyanti et al. [74] reported the catalytic hydrotreatment of PLs from pinewood at 350°C, 200 bar pressure for 4 h using mono- and bimetallic catalysts based on Rh, Pt, and Pd on a zirconia support in a batch set-up (Table 8.6, entry 7). Pd/ZrO_2 showed the highest activity, followed by Rh/ZrO_2. Upgraded oils with the best product properties were obtained with Rh/ZrO_2. For this catalyst, the TGA residue, which is a measure for coking tendency, was 13.4 wt%, and the M_w of the upgraded oil (748 g/mol) was lowest. All noble metal catalysts showed higher activities per gram of metal than the benchmark catalyst CoMo/Al_2O_3. Hydrotreatment reactions at variable batch times were performed using the Rh/ZrO_2 catalyst to gain insights in reaction pathways and showed the involvement of competitive hydrogenation and polymerisation pathways at the initial stage of the process. Temperature-programmed oxidation (TPO) measurements of spent catalysts showed the presence of carbonaceous deposits on the catalyst (2–6 wt%). These deposits may be removed by oxidation at temperatures lower than 600°C without changing the morphology of the catalysts.

Wildschut et al. [50] reported the use of Ru/C (5 wt%) as a catalyst for the hyrotreatment of PLs at 350°C and 200 bar pressure in a batch reactor set-up (Table 8.6, entry 8). Highest oil yields (65 wt%) were obtained after 4 h using a 5 wt% intake of catalyst on PL. Longer reaction times lead to a reduction of the oil yield due to the formation of gas-phase components. A solvent-solvent extraction procedure was applied to gain insights into the molecular transformations during the catalytic hydrotreatment experiments. It appears that specifically the carbohydrate fraction is very reactive. The observations were rationalised by a set of reaction pathways for the various product phases.

Shell [75] patented a process for the catalytic hydrotreatment of PLs derived from forestry residue at 230–340°C for 4 h under 290 bar total pressure using a Ru/C catalyst in a batch autoclave to produce biofuels (Table 8.6, entry 9). At low temperature (230–260°C), a bottom oil in a yield of around 47–49 wt% was obtained with an oxygen content between 24.4 and 28.0 wt% and MCRT values between 9.1 and 11.7 wt%. At higher temperatures (300–340°C), top oils in 48–50 wt% yield were obtained. The lowest oxygen content was 15.5 wt%; the lowest MCRT was 1.8 wt%, obtained at 330°C.

French et al. [76] studied the catalytic hydrotreatment (150–280°C for stabilisation and 340–400°C for hydrodeoxygenation, 69–169 bar) of PLs derived from oak using Pd/C (activated), Pd/char, Pt/char, Ru/char, and NiMo–S in a semibatch reactor to determine the effects of catalyst type on the quality of the product oils (Table 8.6, entry 10). The organic product yield ranged from 24 to 42 wt% (about 51%–77% of the carbon in the feed), and the products showed a low oxygen content (5 wt%) and acidity.

8.8.1.2 Studies in continuous set-ups

Elliott et al. [77] reported the hydroprocessing of PLs from mixed wood in a fixed-bed reactor using a Pd/C catalyst (Table 8.6, entry 11). Reactions were performed at a range of conditions (310 and 375°C, 138 bar) and a LHSV between 0.18 and 1.12 h^{-1}. An oil phase was obtained in a yield of 0.62 g/g dry feed with a H/C ratio of 1.43 at 340°C and a LHSV of 0.25 h^{-1}. The set-up was successfully operated for 8–102 h on stream.

Bi et al. [78] performed catalytic hydrotreatment studies of PLs from forestry residue in two fixed-bed reactors (Table 8.6, entry 12). The reaction temperature in the first reactor was kept at 100°C for stabilisation; the second reactor was maintained at 150–360°C. The pressure was set at 200 bar. Both the PL feed and product oils were analysed by negative-ion ESI FT–ICR MS. The PL feed contained mainly O2–O18 species with double bond equivalent (DBE) values between 0 and 22. After catalytic hydrotreatment, the oxygen numbers reduced to O1–O10, and the DBE values also decreased. The amounts of major oxygen-containing compounds such as carbonyls, carboxylic acids, ethers, carbohydrates, and alcohols considerably changed upon the hydrotreatment process. Hydrogenation of the carbonyls, carbohydrates, and furans occurred below 150°C. Dehydration, hydrodeoxygenation, and dehydration-hydrogenation reactions were observed above 210°C. Decarboxylation and decarbonylation required higher temperatures (>300°C). Lignin polymers were depolymerised after reduction of the carbonyl and methoxy groups.

Venderbosch et al. [5] investigated the catalytic hydrotreatment of PLs in a four-stage packed-bed reactor at different process severities using the benchmark Ru/C catalysts (Table 8.6, entry 13). The results revealed that catalytic hydrogenation and thermal (noncatalytic) polymerisation are parallel reactions, as shown in Scheme 8.8. Polymerisation leads to the formation of higher M_w fragments, especially during the low-temperature stabilisation step as shown in Fig. 8.10. The TG residue, a good indicator for thermal stability of an oil sample, shows the same tendency as the M_w. This clear increase in M_w and TG residue in the stabilisation step points out that the initial activity of Ru/C is not sufficiently high to suppress the polymerisation of thermally labile components, for example, carbonyl-containing compounds and sugars. High M_w fragments with high TG residues will undergo further condensation reactions, ultimately forming char. This char may lead to catalyst deactivation and reactor plugging.

The low initial hydrogenation activity of Ru/C type of catalysts was also confirmed by Elliott and co-workers (Table 8.6, entry 14) [53]. They reported the catalytic hydrotreating of PLs using a two-stage packed-bed reactor with a first catalyst bed composed of Ru/C for low-temperature stabilisation (at around 170°C) and a second bed with sulphided CoMo or NiMo on Al_2O_3 for high-temperature hydrocracking reactions (at around 400°C) (Table 8.6, entry 13). The results revealed that the formation of higher M_w fragments and char in the first bed led to an increase in pressure drop over time, and all tests were terminated in <100 h. In a subsequent study, hydrotreatment experiments were performed in the two-stage packed-bed reactor with a PL feed that was prehydrotreated at 140°C using the same procedures [81]. The run was successful, and a cumulative runtime of 1440 h was reported without major operational issues.

Wang et al. [79] investigated the low-temperature stabilisation of PLs derived from wood in a two-stage packed-bed reactor using a homemade Ru/TiO_2 catalyst at

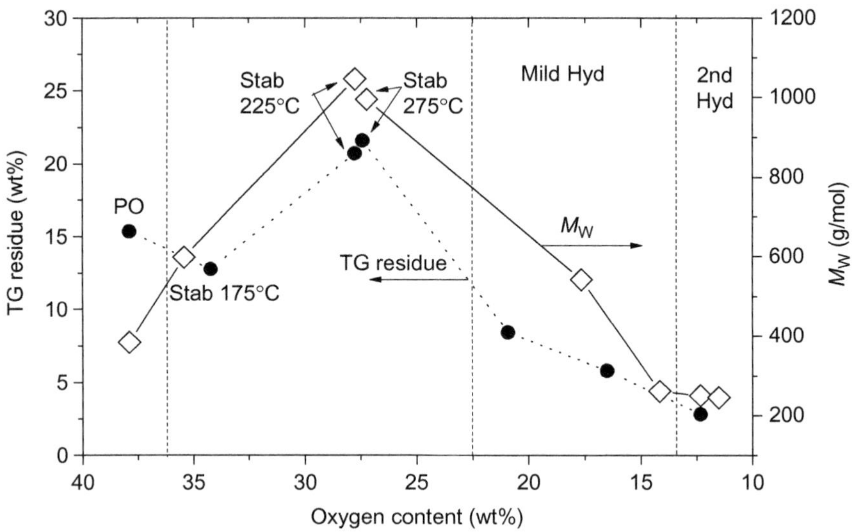

Fig. 8.10 TG and M_w of the organic phases after catalytic hydrotreatment using Ru/C catalyst in four-stage packed-bed reactors [5].

120–160°C (Table 8.6, entry 15). Hydrogen uptake was around 3.8–12.4 mmol/g PL for 48 h tests. Hydrogenation of the sugar fraction was confirmed by ^{13}C NMR; hydrogenation of smaller aldehydes and ketones was shown to take place by performing carbonyl content measurement. Hydrogenation reaction of sugars to sugar alcohols, ketones, and aldehydes to alcohols significantly improved the thermal stability of the PLs and allowed for better performance in the second high temperature, deep hydrotreatment step.

Elliot et al. [80] patented a process for stabilising PLs with hydrogen as the reactant in the presence of Ru/C at low temperature (81–83°C, 84 bar) in a trickle-bed (down flow) reactor configuration (Table 8.6, entry 16). The carbonyl content was reduced slightly from 4.2–7.0 for the feed to 3.1–3.4 (C%) for the product oil. Ageing of the product oils was studied by monitoring the viscosity upon heating at 80°C for 24 h. A viscosity increase of about 5.2% was observed for the product oil compared with 42% for the fresh PL feed.

As such, it can be concluded that noble-metal-based catalysts have been used succesfully for the hydrotreatment of PLs. However, the catalysts are in general not sufficiently active in the initial stabilisation stage of the catalytic hydrotreatment of PLs, and combined with their high prices, the identification of cheap metal-based catalysts with high hydrogenation activity for the thermally labile components in PLs in the low-temperature stabilisation step has been pursued actively.

8.8.2 Nonnoble-metal based catalysts

The first nonnoble-metal-based catalysts explored for the catalytic hydrotreatment of PLs were typical HDS catalysts, namely, sulphided NiMo or CoMo/γ-Al_2O_3 [41,64,76, 82–84]. Close to quantitative deoxygenation levels have been reported [82]. A continuous supply of a sulphur-containing compound is required to keep the catalyst in the sulphided form, which is essential to maintain high catalytic activity. As such, these catalysts are less attractive for the hydrotreatment of sulphur-free biomass streams.

Investigations using nonsulphided noble-metal-free catalysts are mainly focusing on nickel, promoted by Cu, Fe, Pd, or Mo, supported on activated carbon, TiO_2, ZrO_2, Al_2O_3, or SiO_2. An overview of the results on the catalytic hydrotreatment of PLs using such catalysts is given in Table 8.7.

Ardiyanti et al. [85] reported the catalytic hydrotreatment of PLs using bimetallic Ni–Cu catalysts on various supports (CeO_2–ZrO_2, ZrO_2, SiO_2, TiO_2, rice husk carbon, and sibunite) at 350°C, 200 bar total pressure in a batch autoclave (Table 8.7, entry 1). The metal contents ranged from 7.5 to 9.0 (Ni) and 3.1–3.6 wt% (Cu) for the inorganic supports and 17.1–17.8 (Ni) and 7.1–7.8 (Cu) for the carbon supports. After reaction, a product oil was isolated in yields between 35.9 and 45.5 wt% on PL feed. The product

Table 8.7 Catalytic hydrotreatment of PLs using nickel-based catalysts

Entry	PLs	Reactor type	Catalyst	Condition	References
1	Pinewood-derived PLs	Batch	Ni–Cu/TiO_2, Ni–Cu/CeO_2–ZrO_2, Ni–Cu/ZrO_2, Ni–Cu/CRH, Ni–Cu/sibunite, Ni–Cu/δ-Al_2O_3	110 bar H_2, 150°C for 1 h, and 350°C for 3 h	[85]
2	Pinewood-derived PLs	Batch	Ni–Cu/δ-Al_2O_3 with various Ni/Cu ratios	100 bar H_2, 150°C for 1 h, and 350°C for 3 h	[86]
3	Prairie cordgrass-derived PLs	Batch	Ni/AC, Ni–Fe/AC, Ni–Mo/AC, Ni–Cu/AC	34 bar H_2, 350°C, 5 h	[87]
4	Pine sawdust-derived PLs	Batch	Ni/HZSM-5	40 bar H_2, 240°C, 3 h	[88]
5	Pine sawdust-derived PLs	Batch	Reduced Ni–Mo/γ-Al_2O_3	200°C, 2 h, 30 bar	[84]
6	Pine sawdust-derived PLs	Batch	Reduced Ni–Mo/CNT	100–250°C, 2–10 h, 50–60 bar	[89]
7	Pinewood-derived PLs	Batch/continuous	Ni–Cu/SiO_2, Ni–Cu/SiO_2 and Kaolin, Ni–Pd/SiO_2, Ni–Cu/SiO_2–ZrO_2–La_2O_3	120 bar H_2, 150°C for 1 h, and 350°C for 3 h	[38,52]

with the highest H/C ratio (1.43) and lowest TG residue (2.7 wt%) was obtained using Ni–Cu on TiO_2. However, char formation was still high (4.1 wt%), indicating that the rate of hydrogenation in the initial stage of the process is not sufficiently high to suppress thermal polymerisation.

Ardiyanti et al. [86] also performed catalytic hydrotreatment experiments for PLs using bimetallic NiCu/δ-Al_2O_3 catalysts at 350°C in a batch autoclave (Table 8.7, entry 2). The catalysts are characterised by various Ni/Cu ratios (0.32–8.1 w/w) at a fixed total metal intake of about 20 wt%. Hydrotreatment of PLs was performed for 1 h at 150°C followed by 3 h at 350°C, at 100 bar initial pressure and yielded an upgraded oil. Best performance was found for the catalyst with a Ni-to-Cu wt% ratio of eight. For this catalyst, hydrogen consumption was the highest (146 NL/kg PL feed), and the oil had a high H/C ratio of 1.35, a low M_w of 500 g/mol, and a low TGA residue of 6.8 wt%.

Cheng et al. [87] reported the catalytic hydrotreatment of PLs from prairie cordgrass using 34 bar of H_2 at 350°C in a batch autoclave (Table 8.7, entry 3). Supported Ni catalysts on activated carbon including Ni/AC, Ni–Fe/AC, Ni–Mo/AC, and Ni–Cu/AC (6% metal content) were used. All catalysts gave product oils with improved properties compared with the PL feed (water content, pH, oxygen content, higher heating value, and chemical compositions). The oxygen content was reduced from 68.81 for the feed to 21.70%–25.88% for the products. As a result, the higher heating value increased from 11.33 for the PL feed to 29.65–31.61 MJ/kg for the products. Ni/AC gave a product oil with 33% of hydrocarbons (C6–C12) in the gasoline range. For Ni–Mo/AC, a product oil with the highest content of gasoline blending alkyl phenols (38%) was obtained.

Zhang et al. [88] reported the catalytic hydrotreatment of PLs from pine sawdust using a 10 wt% Ni/H-ZSM-5 (Si/Al = 38) catalyst at 240°C in a batch autoclave (Table 8.7, entry 4). A single liquid phase was obtained after reaction, with a water content considerably higher than the PL feed (52.0 vs 38.3 wt%) due to hydrodeoxygenation and polymerisation reactions. The pH of the product oil increased from 2.27 to 4.07, and the hydrogen content increased from 6.28 to 7.01 wt%. The decrease in acidity is advantageous from an application point of view. The spent catalyst was characterised by thermogravimetry analysis (TGA) and revealed the presence of large amounts (up to 36 wt%) of organic residues on the catalyst surface, likely formed by polymerisation of reactive species in the PLs.

Xu et al. [84] reported the upgrading of PLs from pine sawdust using reduced Ni–Mo/γ-Al_2O_3 catalysts at 200°C, 30 bar for 2 h in a batch autoclave (Table 8.7, entry 5). After reaction, a single-phase oil was obtained with a water content of 59.0 wt%, compared with 46.2 wt% for the feed. The pH value increased from 2.16 to 2.84 due to the conversion of organic acids, which was confirmed by GC. The amount of aldehydes and ketones in the product oil was lower than for the PL feed (from 47.0 to 26.1%, GC peak area), whereas a slight increase in alcohol content was observed (23.0 for the feed to 25.6% in the product, GC peak area) The H content of the product oil was higher than the feed (6.93 vs 6.61 wt% for the feed).

Zhou et al. [89] studied the upgrading of PLs from pine sawdust using a reduced Ni–Mo/CNT catalyst at 100–250°C under 5–60 bar hydrogen in a batch autoclave (Table 8.7, entry 6). After reaction, a single liquid product was obtained with a higher water content than the feed (from 42.8 to 51.8%). The pH increased from 2.92 to 3.68, indicative for the conversion of organic acids under the prevailing conditions. The H content of the product (11.26%) was considerably higher than the feed (8.40%).

The use of novel bimetallic Ni–Cu/Pd catalysts for the catalytic hydrotreatment of PLs, referred to as Picula catalysts, was reported by Ardiyanti et al. [52]. The catalysts were prepared using a sol–gel method and are characterised by a high metal loading (i.e. Picula cat D—57.9 wt% Ni, 7 wt% Cu, and 35.1% SiO_2; Table 8.7, entry 7, and Table 8.8 for details).

Catalytic hydrotreament of pinewood-derived PLs in batch autoclaves was performed (Table 8.7, entry 7 for details). The catalysts show clear advantages compared with the benchmark Ru/C catalyst. Examples are a low methane production rate as shown in Fig. 8.11, which has a positive effect on the hydrogen consumption. The best performance was obtained using Picula cat B with Pd as the promoter, giving an oil phase with highest H/C ratio (1.51) and lowest TG residue (0.7 wt%). The results revealed that Pd is a better promoter compared with Cu for the Picula catalysts studied. Further product analyses revealed that the reduced tendency for char formation is likely due to a higher hydrogenation activity of the Picula catalysts for thermal labile components in PLs at low temperature (below 200°C). As such, the rate of polymerisation reactions is considerably suppressed.

Table 8.8 Composition of Picula catalysts for the catalytic hydrotreatment of PLs [52].

Code	Active metal loading (wt%)	Support (wt%)
Picula cat B	Ni 58.3, Pd 0.7	SiO_2 41
Picula cat C	Ni 28.8, Cu 3.7	SiO_2 33.8, kaolin 33.8
Picula cat D	Ni 57.9, Cu 7	SiO_2 35.1
Picula cat E	Ni 36.5, Cu 2.3	SiO_2 12.6, ZrO_2 37.2, La_2O_3 0.9

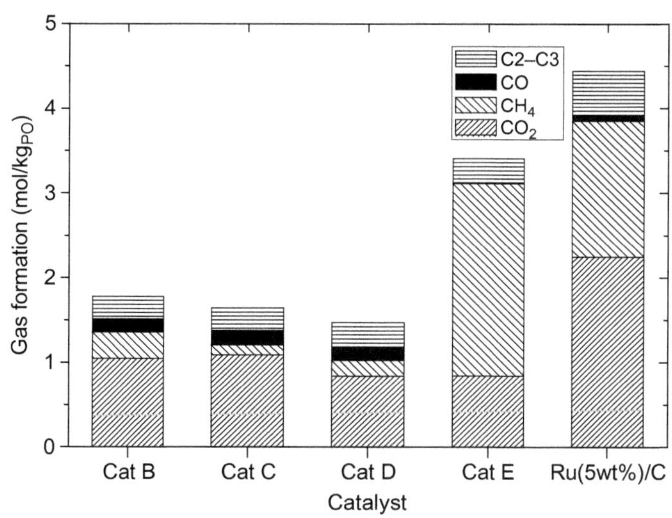

Fig. 8.11 Gas-phase formation for the batch hydrotreatment of PLs using Picula cat B–E.

Further catalytic hydrotreatment studies on PLs using noble-metal-free Picula cat D were performed in a continuous reactor configuration consisting of a packed-bed reactor with four individual reactors arranged in a series configuration. Deactivation of the catalyst and clogging by char formation, a typical phenomenon for Ru/C, was not observed for the Picula cat D at extended times on stream [38].

From the above results, it is clear that the Picula catalysts show considerably better performance for the catalytic hydrotreatment of PLs compared with the benchmark Ru/C catalyst and the Ni-based catalysts with low metal loadings. It was postulated that this good performance is due to a high activity for thermally labile carbonyl-containing molecules and sugars at low temperature (below 200°C), thereby reducing the tendency for char formation.

The stability of the Picula catalysts was investigated for up to 400h time-on-stream in a four-stage packed-bed reactor. Reactor clogging was not observed to a significant extent though some catalyst deactivation occurred [38]. Further characterisation studies on spent and regenerated Picula catalysts (Ni–Pd/SiO_2) revealed that the regenerated catalysts have a considerably lower surface area compared with the fresh catalysts and also show some agglomeration of active metal nanoparticles; see TEM images in Fig. 8.12 [52].

8.8.3 Catalytic hydrotreatment of PLs in solvents

Hydrotreatment of PLs in various solvents using noble-metal- and Ni-based catalysts have been reported, and the results are summarised in Table 8.9. The most commonly

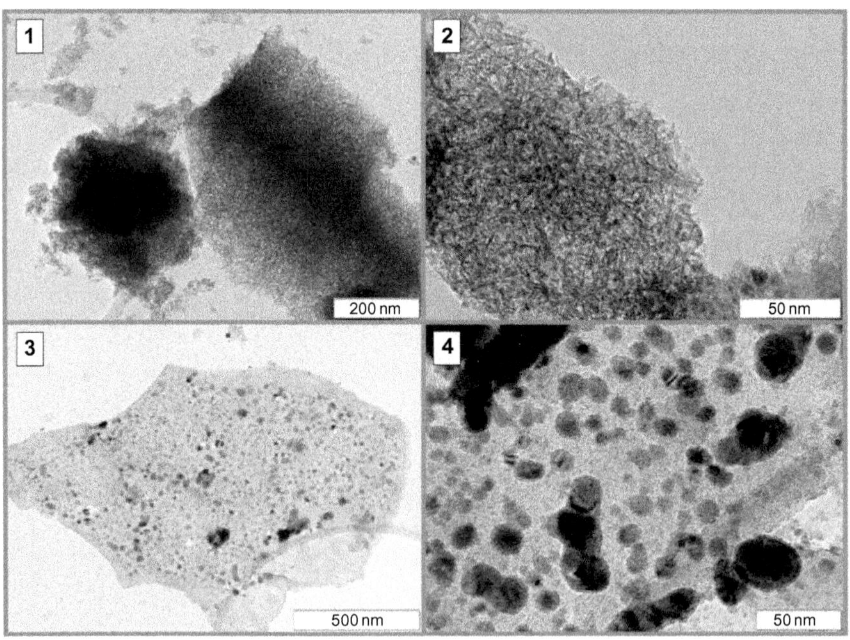

Fig. 8.12 HRTEM images of cat B (NiPd/SiO_2): 1, 2—fresh catalyst, 3, 4—regenerated catalyst.

Table 8.9 Overview of catalytic hydrotreatment of PLs in various solvents using both noble and nonnoble-metal catalysts

Entry	PLs	Reactor type	Solvent	Catalyst	Condition	References
1	Switchgrass-derived PLs	Batch	PLs, polyethylene glycol = 1:9 (v/v)	Ru/C, Rh/C	200–280°C, 20.7–68.9 bar, 5 h	[56]
2	Hardwood (sawdust)-derived PLs	Batch	PLs, glycerol = 2:1 (w/w)	Ru/C	300°C, 3 h, 69 bar H_2, 3 h	[57]
3	Hardwood (sawdust)-derived PLs	Batch	PLs, alcohols (methanol, ethanol, glycerol) = 20:1 (w/w)	Ru/C	300–350°C, 150–200 bar, 3 h	[58]
4	Rice husk-derived PLs	Batch	PLs, ethanol = 1:1–5 (w/w)	Pt/sulphated Zirconia/SBA-15	26–300°C, 5–20 bar H_2, 3 h	[90]
5	Miscanthus sinensis-derived PLs	Batch	PLs, ethanol = 4:1 (w/w)	Pd/C	250–350°C, 30 bar H_2, 0.5–1 h	[91]
6	Rice husk-derived PLs	Batch	PLs, ethanol = 1:2 (w/w)	Ru/HZSM-5, Pt/C, Pd/C, Ru/C	300°C, 20 bar H_2, 5 h	[92]
7	*Pinus sylvestris* L.-derived PLs	Batch	PLs, alcohols (methanol, ethanol) = 1:10 (w/w)	Pt/sulphated zirconia /SBA-15 Pd/sulphated zirconia/SBA-15 Pt/HZSM-5, Pd/HZSM-5	260°C, 20 bar H_2, 3 h	[93]
8	Rice husk-derived PLs	Batch	PLs, methanol = 6:5 PLs; methanol; formic acid = 6:5:2 (w/w)	Raney Ni, Pd/HY, Ru/HY, Pd/SBA-15, Ru/SBA-15	170–200°C, 40 bar H_2, 170–200°C, no hydrogen for the reaction with formic acid, 5–7 h	[60]
9	Rice husk-derived PLs	Batch	PLs, decalin = 3:7 (w/w)	Ni/TiO_2-ZrO_2	300°C, 40 bar H_2, 8 h	[61]
10	Pine (sawdust)-derived PLs	Batch	PLs, butanol = 1:1 (w/w)	Ru/C	25–300°C, 20 bar H_2, 3 h	[59]
11	Pine-derived PLs	Packed bed	PLs, methanol = 1:1 (v/v)	Ni/HZSM-5, Ru/HZSM-5	250°C, 41.4 bar, WHSV = 2.5 h^{-1}	[94]

used solvents are alcohols (methanol, ethanol, glycerol, and higher alcohols). The incentive for the use of solvents is to suppress the repolymerisation of reactive oxygenated molecules [95] and/or to supply hydrogen by using hydrogen-donor solvents like formic acid [60], tetralin [96], or decalin [97]. Most studies were carried out in batch; only one example has been reported in a continuous packed-bed reactor. The results of the various studies will be discussed in the following paragraphs.

8.8.3.1 Reactions in batch

Nan et al. [56] reported the catalytic hydrotreatment of PLs derived from switchgrass in polyethylene glycol (PEG) using Ru/C and Rh/C catalysts in a batch reactor in the temperature range of 200–280°C under hydrogen pressures of 20.7–68.9 bar (Table 8.9, entry 1). A large excess of solvent was applied (PLs to solvent ratio of 1:9). The liquid product yields were in the range of 78.4–95.1 wt% for Ru/C and 84.7–92.2 wt% for Rh/C (including the solvent PEG). Solid yield was low (0.6–2.3 wt% for Ru/C and 0.4–1.1 wt% for Rh/C). Highest activity in terms of deoxygenation level of the product oil was obtained with Ru/C at the most severe reaction conditions (280°C and 68.9 bar). The liquid product samples (280°C and 68.9 bar) were extracted with n-hexane (1:1 v/v) to separate the products from the PEG. The hexane soluble fraction was analysed by GC–MS, and the main products (peak area percentages) were alcohols (5.2%, e.g. propanol and ethylene glycol), cyclic carbonyl compounds (3.8%, e.g. cyclopentanone and cyclohexanone), hydrocarbons (21.2%, e.g. methylcyclopentane and cyclohexane), and phenolics (35.7%, e.g. phenol and 2-methoxy-phenol). Comparison of the product composition by FTIR of runs in PEG with runs in C30 hydrocarbons revealed that the alcohols were formed from the PLs rather than the PEG solvent.

Reyhanitash et al. [57] investigated the hydrodeoxygenation of PLs from hardwood sawdust using a Ru/C catalyst in a batch reactor at 300°C and 69 bar H_2 pressure (Table 8.9, entry 2). Three different feeds were used: (i) PLs pretreated in glycerol in the presence of a solid acid catalyst (Amberlyst 35) at 120°C for 2 h, (ii) a PL-glycerol mixture, and (iii) an undiltued PL only. After reactions, two liquid phases (organic and aqueous) were obtained. The phases were separated for analysis, including the glycerol content (GC-FID). All product oils showed rather comparable H/C and O/C ratios. The H/C ratios and O/C ratios (glycerol and water-free) were considerably lower than for the feed (H/C 1.67, O/C 0.71, glycerol, and water-free), likely due to water rejection reactions. Molecular weight determinations showed that dilution of the PLs by glycerol results in a reduction of the molecular weight of the oil fraction compared with direct catalytic hydrotreatment of PLs. Apparently, the use of a solvent suppresses polymerisation of reactive compounds. The total glycerol amount in the liquid product obtained from an experiment with the glycerol pretreated PL was 15.8 wt%, compared with an initial glycerol intake of 33.7 wt%. Part of this reduction is due to glycerol consumption by esterification of carboxylic acids.

The two-step approach involving pretreatment of the PL with an alcohol (5 wt%, methanol, ethanol, and glycerol) in the presence of a solid acid catalyst (Amberlyst 35) was further investigated by the same group [58] (Table 8.9, entry 3). The pretreatment was conducted at 100°C for 2 h, followed by a catalytic hydrotreatment using

Ru/C at temperatures between 300°C and 350°C under 150–200 bar H_2. After reactions, two liquid phases were obtained: a heavy organic phase and an aqueous phase. The residual alcohol contents were measured by GC-FID. The results demonstrated that the initial pretreatment (esterification of the PLs with a small alcohol in the presence of a solid acid) yielded product oils with similar elemental compositions (after compensation for the alcohol amount) as for the experiments without alcohol addition. However, the molecular weight of the product oils were considerably lower for experiments with alcohols, likely due to a reduction in the rate of polymerisation of reactive oxygenated molecules in PLs. Alcohol consumption by, for instance, gasification and esterification reactions was <15%.

The catalytic upgrading of rice-husk-derived PLs in supercritical ethanol (PL to ethanol ratio of 1:5, w/w) using a Pt/sulphated zirconia/SBA-15 was studied in a batch autoclave at temperatures between 260°C and 300°C and hydrogen pressures from 5 to 20 bar (Table 8.9, entry 4) [90]. After reaction, a single liquid phase with organic compounds (designated as oil), water, and leftover ethanol was obtained. The ethanol and water content in the liquid product were measured by GC and Karl Fischer titration. A high liquid yield was obtained (95.5–99.8 wt%) with an oil fraction between 23.2 and 33.9 wt%. The solid yield was between 0.19 and 4.53 wt%. The heating value (excluding ethanol and water) increased from 21.9 for the PL feed to 26.2–29.6 MJ/kg for the product oils. Around 10–20 wt% conversion of the ethanol was observed. The effect of process conditions on solids formation and relevant product properties was determined. Solid formation was reduced at higher hydrogen pressures and ethanol to PL ratios. Higher ethanol-to-PL ratios increased the yield and heating value of the product oils.

Oh et al. [91] reported the hydrodeoxygenation of *Miscanthus sinensis*-derived PLs in ethanol (PLs to ethanol ratio of 4:1, w/w) in a batch reactor using Pd/C as the catalyst between 250°C and 350°C under 30 bar of H_2 for 0.5–1 h (Table 8.9, entry 5). Two immiscible liquid phases were obtained after reaction, a light fraction (24–41 wt%) and a heavy fraction (13–60 wt%). Here, the yield of each phase is defined based on the sum of the PL and solvent intake on a wet basis, including ethanol. Considerable amounts of char (6–17 wt%) were obtained. The H/C ratio of the heavy oil fractions was between 1.8 and 2.3 compared with 2.2 for the feed (PL and ethanol), while the O/C decreased from 1.0 for the feed to 0.3–0.6 for the products. The higher heating value of the heavy oil fractions was between 25.8 and 30.6 MJ/kg, which is considerably higher than for the PL feed (17.3 MJ/kg). Unfortunately, the elemental analysis and heating value were not compensated for the presence of ethanol and water, and as such, a proper comparison of product properties is not possible. A detailed analysis on molecular level of the heavy oil fraction by GC–MS revealed that oxygen-containing compounds such as acids (2-hydroxy-butanoic acid), aldehydes (furfural and butanedial), and sugars (levoglucosan) were reduced by hydrogenation and deoxygenation reactions. To gain insights in the level of hydrocracking of the pyrolytic lignin fraction, the amounts of pyrolytic lignins in the heavy oil fraction and initial PLs were determined by an extraction method reported by Scholze et al. [98]. The results indicated that approximately 60% of the pyrolytic lignins are cracked into low-molecular-weight compounds during reaction.

The catalytic hydrotreatment of rice-husk-derived PLs in supercritical ethanol (PL to ethanol ratio of 1:2, w/w) in a batch reactor using a number of noble metal catalysts (Ru/HZSM-5, Pt/C, Pd/C, and Ru/C) at 300°C under 20 bar H_2 for 5 h was reported by Chen et al. [92] (Table 8.9, entry 6). After reaction, a single liquid phase was obtained with an oil content between 27.7 and 32.0 wt%, the other main components being ethanol and water. The alcohol content (GC-TCD) in the liquid product was between 42.7 and 50.7 wt%, and the water content (Karl Fischer titration) was between 21.6 and 25.3 wt%. Mass balance calculations revealed that up to 16 wt% of the ethanol is converted during the catalytic hydrotreatment. The heating value (excluding ethanol and water) increased from 21.5 for the PL feed to 29.5–31.0 MJ/kg for the product oils. The pH of the product liquids was in the range of 4.54–5.8 compared with 3.13 for the PLs feed, indicative for a reduction in the amounts of organic acids. The active carbon supported catalysts showed better performance for the catalytic hydrotreatment of PLs in supercritical ethanol in terms of heating value, pH, and yield of the desired product oil compared with Ru/HZSM-5. This difference in performance may be related to differences in the support properties (acidity and pore-size distribution). However, the average Ru nanoparticle size for Ru/HZSM-5 was considerably larger (19 nm, XRD, and TEM) than for the Ru on carbon support (<5 nm), and this may also affect reactivity.

Zhang et al. [93] reported the catalytic hydrotreatment of *Pinus sylvestris* L.-derived PLs in supercritical alcohols (like methanol and ethanol) using Pt/sulphated zirconia/SBA-15, Pd/sulphated zirconia/SBA-15 and Pt/HZSM-5, and Pd/HZSM-5 at 260°C under 20 bar H_2 in a batch autoclave (Table 8.9, entry 7). Three feeds were used as the starting materials: (i) a virgin PL, (ii) a distillate, and (iii) the residue after distillation. The three feeds were mixed with methanol/ethanol (1:10 w/w) before reaction. The product oils were analysed by GC–MS, and relative peak areas were reported. Solid formation on the spent catalysts was determined by TGA/SDTA analysis. The PL feeds were shown to contain acids, aldehydes, alcohols, esters, ketones, phenols, sugars, polycyclic aromatic hydrocarbons (PAHs), and other oxygenated compounds, while esters were the most prominent components in the product oils. Complete conversion of acids and aldehydes in the distillation residue was achieved using sulphated zirconia as the catalyst support, while around 6% of the aldehydes were still present in both methanol and ethanol when using HZSM-5. Yields and elemental compositions of the product oils using the PL feed were not reported. For the runs with the distillation residue, an ethanol conversion of around 25% was reported. Product analysis showed that the catalytic hydrotreatment of the PLs in ethanol using the Pt/sulphated zirconia/SBA-15 gave better results than with Pd/sulphated zirconia/SBA-15 and Pt/HZSM-5, Pd/HZSM-5 and that the use of ethanol is preferred above methanol.

Xiong et al. [60] studied the catalytic hydrotreatment of PLs from rice husk in methanol or methanol combined with formic acid (PL to methanol ratio of 6:5 for methanol only and a PL/methanol/formic acid ratio of 6:5:2 for methanol combined formic acid) using Raney Ni and zeolite-supported Pd and Ru catalysts in a batch autoclave at 170–200°C (Table 8.9, entry 8). For experiments in methanol, 40 bar of external hydrogen pressure was applied, while external H_2 was not used for reactions in methanol/formic acid. Initial experiments with Raney Ni showed that methanol combined with formic acid is preferred and leads to products with the best properties in terms of water content, viscosity, pH, and amounts of low-molecular-weight products. Reactions in methanol and formic acid gave a single liquid phase in yields (based on total liquid intake) between 86.6 and 93.3 wt%. The water content in the liquid phase ranged from 16.5 to 20.4 wt% compared with 14.2 wt% for the initial feed (PL and solvent). The amount of char formed during the experiments was low and ranged from 0.2 to 3.5 wt%. Heating values (excluding methanol) of the product oils were in the range of 17.5–22.6 MJ/kg compared with 17.6 MJ/kg for the PL feed. The pH of the product oils (3.47–4.58) was higher than for the feed (2.16). A detailed analysis of the PL feed and product oils by GC–MS revealed that the amounts of unsaturated components in the PLs were reduced substantially and that the organic acids were converted to esters through esterification with methanol.

Zhang et al. [61] investigated the catalytic upgrading of PLs from rice husk in decalin (PL/decalin ratio of 3:7, w/w) using Ni/TiO_2-ZrO_2 in a batch reactor at 300°C, 40 bar H_2 (Table 8.9, entry 9). After reaction, a top liquid phase was formed containing the organic product and decalin and an aqueous bottom phase. The amount of product oil was calculated by subtracting the weight of the decalin intake from the total organic liquid phase. A dry oil yield of 19.3 wt% was reported in combination with a high amount of char (16.7 wt%). The pH of the organic layer was 4.21 compared with 2.38 for the PL feed, the heating value (excluding decalin) was 25.8 MJ/kg, which is about twice that of the original PL. Both the PL feed and the product oil were analysed by GC–MS, and it was reported that the acids and aldehydes were hydrogenated. Phenolic compounds, especially alkylphenolics, were the main components in the product oil probably from hydrocracking of the pyrolytic lignin fraction of PLs. Examples of typical phenolics are 2-methoxy-phenol, 2-methoxy-4-methyl-phenol, and 3, 5-dimethyl-phenol.

Xu et al. [59] reported the catalytic hydrotreatment of PLs from rice husk in butanol (1:1 w/w) using Ru/C in a batch reactor (250–300°C with 20 bar initial H_2 pressure) (Table 8.9, entry 10). After reaction (3 h), two immiscible liquid phases were present, a light oil (with solvent, 69.4–84.6 wt%) and an aqueous phase. Solids formation was very limited and about 0.2–0.3 wt%. The pH increased from 2.6 for the PL feed to 4.2–4.4 for the product oils, and the heating value increased from 17.9 to 29.5–32.0 MJ/kg (with solvent butanol). GC–MS analysis of the product oil revealed that the amounts of acids, aldehydes, ketones, phenols, and furanics were reduced significantly, whereas the ester and ether contents increased significantly after reaction by esterification, etherification, acetalisation, hydrogenation, and hydrodeoxygenation reactions.

The light oil obtained at 300°C was distilled to separate the solvent from the product oil (60°C, 13 mbar for 40 min in a rotary evaporator). The distillation residue (54 wt% of the total light oil, with 5.4 wt% of butanol, 0.5 wt% of butyl acetate, and 0.4 wt% of water) had a heating value of 35.2 MJ/kg, a pH value 4.0, and a degree of deoxygenation of 64%. The effect of butanol addition was evaluated by performing catalytic hydrotreatment reactions in the absence of butanol. The results revealed that butanol addition effectively suppressed coke formation (2.5–9.9 wt% in the absence of solvent vs 0.2–0.3 wt% in the presence of butanol). It was concluded that butanol not only acts as a solvent and improves the solubility of H_2 in the reaction mixture but also acts as a reactant.

8.8.3.2 Studies in continuous set-ups

Hilten et al. [94] studied the continuous catalytic hydrotreatment of PLs from pinewood in methanol (1:1 v/v) using bifunctional Ni/HZSM-5 and Ru/HZSM-5 catalysts in a packed-bed reactor (Table 8.9, entry 11). Experiments were carried out at 250°C, 41.4 bar, and a WHSV of 2.5 g feed/(g cat h). The TAN was reduced considerably to 66%–76% of the original value by simultaneous esterification and hydrogenation. Methanol conversion and the amounts of esters formed were determined using GC–MS/FID and HPLC. The performance of both catalysts was compared with an iron oxide catalyst with both acid and basic sites (reduced red mud, RRM). Among the three catalysts, the iron oxide catalyst showed better performance than the Ni and Ru/HZSM-5 catalysts in terms of product yield. RRM yielded 11.5 mol% of methyl acetate compared with 4–10 mol% for the other catalysts, and higher levels of long-chain fatty acid methyl esters were also observed in the product. Coke formation was reduced by 60%–80% by using the bifunctional catalysts in the presence of hydrogen. Besides, the iron oxide catalyst yielded a higher phenolic yield (e.g. 2-methoxyphenol), indicating that the pyrolytic lignin in the PL feed was depolymerised to a certain extent.

8.9 Conclusions

Fast pyrolysis of biomass is getting mature, with first commercial plants in operation. The derived PLs contain substantial amounts of monomeric and oligomeric oxygenated molecules, with a wide range of functionalities. Examples include carboxylic acids, aldehydes, ketones, sugars, and lignin-derived monomers and oligomers. The presence of carboxylic acids renders the PLs acidic in nature. High water content and high oxygen content results in a relatively low energy density, and carbonyl-containing molecules and sugars are considered to be the main reason for the limited thermal stability during storage and coke formation when heating PLs.

Catalytic hydrotreatment is a promising technology for the upgrading of PLs. Tuning of the oxygen content of the product is possible by proper selection of process conditions and choice of the catalyst. The use of the product oils for cofeeding in existing FCC refinery units (mixed with crude oil derivatives, e.g. VGO; oxygen content around 10 wt%) or direct use as a transportation fuel (oxygen content <1 wt%) has been considered. However, polymerisation of reactive components and intermediates is a serious issue, leading to the formation of larger M_w fragments, which react further to char, resulting in catalysts deactivation and reactor clogging. As such, it is of high importance to identify catalysts that have a high activity at relatively low temperatures to avoid polymerisation of reactive components. Various types of catalysts have been studied for the catalytic hydrotreatment of PLs in the last decade. Noble-metal-based catalysts are expensive and show relatively low hydrogenation activity, especially in the initial low-temperature stabilisation step, and these drawbacks limit their application as attractive catalysts for PL upgrading. Nonnoble-metal catalysts, mainly Ni-based, have also been tested for the catalytic hydrotreatment of PLs. The results reveal that the activity of nickel-based catalysts with a low metal loading (< 10 wt%) is still insufficient to fully eliminate char formation. Picula catalysts developed by our group in cooperation with BTG, the Netherlands, and the Boreskov Institute of Catalysis (BIC, Russia) are characterised by a high metal loading and small metal nanoparticles and showed higher activity in the low-temperature range compared with noble-metal-based catalysts. Bimetallic catalysts also seem to have good potential, and particularly, Pd is shown to be a good promotor.

The use of solvents has been explored with the incentive to reduce the rate of polymerisation reactions and as such to reduce char formation and the molecular weight of the product oil. Typically, low-molecular-weight alcohols are applied though also some experiments have been reported with hydrogen-donor solvents like tetralin. A main issue is the fact that most solvents are not inert under the reaction condition employed and are partly converted to liquid- and gas-phase components. With the existing data, it is not yet clear whether the use of solvents has a positive effect on product properties. An exception is the observation that the pH of the product oils is always higher than for the feed when using alcohols, explained by the occurrence of esterification reactions between alcohols and the organic acids present in the PL feeds. Further systematic research will be required to assess the benefits of solvent addition, supplemented by technoeconomic evaluations.

References

[1] Chheda J.N., Huber G.W., Dumesic J.A. Liquid-phase catalytic processing of biomass-derived oxygenated hydrocarbons to fuels and chemicals. Angew Chem Int Ed 2007; 46:7164–83.

[2] Serrano-Ruiz J.C., Luque R., Sepulveda-Escribano A. Transformations of biomass-derived platform molecules: from high added-value chemicals to fuels via aqueous-phase processing. Chem Soc Rev 2011;40:5266–81.

[3] Huber G.W., Iborra S., Corma A. Synthesis of transportation fuels from biomass: chemistry, catalysts, and engineering. Chem Rev 2006;106:4044–98.

[4] Venderbosch R.H., Prins W.. Fast pyrolysis technology development. Biofuels Bioprod Biorefin 2010;4:178–208.

[5] Venderbosch R.H., Ardiyanti A.R., Wildschut J., Oasmaa A., Heeres H.J. Stabilization of biomass-derived pyrolysis oils. J Chem Technol Biotechnol 2010;85:674–86.

[6] http://www.btgworld.com/.

[7] Oasmaa A., Kuoppala E., Solantausta Y. Fast Pyrolysis of forestry residue. 2. Physicochemical composition of product liquid. Energy Fuel 2003;17:433–43.

[8] Mettler M.S., Mushrif S.H., Paulsen A.D., Javadekar A.D., Vlachos D.G., Dauenhauer P.J. Revealing pyrolysis chemistry for biofuels production: Conversion of cellulose to furans and small oxygenates. Energy Environ Sci 2012;5:5414–24.

[9] Patwardhan P.R., Brown R.C., Shanks B.H. Product distribution from the fast pyrolysis of hemicellulose. ChemSusChem 2011;4:636–43.

[10] Liu W.-J., Jiang H., Yu H.-Q. Thermochemical conversion of lignin to functional materials: a review and future directions. Green Chem 2015;17:4888–907.

[11] Patwardhan P.R., Brown R.C., Shanks B.H. Understanding the fast pyrolysis of lignin. ChemSusChem 2011;4:1629–36.

[12] Oasmaa A., Elliott D.C., Korhonen J. Acidity of biomass fast pyrolysis bio-oils. Energy Fuel 2010;24:6548–54.

[13] Oasmaa A., Czernik S. Fuel oil quality of biomass pyrolysis oils state of the art for the end users. Energy Fuel 1999;13:914–21.

[14] Czernik S., Bridgwater A.V. Overview of applications of biomass fast pyrolysis oil. Energy Fuel 2004;18:590–8.

[15] Mercader F.d.M., Groeneveld M.J., Kersten S.R.A., Venderbosch R.H., Hogendoorn J.A. Pyrolysis oil upgrading by high pressure thermal treatment. Fuel 2010;89:2829–37.

[16] Oasmaa A., Korhonen J., Kuoppala E. An approach for stability measurement of wood-based fast pyrolysis bio-oils. Energy Fuel 2011;25:3307–13.

[17] van Zandvoort I., Wang Y., Rasrendra C.B., van Eck E.R.H., Bruijnincx P.C.A., Heeres H.J., Weckhuysen B.M. Formation, molecular structure, and morphology of humins in biomass conversion: influence of feedstock and processing conditions. ChemSusChem 2013;6:1745–58.

[18] Oasmaa A., Kuoppala E., Ardiyanti A., Venderbosch R.H., Heeres H.J. Characterization of hydrotreated fast pyrolysis liquids. Energy Fuel 2010;24:5264–72.

[19] Kloekhorst A., Wildschut J., Heeres H.J. Catalytic hydrotreatment of pyrolytic lignins to give alkylphenolics and aromatics using a supported Ru catalyst. Cat Sci Technol 2014;4:2367–77.

[20] Bridgwater A.V., Meier D., Radlein D. An overview of fast pyrolysis of biomass. Org Geochem 1999;30:1479–93.

[21] Mohan D., Pittman C.U., Steele P.H. Pyrolysis of wood/biomass for bio-oil: a critical review. Energy Fuel 2006;20:848–89.

[22] Panigrahi S., Dalai A.K., Chaudhari S.T., Bakhshi N.N. Synthesis gas production from steam gasification of biomass-derived oil. Energy Fuel 2003;17:637–42.

[23] van Rossum G., Kersten S.R.A., van Swaaij W.P.M. Staged catalytic gasification/steam reforming of pyrolysis oil. Ind Eng Chem Res 2009;48:5857–66.

[24] Wang D., Czernik S., Montané D., Mann M., Chornet E. Biomass to hydrogen via fast pyrolysis and catalytic steam reforming of the pyrolysis oil or its fractions. Ind Eng Chem Res 1997;36:1507–18.

[25] Wang Z.X., Dong T., Yuan L.X., Kan T., Zhu X.F., Torimoto Y., Sadakata M., Li Q.X. Characteristics of bio-oil-syngas and its utilization in Fischer- Tropsch synthesis. Energy Fuel 2007;21:2421–32.

[26] Rasrendra C.B., Girisuta B., van de Bovenkamp H.H., Winkelman J.G.M., Leijenhorst E.J., Venderbosch R.H., Windt M., Meier D., Heeres H.J. Recovery of acetic acid from an aqueous pyrolysis oil phase by reactive extraction using tri-n-octylamine. Chem Eng J 2011;176–177:244–52.

[27] Vitasari C.R., Meindersma G.W., de Haan A.B. Glycolaldehyde co-extraction during the reactive extraction of acetic acid with tri-*n*-octylamine/2-ethyl-1-hexanol from a wood-based pyrolysis oil-derived aqueous phase. Sep Purif Technol 2012;95:39–43.

[28] Vitasari C.R., Meindersma G.W., de Haan A.B. Renewable glycolaldehyde isolation from pyrolysis oil-derived aqueous solution by reactive extraction with primary amines. Sep Purif Technol 2012;95:103–8.

[29] Vitasari C.R., Meindersma G.W., de Haan A.B. Conceptual process design of an integrated bio-based acetic acid, glycolaldehyde, and acetol production in a pyrolysis oil-based biorefinery. Chem Eng Res Des 2015;95:133–43.

[30] Bennett N.M., Helle S.S., Duff S.J.B. Extraction and hydrolysis of levoglucosan from pyrolysis oil. Bioresour Technol 2009;100:6059–63.

[31] Chi Z., Rover M., Jun E., Deaton M., Johnston P., Brown R.C., Wen Z., Jarboe L.R. Overliming detoxification of pyrolytic sugar syrup for direct fermentation of levoglucosan to ethanol. Bioresour Technol 2013;150:220–7.

[32] Helle S., Bennett N.M., Lau K., Matsui J.H., Duff S.J.B. A kinetic model for production of glucose by hydrolysis of levoglucosan and cellobiosan from pyrolysis oil. Carbohydr Res 2007;342:2365–70.

[33] Lian J., Garcia-Perez M., Chen S. Fermentation of levoglucosan with oleaginous yeasts for lipid production. Bioresour Technol 2013;133:183–9.

[34] Wang S., Wang Y., Cai Q., Wang X., Jin H., Luo Z. Multi-step separation of monophenols and pyrolytic lignins from the water-insoluble phase of bio-oil. Sep Purif Technol 2014;122:248–55.

[35] Mahfud F.H., Ghijsen F., Heeres H.J. Hydrogenation of fast pyrolyis oil and model compounds in a two-phase aqueous organic system using homogeneous ruthenium catalysts. J Mol Catal A Chem 2007;264:227–36.

[36] Lian J., Chen S., Zhou S., Wang Z., O'Fallon J., Li C.-Z., Garcia-Perez M. Separation, hydrolysis and fermentation of pyrolytic sugars to produce ethanol and lipids. Bioresour Technol 2010;101:9688–99.

[37] Biocoup at a glance. http://www.biocoup.com.

[38] Ardiyanti A.R., Venderbosch R.H., Yin W., Heeres H.J. Catalytic Hydrotreatment of Fast Pyrolysis Oils Using Supported Metal Catalysts. In: Catalytic Hydrogenation for Biomass Valorization, RSC Energy and Environment Series. 2015. p. 151–73. RSC. See: http://pubs.rsc.org/en/Content/eBook/978-1-84973-801-9 [chapter 7].

[39] Venderbosch R.H., Heeres H.J. Pyrolysis oil stabilisation by catalytic hydrotreatment. Biofuel's Engineering Process Technology. 2011. 385. www.intechopen.com.

[40] Elliott D.C., Baker E.G., Piskorz J., Scott D.S., Solantausta Y. Production of liquid hydrocarbon fuels from peat. Energy Fuel 1988;2:234–5.

[41] Elliott D.C., Oasmaa A. Catalytic hydrotreating of black liquor oils. Energy Fuel 1991;5:102–9.

[42] Yin W., Kloekhorst A., Venderbosch R.H., Bykova M.V., Khromova S.A., Yakovlev V.A., Heeres H.J. Catalytic hydrotreatment of fast pyrolysis liquids in batch and continuous set-ups using a bimetallic Ni–Cu catalyst with a high metal content. Cat Sci Technol 2016;6:5899–915.

[43] Busetto L., Fabbri D., Mazzoni R., Salmi M., Torri C., Zanotti V. Application of the Shvo catalyst in homogeneous hydrogenation of bio-oil obtained from pyrolysis of white poplar: new mild upgrading conditions. Fuel 2011;90:1197–207.

[44] Hoffer B.W., Crezee E., Devred F., Mooijman P.R.M., Sloof W.G., Kooyman P.J., van Langeveld A.D., Kapteijn F., Moulijn J.A. The role of the active phase of Raney-type Ni catalysts in the selective hydrogenation of D-glucose to D-sorbitol. Appl Catal A Gen 2003;253:437–52.

[45] Hoffer B.W., Crezee E., Mooijman P.R.M., van Langeveld A.D., Kapteijn F., Moulijn J.A. Carbon supported Ru catalysts as promising alternative for Raney-type Ni in the selective hydrogenation of D-glucose. Catal Today 2003;79–80:35–41.

[46] Makkee M., Kieboom A.P.G., van Bekkum H. Hydrogenation of d-fructose and d-fructose/d-glucose mixtures. Carbohydr Res 1985;138:225–36.

[47] Bindwal A.B., Vaidya P.D. Kinetics of aqueous-phase hydrogenation of levoglucosan over Ru/C catalyst. Ind Eng Chem Res 2013;52:17781–9.

[48] Kobayashi H., Ito Y., Komanoya T., Hosaka Y., Dhepe P.L., Kasai K., Hara K., Fukuoka A. Synthesis of sugar alcohols by hydrolytic hydrogenation of cellulose over supported metal catalysts. Green Chem 2011;13:326–33.

[49] Kobayashi H., Matsuhashi H., Komanoya T., Hara K., Fukuoka A. Transfer hydrogenation of cellulose to sugar alcohols over supported ruthenium catalysts. Chem Commun 2011;47:2366–8.

[50] Wildschut J., Iqbal M., Mahfud F.H., Cabrera I.M., Venderbosch R.H., Heeres H.J. Insights in the hydrotreatment of fast pyrolysis oil using a ruthenium on carbon catalyst. Energy Environ Sci 2010;3:962–70.

[51] Yakovlev V.A., Bykova M.V., Khromova S.A. Stability of nickel-containing catalysts for hydrodeoxygenation of biomass pyrolysis products. Catal Ind 2012;4:324–39.

[52] Ardiyanti A.R., Bykova M.V., Khromova S.A., Yin W., Venderbosch R.H., Yakovlev V.A., Heeres H.J. Ni-based catalysts for the hydrotreatment of fast pyrolysis oil. Energy Fuel 2016;30:1544–54.

[53] Elliott D.C., Hart T.R., Neuenschwander G.G., Rotness L.J., Olarte M.V., Zacher A.H., Solantausta Y. Catalytic hydroprocessing of fast pyrolysis bio-oil from pine sawdust. Energy Fuel 2012;26:3891–6.

[54] Mortensen P.M., Grunwaldt J.D., Jensen P.A., Knudsen K.G., Jensen A.D. A review of catalytic upgrading of bio-oil to engine fuels. Appl Catal A Gen 2011;407:1–19.

[55] Navalikhina M.D., Krylov O.V. Heterogeneous catalysts of hydrogenation. Russ Chem Rev 1998;67:587–616.

[56] Nan W., Krishna C.R., Kim T.J., Wang L.J., Mahajan D. Catalytic upgrading of switchgrass-derived pyrolysis oil using supported ruthenium and rhodium catalysts. Energy Fuel 2014;28:4588–95.

[57] Reyhanitash E., Tymchyshyn M., Yuan Z., Albion K., van Rossum G., Xu C. Upgrading fast pyrolysis oil via hydrodeoxygenation and thermal treatment: effects of catalytic glycerol pretreatment. Energy Fuel 2014;28:1132–8.

[58] Reyhanitash E., Tymchyshyn M., Yuan Z., Albion K., van Rossum G., Xu C. Hydrotreatment of fast pyrolysis oil: Effects of esterification pre-treatment of the oil using alcohol at a small loading. Fuel 2016;179:45–51.

[59] Xu X., Zhang C., Zhai Y., Liu Y., Zhang R., Tang X. Upgrading of bio-oil using supercritical 1-butanol over a Ru/C heterogeneous catalyst: role of the solvent. Energy Fuel 2014;28:4611–21.

[60] Xiong W.-M., Fu Y., Zeng F.-X., Guo Q.-X. An in situ reduction approach for bio-oil hydroprocessing. Fuel Process Technol 2011;92:1599–605.

[61] Zhang X., Long J., Kong W., Zhang Q., Chen L., Wang T., Ma L., Li Y. Catalytic upgrading of bio-oil over Ni-based catalysts supported on mixed oxides. Energy Fuel 2014;28:2562–70.

[62] Boscagli C., Raffelt K., Zevaco T.A., Olbrich W., Otto T.N., Sauer J., Grunwaldt J.-D. Mild hydrotreatment of the light fraction of fast-pyrolysis oil produced from straw over nickel-based catalysts. Biomass Bioenergy 2015;83:525–38.

[63] Elliott D.C., Wang H., Rover M., Whitmer L., Smith R., Brown R. Hydrocarbon liquid production via catalytic hydroprocessing of phenolic oils fractionated from fast pyrolysis of red oak and corn stover. ACS Sustain Chem Eng 2015;3:892–902.

[64] French R.J., Black S.K., Myers M., Stunkel J., Gjersing E., Iisa K. Hydrotreating the organic fraction of biomass pyrolysis oil to a refinery intermediate. Energy Fuel 2015;29:7985–92.

[65] Wang H., Male J., Wang Y. Recent advances in hydrotreating of pyrolysis bio-oil and its oxygen-containing model compounds. ACS Catal 2013;3:1047–70.

[66] Robinson A.M., Hensley J.E., Medlin J.W. Bifunctional catalysts for upgrading of biomass-derived oxygenates: A review. ACS Catal 2016;6:5026–43.

[67] Van Santen R.A., Van Leeuwen P.W.N.M., Moulijn J.A., Averill B.A. Catalysis: An integrated Approach. 2nd, revised and enlarged ed. Elsevier, Amsterdam; 2000.

[68] Wildschut J., Melián-Cabrera I., Heeres H.J. Catalyst studies on the hydrotreatment of fast pyrolysis oil. Appl Catal B Environ 2010;99:298–306.

[69] Elkasabi Y., Mullen C.A., Pighinelli A.L.M.T., Boateng A.A. Hydrodeoxygenation of fast-pyrolysis bio-oils from various feedstocks using carbon-supported catalysts. Fuel Process Technol 2014;123:11–8.
[70] de Miguel Mercader F., Groeneveld M.J., Kersten S.R.A., Geantet C., Toussaint G., Way N.W.J., Schaverien C.J., Hogendoorn K.J.A. Hydrodeoxygenation of pyrolysis oil fractions: process understanding and quality assessment through co-processing in refinery units. Energy Environ Sci 2011;4:985–97.
[71] Mullen C.A., Boateng A.A., Reichenbach S.E. Hydrotreating of fast pyrolysis oils from protein-rich pennycress seed presscake. Fuel 2013;111:797–804.
[72] Capunitan J.A., Capareda S.C. Hydrotreatment of corn stover bio-oil using noble metal catalysts. Fuel Process Technol 2014;125:190–9.
[73] Wildschut J., Mahfud F.H., Venderbosch R.H., Heeres H.J. Hydrotreatment of fast pyrolysis oil using heterogeneous noble-metal catalysts. Ind Eng Chem Res 2009; 48:10324–34.
[74] Ardiyanti A.R., Gutierrez A., Honkela M.L., Krause A.O.I., Heeres H.J. Hydrotreatment of wood-based pyrolysis oil using zirconia-supported mono-and bimetallic (Pt, Pd, Rh) catalysts. Appl Catal A Gen 2011;407:56–66.
[75] J. A. Hogendoorn, S. R. A. Kersten, L. Meesala, F. de Miguel Mercader, A. J. A. Moppi, C. J. Schaverien, Patent EP2504412 A1, 2012.
[76] French R.J., Stunkel J., Black S., Myers M., Yung M.M., Iisa K. Evaluate impact of catalyst type on oil yield and hydrogen consumption from mild hydrotreating. Energy Fuel 2014;28:3086–95.
[77] Elliott D.C., Hart T.R., Neuenschwander G.G., Rotness L.J., Zacher A.H. Catalytic hydroprocessing of biomass fast pyrolysis bio-oil to produce hydrocarbon products. Environ Prog Sustain Energy 2009;28:441–9.
[78] Bi Y., Wang G., Shi Q., Xu C., Gao J. Compositional changes during hydrodeoxygenation of biomass pyrolysis oil. Energy Fuel 2014;28:2571–80.
[79] Wang H., Lee S.-J., Olarte M.V., Zacher A.H. Bio-oil stabilization by hydrogenation over reduced metal catalysts at low temperatures. ACS Sustain Chem Eng 2016;4:5533–45.
[80] D. C. Elliot, A. Oasmaa, Patent EP2707460 A1, 2012.
[81] Olarte M.V., Zacher A.H., Padmaperuma A.B., Burton S.D., Job H.M., Lemmon T.L., Swita M.S., Rotness L.J., Neuenschwander G.N., Frye J.G., Elliott D.C. Stabilization of softwood-derived pyrolysis oils for continuous bio-oil hydroprocessing. Top Catal 2016;59:55–64.
[82] Baldauf W., Balfanz U., Rupp M. Upgrading of flash pyrolysis oil and utilization in refineries. Biomass Bioenergy 1994;7:237–44.
[83] Xu Y., Wang T., Ma L., Zhang Q., Liang W. Upgrading of the liquid fuel from fast pyrolysis of biomass over MoNi/γ-Al_2O_3 catalysts. Appl Energy 2010;87:2886–91.
[84] Xu Y., Wang T., Ma L., Zhang Q., Wang L. Upgrading of liquid fuel from the vacuum pyrolysis of biomass over the Mo–Ni/?-Al_2O_3 catalysts. Biomass Bioenergy 2009;33:1030–6.
[85] Ardiyanti A.R., Khromova S.A., Venderbosch R.H., Yakovlev V.A., Melián-Cabrera I.V., Heeres H.J. Catalytic hydrotreatment of fast pyrolysis oil using bimetallic Ni–Cu catalysts on various supports. Appl Catal A Gen 2012;449:121–30.
[86] Ardiyanti A.R., Khromova S.A., Venderbosch R.H., Yakovlev V.A., Heeres H.J. Catalytic hydrotreatment of fast pyrolysis oil using bimetallic Ni–Cu catalysts on various supports. Appl Catal B Environ 2012;117–118:105–17.
[87] Cheng S., Wei L., Zhao X., Kadis E., Cao Y., Julson J., Gu Z. Hydrodeoxygenation of prairie cordgrass bio-oil over Ni based activated carbon synergistic catalysts combined with different metals. New Biotechnol 2016;33:440–8.

[88] Zhang X., Wang T., Ma L., Zhang Q., Jiang T. Hydrotreatment of bio-oil over Ni-based catalyst. Bioresour Technol 2013;127:306–11.
[89] Zhou M., Tian L., Niu L., Li C., Xiao G., Xiao R. Upgrading of liquid fuel from fast pyrolysis of biomass over modified Ni/CNT catalysts. Fuel Process Technol 2014;126:12–8.
[90] Dang Q., Luo Z., Zhang J., Wang J., Chen W., Yang Y. Experimental study on bio-oil upgrading over catalyst in supercritical ethanol. Fuel 2013;103:683–92.
[91] Oh S., Hwang H., Choi H.S., Choi J.W. Investigation of chemical modifications of micro-and macromolecules in bio-oil during hydrodeoxygenation with Pd/C catalyst in supercritical ethanol. Chemosphere 2014;117:806–14.
[92] Chen W., Luo Z., Yu C., Li G., Yang Y., Zhang H. Upgrading of bio-oil in supercritical ethanol: Catalysts screening, solvent recovery and catalyst stability study. J Supercrit Fluids 2014;95:387–93.
[93] Zhang J., Luo Z., Dang Q., Wang J., Chen W. Upgrading of bio-oil over bifunctional catalysts in supercritical monoalcohols. Energy Fuel 2012;26:2990–5.
[94] Hilten R., Weber J., Kastner J.R. Continuous upgrading of fast pyrolysis oil by simultaneous esterification and hydrogenation. Energy Fuel 2016;30:8357–68.
[95] Oasmaa A., Kuoppala E., Selin J.-F., Gust S., Solantausta Y. Fast pyrolysis of forestry residue and pine. 4. Improvement of the product quality by solvent addition. Energy Fuel 2004;18:1578–83.
[96] Zhang S., Yan Y., Li T., Ren Z. Upgrading of liquid fuel from the pyrolysis of biomass. Bioresour Technol 2005;96:545–50.
[97] Sheu Y.-H.E., Anthony R.G., Soltes E.J. Kinetic studies of upgrading pine pyrolytic oil by hydrotreatment. Fuel Process Technol 1988;19:31–50.
[98] Scholze B., Meier D. Characterization of the water-insoluble fraction from pyrolysis oil (pyrolytic lignin). Part I. PY–GC/MS, FTIR, and functional groups. J Anal Appl Pyrolysis 2001;60:41–54.

Further Reading

[99] Bridgwater A.V. Biomass Bioenergy 2012;38:68–94.
[100] Talmadge M.S., Baldwin R.M., Biddy M.J., McCormick R.L., Beckham G.T., Ferguson G.A., Czernik S., Magrini-Bair K.A., Foust T.D., Metelski P.D., Hetrick C., Nimlos M.R. Green Chem 2014;16:407–53.
[101] Vinu R., Broadbelt L.J. Energy Environ Sci 2012;5:9808–26.
[102] Liu C., Wang H., Karim A.M., Sun J., Wang Y. Chem Soc Rev 2014;43:7594–623.
[103] Zacher A.H., Olarte M.V., Santosa D.M., Elliott D.C., Jones S.B. Green Chem 2014;16:491–515.

Coprocessing of pyrolysis oil in refineries

9

D. Castello, L. Rosendahl
Aalborg University, Aalborg, Denmark

9.1 Introduction

Fast pyrolysis is a fundamental technology in the field of biomass valorisation into energy, and it can be considered as a first upstream treatment of biomass, in view of further processing. Indeed, biomass is generally a solid with relatively low energy density and substantially spread over wide areas. As a result, it is quite uneconomical to handle and transfer it over long distances. Moreover, the fact that biomass is a solid limits the possibilities for its utilisation and further processing. A prior liquefaction step is thus beneficial in order to meet these issues.

Fast pyrolysis is based on reacting biomass at high temperature (usually around 500°C) in the absence of oxygen, with very fast heating rates, in the order of 1000°C/s. Under these conditions, biomass constituents such as cellulose, hemicellulose, and lignin undergo complex cracking, depolymerisation, and oligomerisation reactions, leading to smaller organic compounds to be formed in the vapour phase, along with solids and permanent gases. After condensing such vapours, a liquid product is obtained, which is commonly referred to as 'bio-oil'. It is constituted of a complex mixture of hydrocarbons, anhydrosugars, aldehydes, ketones, phenols, and organic acids, and it normally appears as a brownish emulsion with a relatively high oxygen content: usually around 35% on dry basis. Some authors defined this product as a sort of 'liquid biomass', as its elemental composition resembles that of the original feedstock that was utilised for its production [1].

The high presence of oxygenates in bio-oil makes its utilisation as a fuel quite problematic. It is still possible to burn it into furnaces or industrial boilers, but its usage in internal combustion engines or turbines is extremely complicated, due to its instability. Oxygenated species, indeed, tend to polymerise when heated, even at relatively low temperature. Hence, during the normal operation of an internal combustion engine, they would form undesired polymerisation products, seriously damaging the mechanical parts. Polymerisation may even take place at room temperature, fostered by the high acidity of bio-oil. Storage is indeed a relevant issue. Furthermore, a high content of oxygen implies a high polarity, which makes bio-oil immiscible with conventional fuels. Therefore, it would not be even possible to make blends of fossil fuel and pyrolysis oil, unless proper treatment is carried out.

Better results can be obtained if pyrolysis takes place in the presence of a catalyst, for example, zeolites. Such process, the so-called catalytic fast pyrolysis, is able to produce a liquid with much better properties in terms of oxygen content. The

Direct Thermochemical Liquefaction for Energy Applications. https://doi.org/10.1016/B978-0-08-101029-7.00008-4

molecules in the pyrolysis vapours undergo several reactions, mainly dehydration, decarboxylation, and decarbonylation, which are able to reduce the oxygen content of the produced oil. As a result, catalytic pyrolysis oil (CPO) can have an oxygen content even lower than 10% [2], with usual values around 20%–25% [3]. Although it represents a significant reduction, the values are still too high to be comparable with standard fossil fuels, which hardly have any oxygenates. Again, the obtained liquid must be properly treated in order to be utilised as a fuel.

The most followed strategy for bio-oil upgrading is that of hydrotreatment, that is, processing pyrolysis oil with gaseous hydrogen in the presence of an appropriate catalyst. Such process is able to remove oxygen in the form of H_2O, thus increasing the hydrocarbon content in the fuel. In Chapter 8, hydrotreatment is extensively discussed. According to the severity of the process, that is, temperature, pressure, and, most importantly, the amount of H_2 used, different levels of deoxygenation can be achieved. It is even possible to remove nearly all the oxygen in the feed, thus producing a hydrocarbon mixture, when relevant H_2 amounts are involved.

From the practical point of view, hydrotreating bio-oil to produce oxygen-free liquids would imply the construction of dedicated units, with massive supply of H_2. This perspective presents a number of weak points. First of all, relying on substantial amounts of external hydrogen would result in very high operating costs, which negatively influence the price of the final product. Moreover, as far as capital costs are involved, the size of the upgrading facilities should be large enough to allow reasonable returns and break-even for the investors. Large, centralised, and complex units can be envisaged, which contrasts with the concept of pyrolysis as a relatively simple and distributed technology, useful to comply with the widespread nature of biomass.

Furthermore, the sole hydrotreating upgrading is not sufficient by itself to produce a fuel ready to be distributed and utilised on the market. The product from pyrolysis oil hydrotreating has indeed a relatively large molecular-weight distribution, with a consistent presence of heavy fractions. In order to obtain fuels that could be utilised in current engines, several other operations would be needed, for example, fractionation, cracking, and isomerisation. This would add further complexity (and costs) to a hypothetical biorefinery.

Finally, the construction of new large facilities for the production of upgraded biofuels is deemed to require a considerable amount of time. This would contrast with the general policy addresses, which are setting demanding deadlines for the implementation of a higher share of biofuels on the market. Building specific facilities for biofuels refining is thus not a ready-to-use approach, and it can be only seen as a long-term perspective for the future.

9.1.1 The coprocessing approach: Integration in the existing refineries

A more convenient and ready-to-use approach would consist in utilising the existing fossil oil refineries to perform all or part of the process for the production of biofuels. Many operations required for bio-oil upgrading have indeed many points in common with those carried out in refineries. Bio-oil could be then potentially fed to specific units in the refinery along with traditional fossil streams, without or possibly with only minor modifications to the processes. From a technological point of view, coprocessing bio-oil and traditional oil could represent an important pathway towards biofuels implementation on the market.

In Fig. 9.1, a simplified scheme of the processes taking place in a refinery is depicted. Crude oil first undergoes atmospheric fractionated distillation, in order to separate its different fractions according to their boiling points: liquefied petrol gas (LPG), gasoline, kerosene, diesel, and residue. These are the so-called 'straight-run' products. The produced fractions must normally undergo some conversion treatments in order to increase their quality and comply with regulations. For example, hydrocarbons in the gasoline range need reforming in order to increase aromaticity and thus the octane number. Moreover, all fractions generally need a specific treatment (hydrotreating) in order to remove impurities. The residue from atmospheric distillation is fed to another distillation column operated under vacuum. Low pressure allows lowering the boiling points of the constituents, thus enabling better separation possibilities. The light product of vacuum distillation is denominated vacuum gas oil (VGO), and it is generally cracked into smaller hydrocarbons.

In this scheme, pyrolysis oil, possibly after a mild stabilisation treatment (catalytic pyrolysis or mild hydrodeoxygenation), could be inserted in different locations. For example, it could be fed to the hydrotreater along with the gas oil from the column (straight-run gas oil, SRGO). The crude oil fed to a refinery normally includes other elements than carbon and hydrogen: the so-called heteroatoms. The most abundant one is sulphur, which must be removed due to environmental requirements and for potential corrosion, along with nitrogen and oxygen. Following distillation, such heteroatoms tend to accumulate in the heaviest fractions, and thus, they are found in relevant amounts. Hydrotreating removes such heteroatoms by reaction with H_2 on the surface of a catalyst. Such units could be then used to treat a blend of fossil and bio-oil, as it will be discussed in Section 9.3.

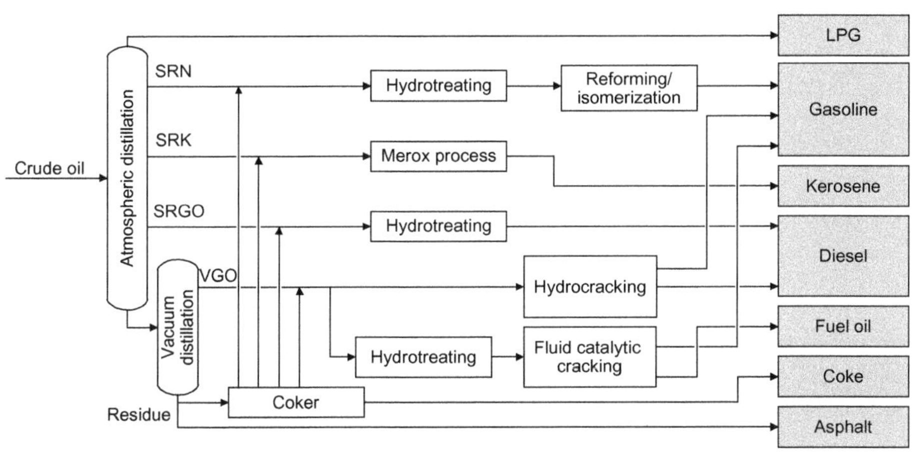

Fig. 9.1 Simplified process scheme of a standard oil refinery.

Another point where pyrolysis oil can be blended is represented by the fluid catalytic cracker (FCC). This unit is utilised to crack VGO into smaller hydrocarbons, in order to enhance the production of gasoline. The reason why FCC is regarded as an interesting unit to coprocess bio-oil with still a quite relevant amount of oxygenates is less evident than for hydrotreating. It is due to the activity of the catalyst used in the FCC process, usually zeolites, which are able to promote deoxygenation reactions. Moreover, in such units, coprocessing is very beneficial, as it can rely on synergetic effects with the fossil stream. In other words, hydrogen transfer from VGO to bio-oil is taking place, thus helping the deoxygenation process. This phenomenon will be more extensively presented in Section 9.2.

Although coprocessing can be evidently competitive in many aspects, it should be also considered that the introduction of a stream still with a relevant fraction of oxygenates can significantly alter the processes, which were actually developed for fossil oil. Indeed, in a coprocessing perspective, the refinery units need to accomplish different purposes than those for which they were designed. For example, while for traditional hydrotreating the main aim is the removal of sulphur, for cohydrotreating the simultaneous oxygen removal becomes relevant as well. This could lead to unsatisfactory performance of the unit. Similar drawbacks could arise in co-FCC, due to extensive coking caused by oxygenates. Such problematics will be also reviewed in the present chapter.

9.2 Coprocessing in fluid catalytic crackers (co-FCC)

The largest part of the available literature in the field of coprocessing has been dedicated to fluid catalytic cracking (FCC). As it was mentioned, the FCC unit is utilised to crack the distillate from the vacuum column (VGO). Such stream is constituted by heavy hydrocarbon compounds with boiling points higher than 300°C. The aim of the FCC treatment is that of converting VGO into smaller hydrocarbons in the gasoline range, that is, C_6–C_8 hydrocarbons. As from the 1940s, FCC has experienced increasing popularity, and nowadays, it is a standard unit, present in a large number of refineries [4]. This has been also caused by the large market request for gasoline, which has represented a strong driving force for the development of this technology.

FCC makes use of acidic microporous catalysts belonging to the class of zeolites, such as ZSM-5 and H-zeolites. Besides cracking reactions, the catalyst is able to promote the aromatisation of hydrocarbons, which results in the production of H_2. This is a beneficial effect in the refinery context, as a higher concentration of aromatics determines a higher octane number in the produced gasoline.

A crucial aspect in FCC operations is to avoid overcracking. Indeed, reactions over zeolite catalysts are very fast, and they tend to consecutively crack the products into smaller and smaller ones. This could result in the production of too small molecules, for example, in the LPG range, and eventually coke and water, which are obviously not useful for fuel applications. Space velocities are therefore kept very high in order to maximise the selectivity towards hydrocarbons in the gasoline range. Despite such short contact time, coke is still produced on the surface of the zeolites.

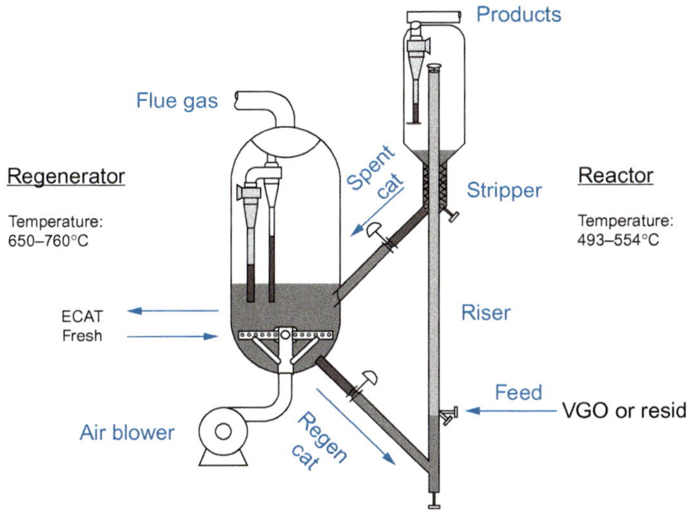

Fig. 9.2 Simplified representation of the fluid catalytic cracking (FCC) process.
From Vogt ETC, Weckhuysen BM. Fluid catalytic cracking: recent developments on the grand old lady of zeolite catalysis. Chem Soc Rev 2015;44(20):7342–70. Available at: http://xlink.rsc.org/?DOI=C5CS00376H—published by the Royal Society of Chemistry.

Coke deposition causes almost instantaneous deactivation of the catalyst, which must be continuously regenerated.

The requirements in terms of short contact time and continuous regeneration of the catalyst are both achieved by utilising a circulating fluidised bed, hence the name of the technology. In Fig. 9.2, a typical configuration for FCC is shown.

The FCC unit involves a riser reactor, where the cracking reactions take place, followed by a regenerator. The feed (VGO) enters from the bottom of the reactor, generally after being preheated. Here, the contact with the catalyst takes place, and both catalyst and hot vapours move co-currently from the bottom to the top of the unit. Reaction takes place at temperatures between 510°C and 565°C and with very short contact times, in the order of 0.1–1 s [5]. The spent catalyst at the top of the riser is separated from the vapours and is conveyed to a regenerator, where an oxidant atmosphere burns the coke on catalyst and makes it available for a new cycle. The thermal energy produced by such combustion heats up the cracking reactor, thus making the process self-sustained.

Different hydrocarbon fractions are produced from the process: light gases, LPG, gasoline, light cycle oil (LCO), and decanted oil (DCO). Gasoline is the preferred product, and it is transferred to the gasoline blending pool. LCO is constituted by hydrocarbons in the diesel range (boiling point: 200–340°C). Due to the effect of the catalyst, LCO has a high degree of aromaticity, which results in a low cetane number. Its applicability for the production of gasoline is therefore limited: it can be blended with SRGO to a minor extent, but it is mostly utilised as a low-grade fuel oil together with DCO.

9.2.1 Catalytic deoxygenation and synergetic effects

The reason why FCC is looked as a coprocessing alternative for oxygenated bio-oil is that the reaction over zeolite catalysts is able to remove the oxygen by means of dehydration, decarbonylation, and decarboxylation reactions (removal of oxygen in the form of H_2O, CO, and CO_2, respectively). The ability of zeolites to promote deoxygenation is quite well-known in the literature. First evidences were obtained for the bio-oil catalytic upgrading process. For example, Adjaye and Bakhshi [6] processed pyrolysis oil through a fixed bed of HZSM-5 and other porous catalysts, and they found that zeolites are very effective for bio-oil deoxygenation. They also proposed a reaction scheme, where thermal and thermocatalytic pathways are taken into account (Fig. 9.3). Similar results were also reported by Vitolo et al. [7] for different kinds of zeolites.

A substantially analogous process is represented by ex situ catalytic pyrolysis, where vapours are contacted with the catalyst in a separate unit downstream the

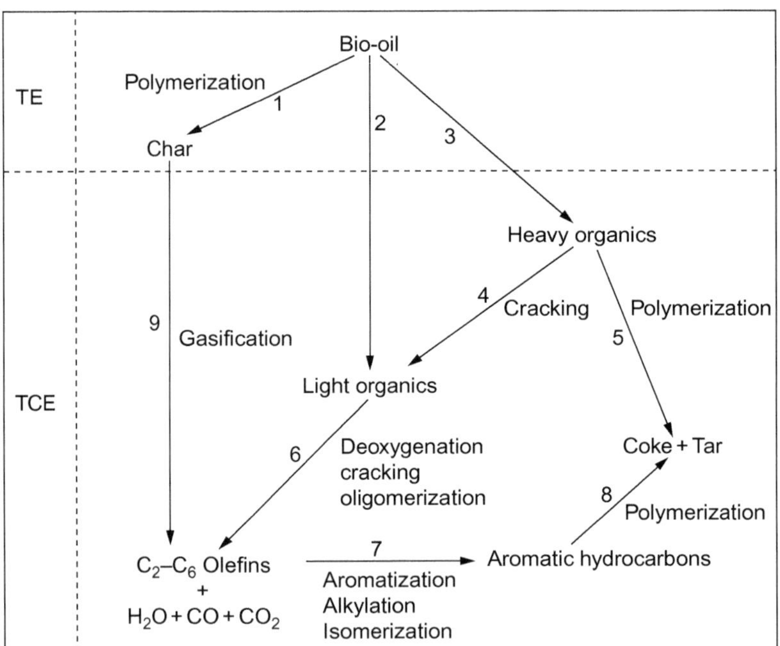

Fig. 9.3 Reaction pathway for the conversion of bio-oil over acidic zeolite catalysts. *TE*, thermal effect; *TCE*, thermocatalytic effect.
From Adjaye JD, Bakhshi NN. Production of hydrocarbons by catalytic upgrading of a fast pyrolysis bio-oil. Part I: conversion over various catalysts. Fuel Process Technol 1995;45(3):161–83. Available at: http://apps.isiknowledge.com/InboundService.do?Func=Frame&product=WOS&action=retrieve&SrcApp=Papers&UT=A1995TG00300001&SID=U2lAJ5eoLkK8NEl5m85&Init=Yes&SrcAuth=mekentosj& mode=FullRecord& customersID=mekentosj& DestFail=http://access.isiproducts.com/custom_i. Copyright 1995, with permission from Elsevier.

pyrolysis reactor. In this field, the largest part of literature studies has been conducted with zeolite catalysts, which proved to be effective for deoxygenation. The experiments highlighted that dehydration and decarbonylation are the preferred pathways for oxygen removal [8].

Compared with the above-mentioned technologies, co-FCC has another significant advantage: the occurrence of hydrogen transfer from the fossil stream to bio-oil. It is indeed known that bio-oil is generally hydrogen-deficient, showing an H/C ratio much lower than hydrocarbons. This can be also put into relation with the concept of effective hydrogen index (EHI), proposed by Chang et al. [9] and reformulated by Chen et al. [10]. EHI is defined as

$$EHI = \frac{\mathrm{H} - 2\mathrm{O} - 3\mathrm{N} - 2\mathrm{S}}{\mathrm{C}} \tag{9.1}$$

where H, O, N, S, and C are the mole fractions of hydrogen, oxygen, nitrogen, sulphur, and carbon, respectively. EHI is an important parameter for the upgrading of an oil, as it corrects the actual amount of hydrogen in the feedstock with the amount of hydrogen that should be spent for the removal of such heteroatoms as H_2O, NH_3, and H_2S. It has been found that, during catalytic upgrading operations, feedstock with EHI lower than or equal to 1 are subject to massive coke production, while this drawback is not observed when EHI is considerably higher than 1 [10]. This can be explained with the low availability of hydrogen, which eventually results in the production of polyaromatic structures and, hence, coke.

For a hydrocarbon refinery stream, the EHI is normally much higher than 1. Indeed, aliphatic hydrocarbons approach values of EHI around 2, while lower values are attained for aromatics. Although some oxygenated species can show high values of EHI (e.g. methanol, EHI = 2), bio-oil has a very low EHI. For fast pyrolysis oil (FPO), values can be lower than 0.5. Higher values can be reached with catalytic pyrolysis oil (CPO) and hydrodeoxygenated oil (HDO): when the oxygen content is around 10%, EHI approaches 1. In any case, the cracking of bio-oils can most likely result into coking and hence in substantial failure.

Processing bio-oil/refinery streams blends instead of pure bio-oil is therefore a way to overcome this limitation. Indeed, the EHI of VGO is around 1.7 [11]. Blending with small amounts of bio-oil, that is, up to 20%, lowers the overall value of the EHI but still keeps it to a reasonable level above. In other words, VGO acts as an internal source of hydrogen to compensate the deficit of the bio-oil. In this way, corefining establishes a useful synergy between renewable and nonrenewable sources. Bio-oil could be then deoxygenated without the need of any external hydrogen source, while being cracked together with VGO into products in the gasoline range.

The possibility to utilise an internal and inexpensive source of hydrogen represents a strong point in favour of the co-FCC technology. Indeed, the H_2 uptake dramatically increases when high objectives in deoxygenation are required. In Fig. 9.4, data concerning H_2 consumption as a function of the degree of deoxygenation of pyrolysis oil are reported [12]. A higher degree of deoxygenation can be obtained by using higher amounts of hydrogen, with an increasing linear trend. However, for deep deoxygenation, the H_2 uptake grows visibly more than the stoichiometric request, as a

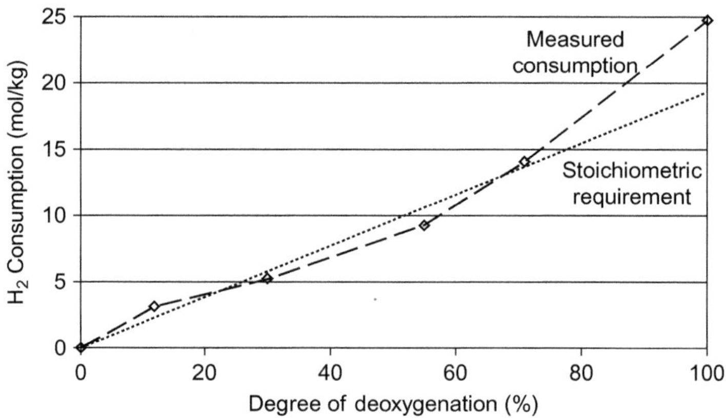

Fig. 9.4 Hydrogen consumption as a function of the degree of deoxygenation, for a bio-oil with an oxygen content of 31 wt%. Experimental data from Venderbosch et al. [12].
From Mortensen PM, et al. A review of catalytic upgrading of bio-oil to engine fuels. Appl Catal A Gen 2010;407(1–2):1–19. https://doi.org/10.1016/j.apcata.2011.08.046. Copyright 2011, with permission from Elsevier.

consequence of the increasing difficulty to convert the most recalcitrant oxygenates. Such relevant H_2 request could result into inacceptable higher costs.

All these aspects can explain the large interest that co-FCC has obtained in the field of drop-in biofuel production. It is indeed regarded as the most promising process for corefining pyrolysis oil, and it has definitely been the most investigated process in this field. There are several literature works dealing with the catalytic cracking of bio-oil with VGO over FCC catalysts, including tests with different kinds of biomass-derived oils, namely, thermal fast pyrolysis oils (FPOs) (i.e. obtained from a noncatalytic process), catalytic pyrolysis oil (CPO), and hydrodeoxygenated oil (HDO).

9.2.2 Co-FCC of different types of bio-oils

The literature in the field of co-FCC accounts for experimental studies carried out in various typologies of experimental apparatuses, with different typologies of bio-oils and blending ratios. In Table 9.1, the main literature works are summarised. For each study, the type of biofeedstock was reported, including its oxygen content on dry basis and the mixing ratio with the refinery stream, usually represented by VGO. In general, each experimental study reports the results of both the co-FCC test and a test with the sole VGO. Such comparison is of great importance, since a major scope of these studies is to understand whether the addition of bio-oil leads to significant changes or drawbacks in the operation of an FCC unit.

9.2.2.1 Hydrodeoxygenated oil (HDO)

In order of time, the first feedstock that was used for co-FCC has been hydrodeoxygenated oil (HDO). One of the first researches was carried out by Samolada et al. [23],

Table 9.1 Summary of the available literature studies on the co-FCC of different bio-oils

Feedstock	Ratio	Oxygen content	Co-FCC yields		FCC yields		Reference
			Gasoline	Coke	Gasoline	Coke	
FPO	10–20	38	36–42	8–10	40–42	7–8	[13]
FPO	5–10	33	37–42	5.6–7.2	40	6.5	[14]
FPO	5–20	32.0	29–35	4.2–5.5	44	5.6	[11]
FPO	20	36.5	17	10	16	5	[15]
CPO	10	27	~55	4–6	~50	2–4	[16]
CPO	15	22	44.4	6.8	44.0	7.1	[17]
CPO	10	19.5	40.7–42	7.5–8.4	40–42	7–8.2	[18]
CPO	20	22.0	19	10	16	5	[15]
HDO[a]	3.75	4.9	42–46	4–6	38–46	2–5.5	[19]
HDO	20	21	45–47	4–7	42–48	2–4	[20]
HDO	10	7–33	30–41	1.2–5.5	25–37	1.2–3.8	[21, 22]
HDO	5	10.0	32.5	4.2	44	5.6	[11]
HDO	20	22.0	18	8	16	5	[15]
HDO[b]	15	4.9	20–26	3–6	16–22	2–3	[23]
HDO	20	17–23	40.2–43.5	5.2–7.2	44	5.9	[24]

Fossil stream is represented by VGO, unless otherwise specified. All values are expressed as percentages. Oxygen content is on dry basis.
[a] A mixture of VGO/LCO/HDO 85/11.25/3.75 was used. FCC yields are referred to VGO/LCO 85/15.
[b] LCO was utilised as fossil stream.

who coprocessed a 15/85 mixture of HDO and LCO in a modified MAT (microactivity test) unit. Their results were quite encouraging, since they managed to obtain even an increase in the gasoline production, with an excellent octane number (around 95).

As it was mentioned, Samolada et al. operated at small scale, by means of the so-called microactivity test (MAT). This is a choice made by large part of literature studies in the field of co-FCC. MAT is a standard procedure in the oil and gas industry [25] that is aimed at testing the activity of a catalyst in view of its utilisation in an FCC unit. The device for MAT includes a feeder and a fixed bed where the catalyst is installed. It is possible to change the flow rate of the feed, in order to modify the catalyst-to-oil ratio. The liquid obtained from the process is then condensed and can be analyzed. Typically, the effluent undergoes a simulated distillation (SimDis) analysis that allows determining the different fractions according to their estimated boiling point. MAT provides a straightforward, standard way to compare different catalysts or feeds.

The next documented test with real bio-oil was carried out by Lappas et al. [19]. Here, the authors prepared a mixture of VGO, LCO, and HDO, where the effective amount of HDO in the overall mixture was only 3.75%. Nevertheless, the results showed very similar conclusions: the amount of gasoline is comparable with the case of 100% VGO, with only a slight increase in coke.

It is worth mentioning that such studies were conducted with a feedstock that already presented a reduced amount of oxygen, as a consequence of a quite strong hydrotreatment. On the other hand, one of the main purposes of coprocessing is that of limiting the extent of hydrotreatment in order to realise significant savings on the operating costs. To this purpose, many of the subsequent works with HDO were carried out with feedstock with a higher oxygen content. In this field, the work by Fogassy et al. [20,26] from IRCELYON (France), carried out in the frame of the EU project 'Biocoup', is very significant. Here, the authors ran experiments with a HDO at 21% oxygen, managing to coprocess it at 20/80 ratio with VGO. What they found is that the yields of gasoline do not seem to be affected by the introduction of oxygenates. As expected, coke yields were higher, going from 4% of pure VGO to 7%.

An important aspect they highlighted is the competition for hydrogen. Indeed, the authors found that hydrogen transfer from the refinery stream to bio-oil is the preferred mechanism for deoxygenation, and it is more relevant than decarboxylation and decarbonylation. As a result, they could appreciate lower yields of H_2 in the process gas than those observed with pure VGO. They also observed a higher degree of aromaticity in the products, which is consistent with the more aromatic nature of bio-oil.

Another significant work in co-FCC was carried out by de Miguel Mercader et al. [24] from the University of Twente (the Netherlands), as part of the same 'Biocoup' project. Here, the authors produced different types of hydrodeoxygenated pyrolysis oils at increasing severity conditions, obtaining liquids with oxygen concentrations from 28% to 17%. Such oils were then coprocessed with long residue in an MAT reactor at 20% blending ratio, obtaining similar gasoline yields as with the pure hydrocarbon feedstock. An increase in coke yields was observed for those oils with a higher oxygen content, while for the others, the variations were minimal. The authors also performed FCC runs with undiluted hydrotreated oils (those with the lowest oxygen content) and compared the results with those extrapolated from the coprocessing tests. Interestingly, coke yields from extrapolation were far more reduced than the measured ones. This is a strong argument for the hydrogen transfer synergetic effects, taking place during co-FCC operations.

9.2.2.2 Catalytic pyrolysis oil (CPO)

Although many studies have been carried out with HDO, in the last few years, some experimental activities were performed with CPO. Through this process, it is indeed possible to completely avoid the usage of external H_2 by contacting the hot vapours with a catalyst, usually represented by zeolites. In the literature, a few works are reported, where blending ratios ranging from 10% to 20% where achieved. In general, results are very similar to the case of HDO. Gasoline productions are comparable with those obtained from VGO. Coke yields are usually higher than the VGO case, but not in a very relevant way. In a few studies, they were even found to be equal or lower.

The first documented experimental work in this field was carried out by Agblevor et al. [17], as a result of a collaboration among Utah State University, Virginia Polytechnic Institute and State University, and BASF Inc. (the United States). In this study, the authors utilised an oil derived from a fractional catalytic pyrolysis process [27], with an oxygen content of 22% on dry basis. They tested blending at 15% with VGO and found that such oils are stable enough for coprocessing, observing a quite similar gasoline production with respect to the pure VGO case. Surprisingly, lower coke yields were obtained when processing the blend. The authors attributed this effect to the presence of steam and hydrogen generated by the process, which could moderate coke formation.

Thegarid et al. [16] compared the co-FCC of HDO and CPO with similar oxygen content. They found that, in general, the overall performance of these two feeds is quite modest, but significant differences can be found for the product quality. CPO is able to yield a higher amount of gasoline than both HDO and pure VGO. Conversely, a lower amount of hydrogen is produced. They concluded that the overall bio-oil yield of catalytic pyrolysis followed by co-FCC is higher than that obtained from HDO followed by co-FCC: around 30% vs 24%. This testifies the importance of the process pathway based on catalytic pyrolysis, which can be considered a promising approach for future developments of the biorefinery concept. In Fig. 9.5, the obtained yields of the main products from the co-FCC process are shown as a function of the catalyst-to-oil ratio. A blending ratio of 10% was used, for both HDO and CPO.

Recently, Wang et al. [18] from the Beijing University of Chemical Technology performed a co-FCC study of a CPO obtained with Fe/ZSM-5 catalyst. For their experimental campaign, they utilised a pilot-scale riser (7 mm inner diameter and 9 m height). Mixtures of CPO/VGO up to 10% were successfully processed, showing similar yields as pure VGO. Interestingly, they found out that the CPO/VGO blend is more crackable than pure VGO. The authors reported that, when blending with 15% CPO, blockage due to coking was observed. However, they also reported that this limitation could be overcome in real refineries due to the larger diameters of the riser units and to the possibility of a separate feeding line for bio-oil.

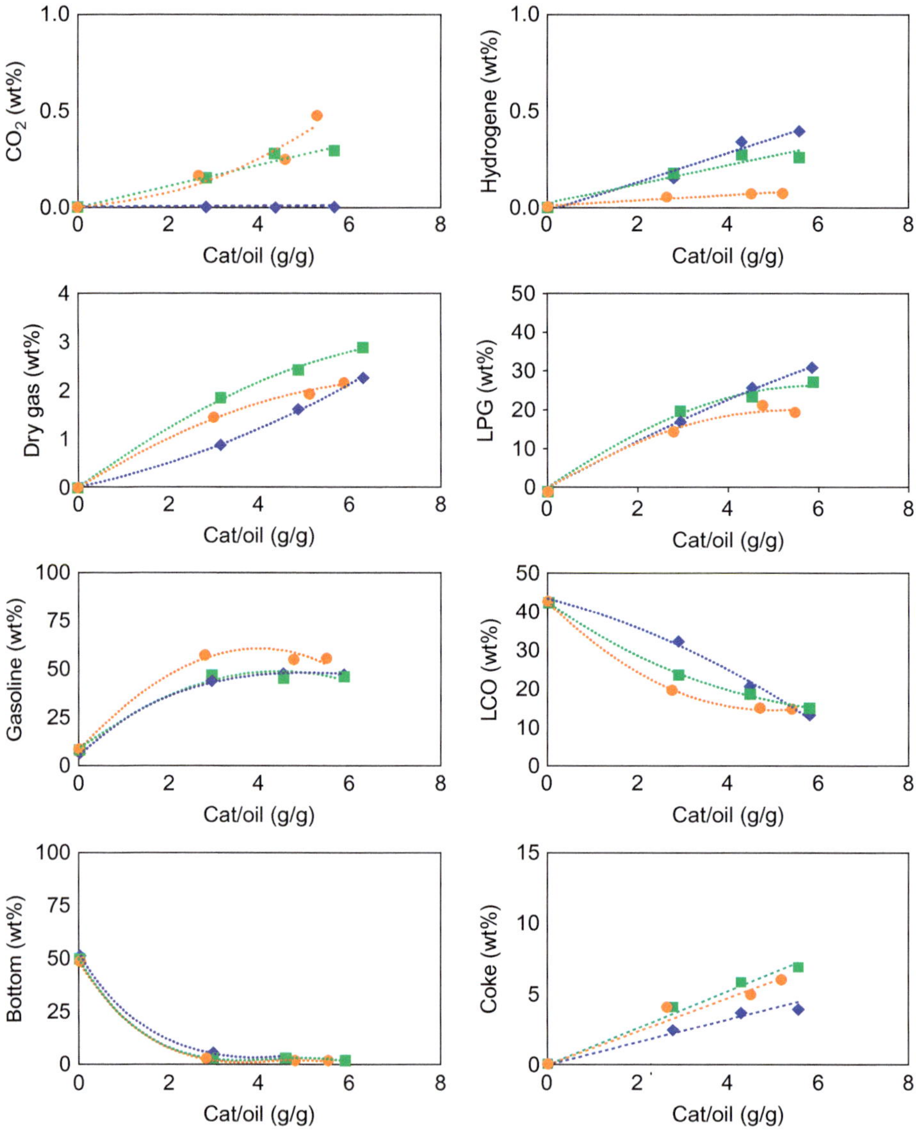

Fig. 9.5 Product distribution as a function of the catalyst-to-oil ratio for (◆) VGO processing, (■) HDO/VGO processing, and (●) CPO/VGO processing.
From Thegarid N. et al., 2014. Second-generation biofuels by co-processing catalytic pyrolysis oil in FCC units. Appl Catal B Environ 2014;145:161–66. https://doi.org//10.1016/j.apcatb.2013.01.019. Copyright 2013 with permission from Elsevier.

9.2.2.3 Thermal fast pyrolysis oil (FPO)

Although co-FCC is often considered for a feedstock that has already received a first gross deoxygenation treatment (hydrotreating or catalytic pyrolysis), some studies in the literature report experiments performed with thermal FPO, that is without any prior treatment. In this way, the possibility of a direct deoxygenation through co-FCC is explored. While for HDO and CPO the oxygen content never exceeds ~20%, in the case of untreated pyrolysis oil such percentages are definitely higher: between 32% and 38% on dry basis. This is due to the high presence of carbohydrates, aldehydes, and ketones that are commonly derived from the cellulosic fraction of biomass. Generally, for such feedstock, lower blending ratios are adopted, although results with a mixture at 20% are available.

The most relevant experimental activities with FPO have been carried out by the group of Pinho et al. from Petrobras (Brazil). Their experimental work is relevant also because they made use of a demonstration-scale FCC unit: an 18 m-high riser with a throughput of 150 kg/h and an amount of catalyst in the system of 450 kg. Their results are therefore very interesting in view of the industrial exploitation of this technology, since they were obtained on a scale comparable with practical applications.

An important issue of coprocessing FPO is that such liquid is practically immiscible with VGO, due to its high oxygen content and thus polarity. Moreover, it is quite thermally unstable, leading to extensive unwanted polymerisation once heated up even to quite low temperatures (50°C). Pinho et al. elegantly solved this issue by providing two separate feeding lines for VGO and bio-oil. In this way, they have the possibility to heat up the two liquids at two different temperatures, avoiding extensive coking. Bio-oil is thus injected through a nozzle below the injection point for VGO (see Fig. 9.6) at a temperature lower than 50°C. The possibility to preheat VGO at temperature higher than 200°C lowers its viscosity and improves the dispersion of bio-oil itself.

In their experimental work, de Rezende Pinho et al. [13] reported that, until 10% mixing, the process could be considered successful, yielding similar amounts of gasoline as with pure VGO. However, a blend at 20% resulted in a significant drop, with severe coke production. Unlike the experimental works conducted with HDO and CPO, the authors found out that, with FPO, more CO than CO_2 is produced, which indicates that decarbonylation is the preferred deoxygenation pathway.

In a subsequent work [14], the authors performed several coprocessing runs in order to prove the feasibility of the process, totalising around 400 h of operations over 18 months. However, the authors also reported that phenols were detected in the gasoline. Such amount increases with the increasing share of renewable feedstock that was processed. For pure VGO, measured total phenols were 2265 ppm, while for the 5/95 mixture, they were 3321 and 6657 ppm for the 10/90 blend. Despite being still quite low values, the overall effect of phenols presence in bio-based fuels should still be fully investigated. Indeed, on the one hand, such presence can be beneficial, as phenols slightly increase the octane number and act as antioxidant additives. On the other hand, they negatively affect the stability of gasoline, and some phenolic compounds may have boiling points even higher than the maximum allowed for gasoline (225°C). All these issues should be further investigated.

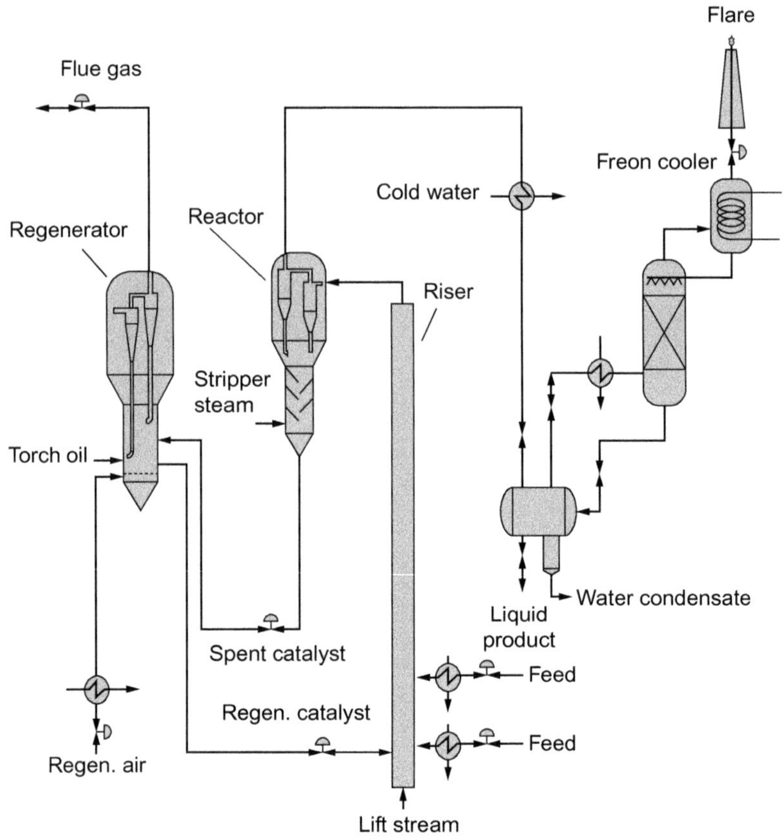

Fig. 9.6 Schematic representation of the demonstration-scale unit utilised by de Rezende Pinho et al. [13], with the two different lines for VGO and bio-oil feeds.
From Pinho ADR, et al. Co-processing raw bio-oil and gasoil in an FCC Unit. Fuel Process Technol 2015;131:159–66. https://doi.org//10.1016/j.fuproc.2014.11.008. Copyright 2015, with permission from Elsevier.

FPO was also utilised by Naik et al. [11], who tested mixtures from 5% to 20%. It should be pointed out that, in this work, FPO consists in the organic phase from the fractionate condensation of pyrolysis oil, where the aqueous phase was not collected. Despite the addition of FPO results in lower gasoline yields, coke yields are similar and even lower. However, they conclude that FPO is not suitable as a feed, since the FCC catalyst is not able to crack the lignin-derived oligomers. On the other hand, for lower blending ratios (5%–10%), FPO could be utilised for the production of light olefins, mainly in the LPG range.

An interesting comparison among FPO, CPO, and HDO was proposed by Lindfors et al. [15] from VTT, Finland. Here, the authors compared FPO, CPO, and HDO samples with similar oxygen contents (around 20%), operating with 20/80 oil/VGO mixtures. Also in this case, FPO was represented by the dry, organic-rich phase of pyrolysis oil, obtained by the partial condensation of the vapours at 66°C. They found

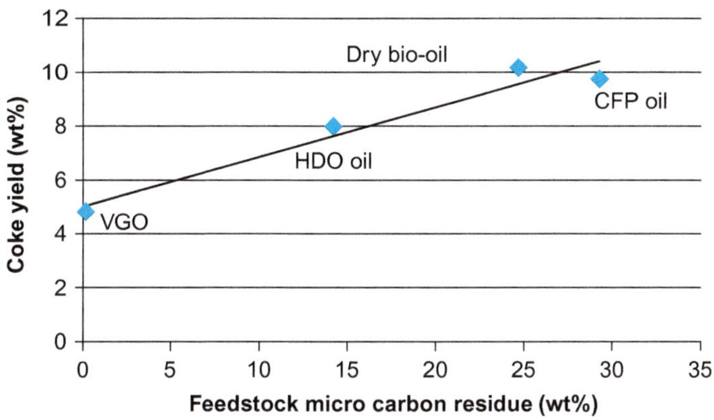

Fig. 9.7 Coke yield as a function of the microcarbon residue of the feedstock for mixture at 20% [15].
From Lindfors C, et al. Co-processing of dry bio-oil, catalytic pyrolysis oil, and hydrotreated bio-oil in a micro activity test unit. Energy Fuels 2015;29(6):3707–14. Copyright 2015 American Chemical Society.

out that, in general, the three feedstocks give similar gasoline yields. However, they highlighted a clear dependence of the coke production on the microcarbon residue (MCR) of the feedstock (Fig. 9.7). MCR is a standard test [28] utilised in the oil industry to determine the amount of solid residue that is obtained once the feedstock is slowly evaporated under inert atmosphere. For VGO, this value is generally close to zero, while for oxygenated feedstock it can be relevant, due to the occurrence of polymerisation reactions. MCR is probably a more comprehensive indicator for the FCC processability of a feedstock than the mere oxygen content. The authors concluded that CPO and HDO are preferable to FPO, since higher liquid yields can be obtained. Nevertheless, the chemical composition of the liquid products was found to be almost identical.

9.2.3 The amount of renewable carbon in the co-FCC product

An interesting question to be answered about corefining is the determination of the amount of renewable feedstock that is actually transferred in the final fuel product. Indeed, the produced fuel should incorporate a significant share of renewable feedstock. In other words, it should be ensured that the utilised cofeed (FPO, CPO, HDO) actively contributes to the production of gasoline and that it neither is merely converted into coke, water, and light gases nor remains largely unconverted, thus reducing the efficiency of the FCC unit. The quantification of the effective share of renewables in the final product is also fundamental in order to assess if the produced fuel can fulfil the requirements of environmental laws. Different regulations impose that a certain share of fuel should be constituted by renewable sources: for example, the European Union [29] imposes a target of 10% by 2020.

Such quantification can be roughly obtained through calculations, by assuming linear mixing, once known the output from the pure VGO tests. Though simple and straightforward, this method is not reliable. Indeed, the interactions and synergetic effects between the two streams result into a relevant degree of nonlinearity, which can make the calculated share of renewables in the product much different from the actual one. Therefore, this parameter should be determined experimentally.

The determination of the amount of renewable carbon in the product can be obtained by means of the ^{14}C analysis, also known as radiocarbon method. Carbon is naturally available in different isotopes. ^{14}C, that is, carbon with a mass number $Z = 14$, is quite abundant in nature. However, it naturally tends to decay in ^{14}N during a decay time of 5730 years. As a consequence, the carbon in fossil oil has a negligible amount of ^{14}C, as it dates back to millions of years, while the carbon in biomass presents an appreciable share of ^{14}C. Therefore, by measuring such carbon isotope in the final products from FCC, it is possible to obtain an accurate estimate on the actual amount of renewable carbon that was converted into usable fuel.

For co-FCC, the ^{14}C method was first applied by Fogassy et al. [30]. In this study, the authors performed co-FCC tests using a blend of VGO and HDO at 90/10, carried out in a fixed-bed quartz reactor. Not only they quantified ^{14}C in the liquid products, but also they measured it into all the products, thus being able to retrieve the distribution of carbon in each fraction. A major result was that, compared with a biocarbon content of 8.5% in the feeding mixture, the obtained gasoline had a renewable carbon content of 7.2%, therefore slightly lower. On the other hand, a significant higher share of biocarbon was found in the gas (10.6%) and especially in the coke (15.8%).

The radiocarbon technique can be also useful to obtain more insight in the production of coke, whose yield increases considerably when oxygenated species are utilised. Gueudré et al. [21] performed a ^{14}C analysis on the solid products of co-FCC of a 9/91 HDO/VGO mixture, and they obtained that not all the increase of coke can be attributed to the oxygenates. Indeed, while coke yields increase from 2 to around 3.8%, only 0.5% is represented by biocarbon (Fig. 9.8). The ^{14}C technique was thus useful to improve the understanding of the synergetic effects taking place during co-FCC.

The technique has been also applied to the co-FCC of thermal pyrolysis oil. De Rezende Pinho et al. [13,14] from Petrobras found that the amount of renewable carbon in the liquid product can be up to around 5% for a FPO/VGO mixture at 20%. However, it should be considered that, unlike the other literature works, here, FPO has a significant water content (around 32%). Hence, the actual amount of cofed

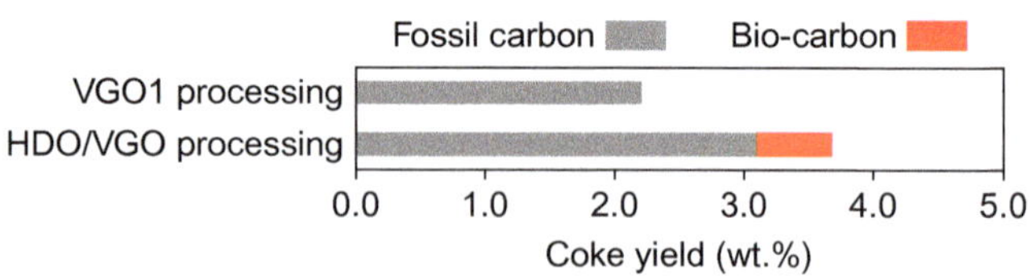

Fig. 9.8 Distribution of biocarbon and fossil carbon in the coke formed during the coprocessing of a blend HDO/VGO 9/91.
From Gueudré L, et al. Coke chemistry under vacuum gasoil/bio-oil FCC co-processing conditions. Catal Today 2015;257(P2):200–12. https://doi.org/10.1016/j.cattod.2014.09.001.

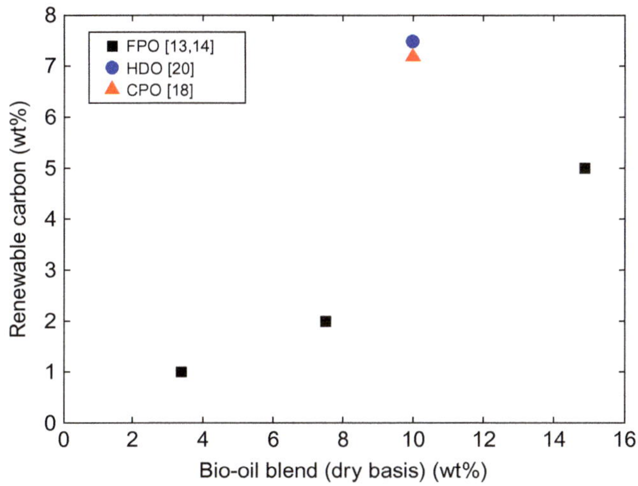

Fig. 9.9 Renewable carbon in the liquid product (obtained with the ^{14}C technique) as a function of the mixing percentage of bio-oil in the feed.

bio-organics is lower, compared to studies where a dry feedstock is utilised. Nevertheless, the authors estimated that around 30% of the fed biocarbon is transferred to gasoline, for mixtures of pyrolysis oil and VGO from 5/95 to 20/80.

As far as CPO is concerned, Wang et al. [18] applied the radiocarbon technique to the product of a co-FCC process with CPO/VGO at 10/90. Here, the authors obtained a renewable carbon content in the gasoline product of about 7%, thus, in line with the experiments using HDO [30].

The radiocarbon technique can thus be considered an interesting approach for quantifying the effectiveness of the co-FCC treatment and, in general, of the coprocessing technologies. The available data show that, when processing HDO and CPO, a higher share of biocarbon can be actually transferred in the final liquid product, with respect to FPO (Fig. 9.9). On the other hand, it should be also taken into account that the yields of HDO and CPO from raw biomass are much lower, as they come from upgrading operations. Therefore, a comprehensive evaluation should be conducted. Further experimental studies are needed in order to quantify the amount of renewable carbon when higher blending ratios are involved for HDO and CPO.

9.3 Cohydrotreating

Another way for coprocessing is represented by cohydrotreating. In this case, the idea is that of integrating the deoxygenation step in the hydrotreating units normally present in a refinery. In common hydrocarbon refineries, hydrotreating is indeed utilised for the removal of heteroatoms, namely, sulphur and nitrogen, which are removed as H_2S and NH_3. Moreover, it can be also utilised to reduce the aromatic content of the oil, by saturating the aromatic bonds. This is a useful operation for diesel production, because the reduction of the aromatics increases the cetane number and hence the quality of the fuel. In refineries, the removal of oxygen has a minor importance, compared with denitrification and desulphurisation. This is due to the very small amount of oxygen present in crude oil, while sulphur concentrations can be very relevant. The presence of sulphur in the oil represents a major issue. Indeed, due to the concern about air pollution, regulations impose very strict limits for this element in the fuels [31].

Coprocessing in hydrotreaters aims at utilising such units but adding the need for deoxygenation. The success of cohydrotreating relies in the simultaneous achievement of both deoxygenation and desulphurisation. As in the case of co-FCC, the process is thus aimed at a slightly different purpose than the one it was originally conceived for.

Compared with co-FCC, cohydrotreating has received much lower attention, at least for pyrolysis oil. Some more effort was done for cohydrotreating with vegetable oil, for the production of fuels in the diesel range. A comprehensive review of the studies in this field can be found in [32]. Apart from vegetable oil, some studies in the literature deal with model compounds, in order to understand the functioning of the process. A few experimental data are also available for pyrolysis oil, and they will be presented in Section 9.3.2.

9.3.1 Process understanding: Tests with model compounds

As mentioned, hydrotreating in standard refineries is aimed at the removal of heteroatoms by means of hydrogen. The most relevant species to be removed by hydroprocessing are nitrogen and especially sulphur. On the other hand, the most abundant element to be removed from pyrolysis oil is represented by oxygen. Therefore, when coprocessing is carried out, it should be ensured that the deoxygenation pathway does not hamper desulphurisation, thus leading to an out-of-specification product or to an increased H_2 consumption, which would negatively influence the feasibility of the process.

Studies with model compounds can be very useful to understand the functioning of the process and its bottlenecks. Coprocessing of guaiacol and straight-run gas oil (SRGO) was carried out by Bui et al. [33]. Guaiacol is often adopted as a model compound for pyrolysis oil, as it is indicative of lignin-derived species, which are common in this kind of liquids. The presence of guaiacol results in an inhibition of desulphurisation. However, it was observed that this effect takes place until all guaiacol is deoxygenated; desulphurisation takes place normally afterwards.

An important study was carried out by Pinheiro et al. [34], who tested the cohydrotreating of several types of oxygenates with SRGO, using a $CoMo/Al_2O_3$ catalyst. Interestingly, they found out that the oxygenated species behaved in a very different way. In particular, a whole class of compounds, namely, those following a deoxygenation route through dehydration, did not show any significant inhibition effect on desulphurisation and denitrification. In this group, the tested compounds were 2-propanol, cyclopentanone, anisole, and guaiacol. On the opposite way, those compounds whose degradation follows decarboxylation or decarbonylation, hence producing CO_2 and CO, did show an inhibition effect. In this class, the tests were conducted with ethyldecanoate and propanoic acid.

Such finding was furtherly proved by a subsequent study by the same authors [35]. Here, they added CO and CO_2 to the reacting stream and clearly visualised the inhibition effect played by such species. According to their explanation, CO and CO_2 compete with some of the intermediates from the hydrodesulphurisation process for adsorption on the active sites of the catalyst. Such competition does not concern the oxygenated compounds themselves, but only CO and CO_2 arising from their conversion. Therefore, they suggest adopting biomass conversion processes able to reduce the amount of the inhibiting compounds or developing a catalyst able to enhance hydrogenation reactions over decarboxylation and decarbonylation.

9.3.2 Cohydrotreating of pyrolysis oil

In the literature, cohydrotreating of bio-oil produced by a thermochemical process is quite rare. To the best of our knowledge, the only study in this field was carried out by de Miguel Mercader et al. [36] in the frame of the EU project 'Biocoup'. Here, the authors obtained an aqueous and an organic phase from pyrolysis oil by diluting it with water, namely: AFWA (aqueous fraction water addition) and OFWA (oil fraction water addition). Then, each fraction underwent a mild hydrotreatment, and two different oils were produced, respectively, $AFWA_{oil}$ and $OFWA_{oil}$. Each of them was then mixed with SRGO, after being diluted with isopropanol (IPA) in order to reduce viscosity. The mixture was then fed to a laboratory-scale hydrotreater, in the presence of a catalyst aimed at desulphurisation.

The results are shown in Table 9.2. It can be observed that the presence of bio-oil generally reduces the extent of desulphurisation, compared with the pure SRGO case. Such effect is much more evident for the $AFWA_{oil}$, whose oxygen content is higher and which was utilised with a higher blending ratio. In this case, desulphurisation rate dropped from 99% to around 79%. Better results were achieved with $OFWA_{oil}$. In this case, desulphurisation efficiency was still reduced but to a minor extent (375 vs 136 ppm of pure SRGO). Such feedstock was however used with a higher dilution rate with respect to $AFWA_{oil}$.

Interestingly, the authors also performed hydroprocessing tests with pure SRGO after the coprocessing runs, finding out that the concentrations of sulphur in the product go back to the previous values. Hence, the reduction of the desulphurisation

Table 9.2 Summary of the experimental results from the cohydroprocessing tests by de Miguel Mercader et al. [36]

Feedstock	Ratio (%)	Oxygen bio-oil (%)	Sulphur in (ppm)	Sulphur out (ppm)	Sulphur removal (%)
$AFWA_{oil}$/IPA/SRGO	70/20/10	18.7	9450	~2000	78.8
$OFWA_{oil}$/IPA/SRGO	70/15/15	13.7	9450	376	96.0
SRGO	–	–	13,500	136	99.0

Oxygen in bio-oil is expressed on dry basis. It is assumed that bio-oil is sulphur-free.

efficiency is reasonably caused by competition effects, rather than catalyst deactivation. Competition between deoxygenation and desulphurisation is thus a serious limitation for such technology. Even though in the case of $OFWA_{oil}$ the decrease of desulphurisation removal appears quite small, it should be mentioned that the requirements on sulphur in diesel are quite strict: 10 ppm in 2009. Therefore, even a small percentage loss in desulphurisation efficiency may result in an off-specification product, and thus, it could imply the adoption of more severe reaction conditions. Additionally, the authors also found the presence of phenols in the product.

Cohydroprocessing of pyrolysis oil is therefore a technological approach for which the integration in existing refineries is more problematic, at least at the current state of knowledge. On the other hand, research work performed in this field is definitely smaller compared with the experiences with co-FCC. Much more experimental effort is needed to evaluate possible optimal process conditions, especially for what concerns the choice of feedstock, blending ratio, and type of catalyst.

9.4 Challenges and perspectives

Coprocessing is an interesting option for the conversion of biomass feedstock into valuable fuels. However, this is a relatively less investigated field, and still, much research is needed in order to improve the existing technologies and, possibly, to find new ones. In this section, some aspects needing to be focused in future research activities are presented. They include the extent of the pyrolysis oil pretreatment prior to the upgrading operations, the development of coprocessing routes for the production of hydrocarbons in the diesel range, and the potential represented by hydrothermal liquefaction (HTL) oils.

9.4.1 The extent of pretreatment

An important question to be properly answered resides in the extent of any treatment to be performed on pyrolysis oil before its coprocessing. Although coprocessing of untreated pyrolysis oil has been documented (see Section 9.2.2.3), it is more likely that a prior deoxygenation process (i.e. hydrotreating or catalytic pyrolysis) would be required. This would allow plainer operations and the possibility of higher blending ratios.

As it has been explained, co-FCC also serves as a concurrent approach with respect to deep dehydrogenation. Hydrogen transfer, which takes place from fossil oil to bio-oil, can greatly reduce the need for an external source of hydrogen, with beneficial effects for the overall cost of the operation. On the other hand, deeper hydrotreatment before co-FCC could increase the actual amount of bio-oil that could be blended and, thus, the share of renewable carbon in the final product.

Therefore, there should be an optimal point between these two contrasting trends, which has to be found for each feedstock and under different process conditions. Similar considerations can be traced for the alternative pathway, involving catalytic pyrolysis instead of hydrotreatment. In this case, the achievement of a higher degree of deoxygenation usually implies a reduction in the oil yields [3]. Reducing the severity of the process to the minimum deoxygenation level required by the following upgrading would then result in higher yields and better profitability of the entire process.

Clear and exhaustive information about the extent of the prior deoxygenation step is currently missing. Nevertheless, understanding this aspect, also by means of technoeconomic analyses, would lead to a significant step forward in order to design a possible biofuel industrial chain. Future studies should also report the final yields of fuel on biomass basis, thus allowing to compare the different process alternatives also on quantitative basis. This is quite difficult to achieve right now, since the studies on coprocessing do not include extensive information about the production of the utilised feedstock. A more comprehensive view on the overall process from biomass to the final fuel would constitute a precious piece of information.

9.4.2 Enhancing the production of middle distillates

Middle distillates, that is, jet-fuel (kerosene) and diesel, are of great importance for some specific fields. Indeed, when aviation, maritime or long-range terrestrial transportations are involved, such middle distillates represent the preferred products. The possible future expansion of electric vehicles for short-range mobility could greatly reduce the market request for gasoline, while middle distillates would still have considerable importance.

As it has been shown, most research in coprocessing has been performed with co-FCC, which is mainly aimed at the production of hydrocarbons in the gasoline range. Nevertheless, co-FCC also produces heavier fractions, such as light cycle oil (LCO) and decanted oil (DCO), which could be recycled at some points in the refinery. It would be interesting to understand the actual degree of deoxygenation of such fractions. Indeed, if co-FCC is able to remove a substantial amount of oxygen also from such heavier compounds, it could be still considered as a sort of preliminary treatment for other units in the refinery.

However, the most suitable coprocessing processes for the production of middle distillates are those based on reactions with H_2, such as hydrotreating. In this field, there are very few experimental results available, and therefore, much work should be performed. From the few available data, it was shown that relevant inhibition of desulphurisation is observed when oxygenates are cofed. A perspective could be that of developing specific catalysts able to reduce decarboxylation and decarbonylation reactions (see Section 9.3.1). Also in this case, the extent of the prior dehydrogenation treatment should be investigated.

A novel perspective that could be explored is that of coprocessing in hydrocrackers. With respect to hydrotreaters, hydrocrackers involve much more severe reaction conditions in terms of both temperature and pressure. They are thus able to remove heteroatoms and, at the same time, to crack larger molecules into smaller ones. Hydrocracking has been used in order to process low-grade crudes and to convert the heavier fractions produced in refineries, such as the residues from distillation and the heavy fractions refractory to FCC. Its application to bio-oil blends, especially to those bio-oils with high molecular weight, could be then potentially beneficial for the production of middle distillates.

9.4.3 The potential of hydrothermal liquefaction oils

As reported in the previous sections, the whole work concerning coprocessing has been essentially carried out on pyrolysis oil or with the products derived from its upgrading. Further experimental work is available with vegetal oils, even though such materials are out of the scope of this publication.

A completely new field for the investigation in coprocessing would be represented by the oils from hydrothermal liquefaction (HTL). HTL can be considered as the wet equivalent of pyrolysis. It involves biomass processing in hot compressed water, either in the subcritical or in the supercritical range. The presence of water as a hydrogen donor allows obtaining a lower oxygen content in the produced liquid. Moreover, it paves the way to the utilisation of nonconventional biomass feedstock such as those characterised by a very high water content, which cannot be utilised for other thermochemical conversion processes. Municipal waste, agricultural residues, and algae are only some of the feedstocks that could be processed and that are definitely not suitable for pyrolysis, due to their very high moisture [37].

The oil from the HTL process is generally referred as 'bio-crude'. It has much better characteristics than raw pyrolysis oil. Indeed, its oxygen content is already quite reduced. Common values are in the range 10%–20% [38], which is lower or comparable with that of many feedstock utilised in co-FCC studies. Bio-crude could be then a very good candidate for coprocessing operations, with the potential of achieving even higher blending ratios than the attested state of the art. The complete lack of experimental data for the co-FCC or cohydrotreating of bio-crude should be definitely overcome in the next future.

Bio-crude with a low oxygen content can successfully undergo fractional distillation along with crude oil [39]. Such perspective is hardly possible for pyrolysis oil, due to its large instability that results in the production of massive amounts of coke when heated. Fractional distillation enables separating the different fractions of bio-crude, in order to utilise each of them for the most suitable upgrading process. Bio-oil would thus follow the same processing route as for standard crude oil. The residual presence of oxygen in the produced fractions could be then removed by using the standard units for heteroatoms removal, such as the hydrotreater. The very low amount of oxygen found in the distilled fractions from bio-crude (around 2%, as in [39]) is expected not to cause relevant inhibition drawbacks to desulphurisation, as observed for other bio-oils.

Distillation of bio-crude results in a consistent fraction of residues, which should be treated accordingly. Residues could undergo hydrocracking together with the heavy residues from standard oil. This potentially opens another coprocessing perspective to be investigated.

9.5 Conclusions

Coprocessing is a potential ready-to-use way for the concrete implementation of biofuels in the commercial field. Such approach allows utilising the existing infrastructure for refining and distribution, and hence, it would represent a drop-in possibility for the practical utilization of biomass-derived fuels.

The largest part of research activities in the field of coprocessing has been performed with co-FCC. Doubtless, this is a very flexible and promising approach. Evidence of pyrolysis oil treatment up to a blending ratio of 10% has been documented even in large-scale plants, while laboratory tests have shown that higher blending can be achieved with partially deoxygenated bio-oils. At the moment, co-FCC is the only coprocessing approach that has proved its technical feasibility. Despite some drawbacks concerning the residual amount of oxygenates in gasoline, the actual implementation of co-FCC in a real refinery appears to be quite concrete.

On the contrary, coprocessing operations for diesel production have not been developed to sufficient extent. An evident decrease in the desulphurisation efficiency has been observed during cohydrotreating tests, which appears to be a strong drawback for such coprocessing approach. Although the available experimental data are scarce, it can be foreseen that such process is not going to be effective unless substantial modifications of the catalyst or feedstock pretreatment are carried out.

A perspective for the production of diesel hydrocarbons could be given by cohydrocraking, which could also involve the residual fractions from co-FCC. Another interesting field to explore is represented by hydrothermal liquefaction (HTL) oils, which are much less oxygenated and could be potentially codistilled. However, a fundamental question to be answered is the extent of pretreatment of biomass-derived oils prior to corefining. A trade-off between pretreatment severity and oil yields should be found, in order to perform comprehensive technoeconomic analyses and give precious information to potential stakeholders.

References

[1] Oudenhoven S.R.G., Kersten S.R.A. Thermochemical conversion: an introduction to fast pyrolysis. In: De Jong W., Van Ommen J.R., editors. Biomass as a sustainable energy source for the future: fundamentals of conversion processes. John Wiley & Sons, Inc., Hoboken, NJ. 2014. p. 359–87. https://doi.org/10.1002/9781118916643.ch11.

[2] Iliopoulou E.F., et al. Pilot-scale validation of co-ZSM-5 catalyst performance in the catalytic upgrading of biomass pyrolysis vapours. Green Chem 2014;16(2):662–74. Available at: http://xlink.rsc.org/?DOI=C3GC41575A.

[3] Venderbosch R.H. A critical view on catalytic pyrolysis of biomass. ChemSusChem 2015;8:1306–16.

[4] Vogt E.T.C., Weckhuysen B.M. Fluid catalytic cracking: recent developments on the grand old lady of zeolite catalysis. Chem Soc Rev 2015;44(20):7342–70. Available at http://xlink.rsc.org/?DOI=C5CS00376H.

[5] Talmadge M.S., et al. A perspective on oxygenated species in the refinery integration of pyrolysis oil. Green Chem 2014;16(2):407–53. Available at http://pubs.rsc.org/en/content/articlehtml/2014/gc/c3gc41951g.

[6] Adjaye J.D., Bakhshi N.N. Production of hydrocarbons by catalytic upgrading of a fast pyrolysis bio-oil. Part I: conversion over various catalysts. Fuel Process Technol 1995;45(3):161–83. Available at: http://apps.isiknowledge.com/InboundService.do?Func=Frame&product=WOS&action=retrieve&SrcApp=Papers&UT=A1995TG003000001&SID=U2lAJ5eoLkK8NEl5m85&Init=Yes&SrcAuth=mekentosj&mode=-FullRecord&customersID=mekentosj&DestFail=http://access.isiproducts.com/custom_i.

[7] Vitolo S., et al. Catalytic upgrading of pyrolytic oils to fuel over different zeolites. Fuel 1999;78(10):1147–59.
[8] Samolada M.C., Papafotica A., Vasalos I.A. Catalyst evaluation for catalytic biomass pyrolysis. Energy Fuels 2000;14(6):1161–7. Available at: http://dx.doi.org/10.1021/ef000026b.
[9] Chang C., Lang W., Silvestri A. Manufacture of gasoline. Patent No. 3,998,898. United States; 1976.
[10] Chen N.Y., Walsh D.E., Koenig L.R. Fluidized bed upgrading of wood pyrolysisliquids and related compounds. In: Soltes E.J., Milne T.A., (Eds.), Pyrolysis oils from biomass, Washington DC, American Chemical Society. 1988. p. 277–89. https://doi.org/10.1021/bk-1988-0376.ch024.
[11] Naik D.V., et al. Catalytic cracking of jatropha-derived fast pyrolysis oils with VGO and their NMR characterization. RSC Adv 2015;5(1):398–409. https://doi.org/10.1039/C4RA08128E.
[12] Venderbosch R.H., et al. Stabilization of biomass-derived pyrolysis oils. J Chem Technol Biotechnol 2010;85(5):674–86.
[13] de Rezende Pinho A., et al. Co-processing raw bio-oil and gasoil in an FCC unit. Fuel Process Technol 2015;131:159–66. https://doi.org/10.1016/j.fuproc.2014.11.008.
[14] de Rezende Pinho A., et al. Fast pyrolysis oil from pinewood chips co-processing with vacuum gas oil in an FCC unit for second generation fuel production. Fuel 2017;188:462–73.
[15] Lindfors C., et al. Co-processing of dry bio-oil, catalytic pyrolysis oil, and hydrotreated bio-oil in a micro activity test unit. Energy Fuels 2015;29(6):3707–14.
[16] Thegarid N., et al. Second-generation biofuels by co-processing catalytic pyrolysis oil in FCC units. Appl Catal Environ 2014;145:161–6. https://doi.org/10.1016/j.apcatb.2013.01.019.
[17] Agblevor F.A., et al. Co-processing of standard gas oil and biocrude oil to hydrocarbon fuels. Biomass Bioenergy 2012;45:130–7. https://doi.org/10.1016/j.biombioe.2012.05.024.
[18] Wang C., Li M., Fang Y. Coprocessing of catalytic-pyrolysis-derived bio-oil with VGO in a pilot-scale FCC riser. Ind Eng Chem Res 2016;55(12):3525–34.
[19] Lappas A.A., Bezergianni S., Vasalos I.A. Production of biofuels via co-processing in conventional refining processes. Catal Today 2009;145(1–2):55–62.
[20] Fogassy G., et al. Biomass derived feedstock co-processing with vacuum gas oil for second-generation fuel production in FCC units. Appl Catal Environ 2010;96(3–4):476–85. https://doi.org/10.1016/j.apcatb.2010.03.008.
[21] Gueudré L., et al. Coke chemistry under vacuum gasoil/bio-oil FCC co-processing conditions. Catal Today 2015;257(P2):200–12. https://doi.org/10.1016/j.cattod.2014.09.001.
[22] Gueudré L., et al. Optimizing the bio-gasoline quantity and quality in fluid catalytic cracking co-refining. Fuel 2017;192:60–70.
[23] Samolada M.C., Baldauf W., Vasalos I.A. Production of a bio-gasoline by upgrading biomass flash pyrolysis liquids via hydrogen processing and catalytic cracking. Fuel 1998;77(14):1667–75. Available at http://www.sciencedirect.com/science/article/pii/S0016236198000738.
[24] de Miguel Mercader F., et al. Production of advanced biofuels: co-processing of upgraded pyrolysis oil in standard refinery units. Appl Catal Environ 2010;96(1–2):57–66. https://doi.org/10.1016/j.apcatb.2010.01.033.
[25] ASTM D3907/D3907M-13. Standard test method for testing fluid catalytic cracking (FCC) catalysts by microactivity test. 2013.
[26] Fogassy G., et al. From biomass to bio-gasoline by FCC co-processing: effect of feed composition and catalyst structure on product quality. Energy Environ Sci 2011;4(12):5068.

[27] Agblevor F.A., et al. Production of stable biomass pyrolysis oils using fractional catalytic pyrolysis. Energy Fuels 2010;24(7):4087–9.
[28] ASTM D4530-15. Standard test method for determination of carbon residue (micro method). 2015.
[29] European Parliament. Directive 2009/28/EC of the European Parliament and of the council of 23 April 2009. 2009.
[30] Fogassy G., et al. The fate of bio-carbon in FCC co-processing products. Green Chem 2012;14(5):1367–71. Available at http://xlink.rsc.org/?DOI=c2gc35152h%5Cnhttp://dx.doi.org/10.1039/C2GC35152H.
[31] European Union. Directive 2009/30/EC of the European Parliament and of the council. 2009.
[32] Al-sabawi M., Chen J. Hydroprocessing of biomass-derived oils and their blends with petroleum Feedstocks a review. Energy Fuels 2012;26:5373–99.
[33] Bui V.N., et al. Co-processing of pyrolisis bio oils and gas oil for new generation of bio-fuels: hydrodeoxygenation of guaiacol and SRGO mixed feed. Catal Today 2009;143(1–2):172–8.
[34] Pinheiro A., et al. Impact of oxygenated compounds from lignocellulosic biomass pyrolysis oils on gas oil hydrotreatment. Energy Fuels 2009;23(2):1007–14.
[35] Pinheiro A., et al. Impact of the presence of carbon monoxide and carbon dioxide on gas oil hydrotreatment: investigation on liquids from biomass cotreatment with petroleum cuts. Energy Fuels 2011;25(2):804–12.
[36] de Miguel Mercader F., et al. Hydrodeoxygenation of pyrolysis oil fractions: Process understanding and quality assessment through co-processing in refinery units. Energy Environ Sci 2011;4:985.
[37] Toor S.S., Rosendahl L., Rudolf A. Hydrothermal liquefaction of biomass: a review of subcritical water technologies. Energy 2011;36(5):2328–42. https://doi.org/10.1016/j.energy.2011.03.013.
[38] Peterson A.a., et al. Thermochemical biofuel production in hydrothermal media: a review of sub- and supercritical water technologies. Energy Environ Sci 2008;1(1):32.
[39] Hoffmann J., Jensen C.U., Rosendahl L.A. Co-processing potential of HTL bio-crude at petroleum refineries—part 1: fractional distillation and characterization. Fuel 2016;165:526–35. Available at: http://www.sciencedirect.com/science/article/pii/S0016236115011084.

Further reading

[40] Mortensen PM, et al. A review of catalytic upgrading of bio-oil to engine fuels. Appl Catal A Gen 2011;407(1–2):1–19. https://doi.org/10.1016/j.apcata.2011.08.046.

Hydrofaction™ of forestry residues to drop-in renewable transportation fuels

10

C.U. Jensen*, J.K.R. Guerrero†, S. Karatzos†, G. Olofsson*, S.B. Iversen*
*Steeper Energy ApS, Hørsholm, Denmark, †Steeper Energy Canada Ltd, S.W. Calgary, AB, Canada

10.1 Introduction

On a global and unprecedented scale, climate change is threatening humanity through changing weather patterns and rising sea levels. An increasing frequency of catastrophic droughts and floodings affects essential food and water supply, forcing so-called climate migrants to move due to sudden or gradual changes in the natural environment [1,2]. The Paris agreement (in force November 2016, signed by 193 countries) addresses the urgent and potentially irreversible threat of climate change and the necessity to reduce greenhouse gas (GHG) emissions from the energy sector [2,3]. In the World Energy Outlook 2016 [3], IEA projects the future energy sector and how existing technologies for renewable heat and power generation will be deployed to approach emission targets. This includes a significant increase in electricity generation from, for example, solar and wind, which indirectly introduces renewables in the transportation sector, where electric cars are considered momentous for urban and short-distance transport [3]. However, the same report emphasises that the lack of renewable fuel alternatives for heavy transportation such as trucks and aviation will account for the continued growth in global consumption of fossil and carbon-intensive petroleum [3]. The European Commission also highlights the need for low-emission fuels for heavy transportation in order to meet the emission targets for the transportation sector, which currently represents about 25% of EU's total GHG emissions [4]. Steeper Energy is addressing this gap by commercializing its proprietary hydrothermal liquefaction (HTL) technology, Hydrofaction™, which facilitates the production of sustainable drop-in biofuels targeting heavy transportation.

The competitive strengths of HTL begin from the ability to process a wide range of wet and dry low-value nonfood biomass residues, thereby mitigating feedstock supply risks and interference with food supply. Secondly, HTL is currently considered a competitive and resource-effective thermochemical pathway to advanced biofuels, due to its high energy and carbon efficiency [5,6]. A 2014 study commissioned by the US DoE estimated a 70% GHG emission reduction from hydrotreated HTL biofuel compared with the 2005 petroleum baseline [5]. Finally, the hydrotreated HTL products are fully compatible with the existing petroleum infrastructure (pipelines, stations, and

Direct Thermochemical Liquefaction for Energy Applications. https://doi.org/10.1016/B978-0-08-101029-7.00009-6

engines) enabling drop-in without blend-wall limitations. This facilitates a gradual phase-in where production capacity and cost-efficiency can be balanced during the transition to fossil independence [7].

10.1.1 Forestry residues as feedstock

Steeper Energy is currently focused on forestry residues and mill wastes as Hydrofaction™ feedstock due to the global abundance of these resources and the industry's need for cost- and carbon-efficient utilisation of their wastes.

Forests cover around one-third (30.7% in 2010) [8] of the global land area and are an essential part of life. The significant amounts of carbon that is sequestered both below- and aboveground by photosynthesis also make forests essential for climate-change adaption and mitigation [8]. As an example, afforestation and reforestation were approved as GHG mitigating strategy under the Kyoto Protocol [9]. Carbon sequestration can be optimised by agroforestry and sustainable forest management, thus enhancing the rate of carbon absorption [8–11]. In other words, since a tree grows slower as it gets older, there is an optimum between the harvest rotation cycle and the rate of carbon uptake. In fact, [11] states that the shorter the harvest rotation age, the more favourable the carbon balance becomes over time. This however implies that the sequestered carbon (wood) is used to substitute fuel-intensive building materials such as steel or concrete, while the manufacturing residues (bark, branches, and sawdust) are used to substitute fossil fuels [11]. Such efficient use of the forestry residues is however not granted, as the industry currently wastes up to 40% of the harvested material [12].

The revenue from residuals is also crucial for the forest industry. This was concluded in a recent study on the biomass supply and utilisation chain in three Canadian provinces [13]. The study showed how sawmills were running at a small loss on their core process, the lumber output, but stayed in business due to revenue from residuals as inputs to, for example, electricity generation. Similarly, if the price of lumber would decrease, only the sawmills located close to residual offtakers (e.g. pulp mill) would stay in operation from the value of the residual products [13]. These residuals (wood chips, sawdust, planer shavings, and residual bark) are mainly produced at the mill during the primary processing. However, the major fraction of forestry residuals is the so-called slash, produced during the preliminary processing at the felling site. These have been estimated to account for more than 20% of the total harvested volume in Canada [12].

As mentioned previously, the forestry residuals need to be used as bioenergy to replace fossil fuels in order for agroforestry to be a carbon-efficient GHG mitigation strategy [11]. As such, the slash can be chipped and transported at a cost around US$ 30–60/ODT (oven-dried tonne) [12–14]. Due to the costs of processing and transportation, a significant amount of such slash is either disposed in stockpiles or burnt on site instead of being efficiently utilised. Wood waste stockpiles decompose over time releasing significant methane emissions from anaerobic digestion, which has a significant GHG footprint (1 kg CH_4 = 25 kg CO_2e) [15–17]. A portion of the forest residuals are currently used for electricity and heat types of bioenergy [3,13]. However, in the longer run, this is unlikely to be a cost-competitive solution to alternative renewable sources such as solar PV and wind power, as IEA projects an average cost reduction on these by 2040 of 40%–70% and 10%–25%, respectively [3]. Furthermore, the pulp industry, which is a major offtaker of forestry residuals, is currently struggling in Europe and North America with a decreasing demand and increasing competition from Asian and South American mills [18,19]. This emphasises the need for finding alternative uses of forestry residuals, such as a feedstock for thermochemical production of biofuels. In fact, the transformation of pulp and paper mills into biorefineries in Europe and North America has been mentioned as a potential cornerstone of a green economy [20]. This is where the Hydrofaction™ technology becomes relevant by bridging the gap between a globally abundant waste product that needs to be cost-effectively and carbon effectively utilised and the heavy transportation sector that needs a renewable fuel alternative. A life-cycle GHG emission analysis of Hydrofaction™ products produced in Alberta, Canada, from forestry residuals (slash) as feedstock is presented later in Section 10.3.

10.2 The Hydrofaction™ process

Hydrofaction™ is a proprietary thermochemical technology platform that enables conversion of wet (as harvested) biomass residues directly to renewable drop-in biofuel indistinguishable from its petroleum equivalent. The reaction pathway includes two major subprocesses, which are supercritical HTL (first stage) and subsequent upgrading through hydrotreating (second stage). During the first stage, Hydrofaction™ utilises a combination of the following features to efficiently convert the biomass to an energy-dense renewable crude oil [21]:

- Operating pressures (300–350 bar) and temperatures (390–420°C) above the critical point of water
- Recirculation of water-soluble organics and Hydrofaction™ oil for improvements in feed characteristics, energy balance, oil quality, and yields
- Use of homogeneous alkali metal catalyst (K_2CO_3 and NaOH) for desired catalytic effects and control of pH to minimise corrosion
- Recovery and recycling of alkali metal catalysts for improved process economics and reduced footprint

The resulting biocrude is characterised by relatively high aromaticity and an oxygen content around 10 wt%, which strongly affects the physiochemical properties (density, viscosity, volatility, polarity, and hydrophobicity) that determine the biocrude compatibility with petroleum equivalents. Therefore, as any petroleum crude, the renewable crude oil needs further refining to meet the various fuel specifications. Whereas hydrodesulphurisation (HDS) and hydrodenitrogenation (HDN) are the major reaction steps in refining petroleum crudes, upgrading of wood-derived HTL crude oil is mostly focused on hydrodeoxygenation (HDO) and hydrodearomatisation (HDA). During Hydrofaction™, this is achieved by hydrotreating on commercial catalysts, which is a standard and widely used refinery operation [22]. Finally, the hydrotreated product is fractionated by boiling point to renewable fuels for drop-in in the diesel and marine fuel pools. Thereby, forestry residues can be used to substitute fossil transportation fuels with a renewable alternative.

The Hydrofaction™ subprocesses are schematised in Fig. 10.1 including major interconnected operations, such as recirculation streams, water treatment, and utilisation of gaseous products in a fired heater. Fig. 10.1 is divided into an HTL and an upgrading part; accordingly, the description and presentation of experimental data for the first and second stage are divided into Sections 10.2.1 and 10.2.2, respectively.

10.2.1 Supercritical HTL (first stage)

Characteristics of Hydrofaction™ HTL are the use of high-density, supercritical water; homogenous alkali metal catalysts; and the recycling of aqueous and oil products back into the process. The Hydrofaction™ HTL stage applies operating conditions (300–350 bar and 390–420°C) that are above the critical point of water, which is higher than most other HTL processes reported in literature [23–28]. The chemical properties, such as the dielectric constant and ionic product, that make near-critical water an appealing reaction medium for HTL are a direct function of density [29]. Thus, by ensuring a high pressure, key thermodynamic properties of water can be maintained at the same order of magnitude as for the subcritical conditions while taking advantage of improved kinetics, mass, and heat transfer at higher temperatures. As an example, the ionic product of water is an important parameter within hydrothermal processing, because it reflects whether a medium favours ionic or radical reactions [30,31]. Due to the relatively high

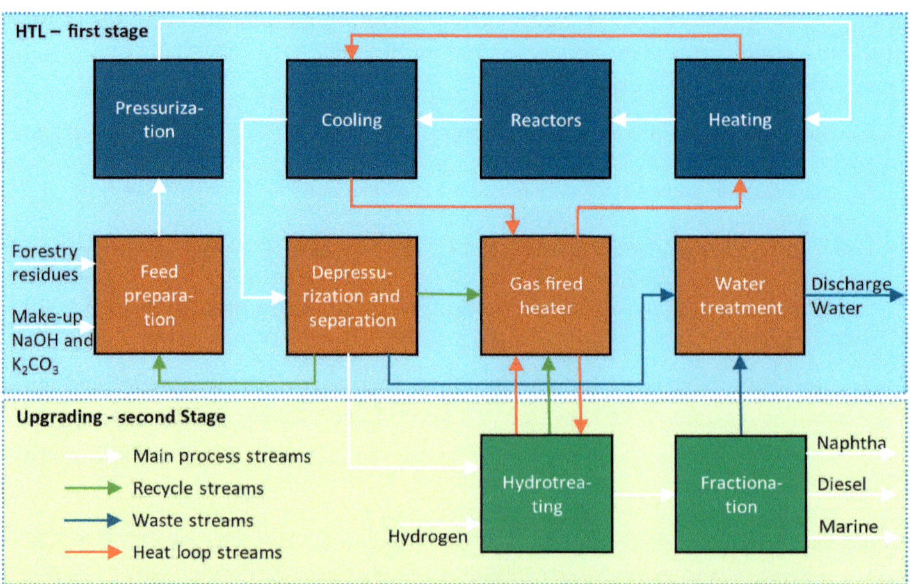

Fig. 10.1 Schematic of the Hydrofaction™ process consisting of supercritical HTL and subsequent hydrotreating. Besides the major heat loop streams indicated, the process utilises additional heat integration between the stages.

pressure, the ionic product of water at Hydrofaction™ conditions is up to several orders of magnitude higher than at the critical point. Similarly, the higher pressure ensures that density and derived properties are less dependent on temperature changes, which is beneficial during plant operations to ensure a smooth transition of process parameters such as fluid velocities, Reynolds numbers, specific heat capacity, heat-transfer coefficients, and retention times. Consequently, due to the supercritical temperature, it is crucial to maintain a high pressure during the Hydrofaction™ HTL stage.

At near-critical conditions, the dielectric constant of water is reduced more than 90% as compared with ambient conditions. Thereby, water dissolves nonpolar compounds, such as biomass and biocrude molecules that are hydrophobic at ambient conditions. This accelerates conversion through, for example, hydrolysis and solvolysis reactions by reducing mass-transfer limitations. Such ionic reactions are further promoted by water dissociation, which is maintained high in relatively high-density supercritical water [29].

The recirculation of produced organic compounds in the form of both water-soluble organics and oil has a number of benefits. It simplifies feed preparation and reduces the heat capacity of the process fluid, and it appears that the presence of water-soluble organics in the feed inhibits further formation of water-soluble organics, thereby increasing the relative amount of oil produced. Ref. [21] provides further details on the characteristics of Hydrofaction™ HTL, the reasoning behind the operating conditions, and the considerations around the major chemical reactions occurring during the HTL stage.

10.2.1.1 Feed preparation

The first step of the HTL stage in Fig. 10.1 is to prepare a pumpable feed of the different reactants including the forestry residues. The feed is prepared from size-reduced forestry residues, recycled oil, and aqueous products as well as make-up homogenous catalysts in the form of potassium carbonate and sodium hydroxide. Recirculation of both the aqueous and oil products is beneficial during feed preparation, because the water-soluble organics and phenolic oil compounds introduce a partial dissolution mechanism, which increases homogeneity and improves rheological properties of the feed. Similarly, the presence of alkali salts gives rise to carbonate/bicarbonate, which accelerates the depolymerisation by hydrolysis during pretreatment. The homogeneous catalyst in the form of potassium is mainly introduced by recirculation of the aqueous products, but make-up is required because a small fraction of the aqueous products are discharged to prevent build-up of trace compounds. Sodium hydroxide is added for control of the pH to alkaline conditions throughout the process.

A high biomass dry-matter loading in the feed is important to drive down both capital and operational expenses per fuel volume produced. Wood feeds with a dry-matter content of 17–25 wt% are used during pilot testing of Hydrofaction™ HTL [21,32,33].

10.2.1.2 Pressurization and heating

The feed mixture is pressurised to a pressure in the range of 300–350 bar and subsequently rapidly heated to a reaction temperature in the range of 390–420°C. The heating is to a large degree performed by heat recovered from the outgoing product stream.

This is performed by indirect heat exchange using high-pressure water as heat-transfer medium in a loop from the cooling section to the heating section.

It should be noted that several benefits are provided by the supercritical operating pressure and temperature applied in the Hydrofaction™ process. This is related to the fact that Hydrofaction™, despite being supercritical, is operating below the so-called pseudocritical line (PCL), which is defined as the temperature at a given pressure where the specific heat capacity of water is at a maximum [29]. This means, for example, that the enthalpy of water at, for example, 335 bar and 400°C is similar to the specific enthalpy at the critical point of water [29]. The specific enthalpy of the actual mixture is even lower as water typically only comprises around 50 wt% of the feed mixture due to the high dry-matter and organic content. Recirculation of organics through both the aqueous and oil product reduces the overall heat-transfer requirement, due to the lower specific heat capacity of organics compared with water. This reduces the overall heat requirement of the process, which is emphasised with Fig. 10.2 that shows how the power consumption of the two serially connected heaters in the pilot plant decrease once the process fluid is switched from water to feed at a constant mass flowrate. Note that the pilot does not facilitate heat exchange and thus it has a slightly different configuration than that given for the first stage in Fig. 10.1. The pilot plant is described in Section 10.2.1.6.

Heat exchange is an obvious feature that improves the cost and energy efficiency of a hydrothermal process that utilises near-critical water as a reaction medium. The relatively high pressure applied in Hydrofaction™ ensures that the heat-transfer coefficient of water does not drop above the critical point. Likewise, the higher temperature, compared with subcritical, results in a high thermal effusivity and a more favourable driving force for

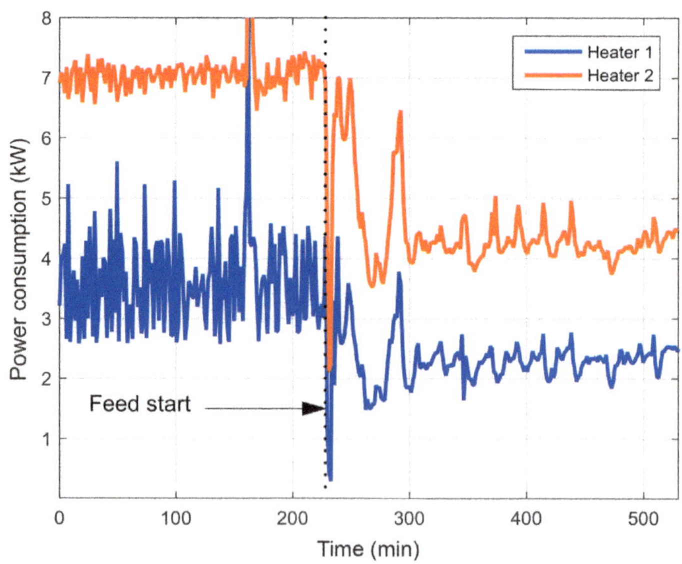

Fig. 10.2 The effect of switching from water to feed on power consumption of the pilot plant heaters.

heat exchange, which improves the heat recovery potential. Typically, 50%–75% of the heat required for heating to reactor temperature is recovered through heat exchange. The remaining heat load of the process is supplied by a fired heater, which at steady state is fired with gas from the process. This stream possesses more than sufficient calorific value, even if the hydrogen fraction is extracted (for upgrading use) before combustion. In fact, this makes the Hydrofaction™ process self-sustained with thermal energy at steady state.

10.2.1.3 Reactors

The heated feed/product intermediate enters the reactors, where it is maintained at reaction temperature and pressure for a certain period of time, typically 10–15 min. A proposed set of major chemical reactions occurring during Hydrofaction™ is presented in Ref. [21].

10.2.1.4 Cooling and pressure reduction

The product mixture from the reactors is cooled to about 150°C by heat exchange with the incoming feed via the high-pressure water loop described above, before the pressure is reduced to separation pressure in the pressure let-down system.

10.2.1.5 Separation and water treatment

In the separation part of the process, the cooled and depressurised products are separated into a gas stream, a water stream containing dissolved salts, a crude oil stream, and optionally a solid stream. This is performed by a combination of filtration and gravimetric phase separation.

The gas phase produced is mixed with product gas from the upgrading section and used to fuel the fired heater and heat the high-pressure water loop as described above. Optionally, the first-stage gas phase may be further separated to extract hydrogen and/or CO_2 before combustion (see Section 10.3 for details).

The aqueous phase is split into a concentrate and a distillate phase in a recovery unit. The distillate is purified to allow disposal and discharged in amounts that balances the water deducted from the feedstock biomass. The concentrate, which contains most of the total organic carbon (TOC) and alkali metal catalysts, is recycled back to the feed preparation step, except a bleed of 5%–15% that is withdrawn to prevent build-up of trace compounds such as chloride. A fraction of the oil phase is also recycled back to the feed preparation step, while the remainder is conditioned through desalting to recover potassium and sodium alkali catalysts and reduce the ash and water content prior to hydrotreating. The first-stage product is a renewable crude oil that similar to fossil equivalents requires further upgrading to meet transportation fuel specifications.

10.2.1.6 Summary of HTL pilot operation data (first stage)

A continuous pilot facility dedicated to supercritical HTL was commissioned at Aalborg University, Denmark, in 1Q2013. Since then, the pilot has completed >1600 oil production hours and constituted a centrepiece in the generation of data from the first stage (HTL) of Hydrofaction™.

The pilot is a semicontinuous HTL plant; feeds are prepared in 100 kg batches and then processed continuously at a flowrate around 20 kg/h. In standby mode between batches, the plant circulates pressurised deionised water, which is also the medium used for heat-up and cool-down of the plant. The batchwise processing of feeds complicates the recirculation of aqueous and oil products, that is, characteristic to Hydrofaction™ HTL. The recirculation is however demonstrated in the pilot by using a start-up oil and recycling the products a number of times until the start-up oil is sufficiently diluted/converted and a steady-state product is produced. Using this approach, it has been observed that recirculation of the aqueous product enhances oil yield, because the TOC content of the aqueous product is stable at steady state [21,33].

Table 10.1 lists key figures on the operating conditions, the feed characteristics, and the product yields observed during an extensive oil production campaign that is published in Ref. [21]. A 50:50 mix of spruce and pine wood chips was used as Hydrofaction™ feedstock to produce >150 kg of renewable Hydrofaction™ crude oil. The total mass balance at steady state was on average 100.3 wt%, with no detectable production of char or filter retentate products. The yield of renewable Hydrofaction™ crude oil from dry ash-free woody biomass was on average 45.3 wt%, and similarly, the gas yield was 41.2 wt%. Based on closure of the carbon, hydrogen, and oxygen elemental balances, the remaining product mass was suggested to be water synthetically produced during the biomass conversion.

10.2.1.7 Effect of feedstock on renewable crude oil quality (first stage)

Table 10.2 lists basic physiochemical properties, and Fig. 10.3 depicts distillation profiles of six different renewable Hydrofaction™ crude oils that were produced from three different woody biomass types and using two different start-up oils, crude tall oil (CTO) and distilled tall oil (DTO). The results indicate that the different Hydrofaction™ oils appear to have very similar key characteristics, which emphasises that the start-up oil has little effect on the final product quality because it seems to be diluted sufficiently after 4–6 recirculation cycles. It needs mentioning that the IBP-130°C volatiles were lost during the production of oil A, which increased the viscosity relative to the other oils. The H/C ratio of oil A and the volatility in Fig. 10.3 are also slightly affected by this.

The effect of hardwood (birch) versus softwood (pine/spruce) was also tested at the pilot at the same operating conditions as those given in Table 10.1, but Table 10.2 and Fig. 10.3 indicate only little noticeable difference between the key quality characteristics of the Hydrofaction™ oil produced from the two different types of woody biomass. Though it seems that birch wood results in a slightly higher nitrogen content compared with pine/spruce mixtures. A 10% bark in a pine/spruce feedstock was also tested, based on the motivations previously mentioned for using forestry residues in which bark will be a significant constituent. Again, the quality of the resulting crude oil is comparable, and thus, it could be suggested that a renewable Hydrofaction™ crude oil produced from woody biomass is expected to be relatively viscous; contains roughly 10 wt% oxygen, 1000–2000 ppm nitrogen, and 100–200 ppm sulphur; and possesses an HHV around 38 MJ/kg and a H/C ratio around 1.3–1.4.

Table 10.1 **Key figures on the first-stage Hydrofaction™ HTL are based on Ref. [21]**

Feed characteristics			**Operating conditions**			**Yield figures**		
Feedstock		50:50 Spruce/pine	Feed flowrate	kg/h	20–22	Mass balance	wt%	100.3
Dry-matter content	wt%	17–20	Reactor temperature	C	390–410	Oil yield	wt%	45.3
Oil-to-wood ratio	kg/kg	0.8–1.0	Reactor pressure	bar	300–320	Gas yield	wt%	41.2
Particle size	mm	<2	Heating rate	C/min	<350–450	Water yield	wt%	13.8

Table 10.2 **Basic properties of renewable crude oils produced by Hydrofaction™ of various woody biomasses using different start-up oils**

Oil	Cycle	Start-up oil	Feedstock	Viscosity[a] (cP)	HHV[b] (MJ/kg)	MCR[b,c] (wt%)	C[b] (wt%)	H[b] (wt%)	N[b] (ppm)	S[b] (ppm)	O[d] (wt%)	H/C
Oil A [21]	6	CTO	Pine/spruce	17,360	38.6	21.0	81.4	8.6	1124	100	9.8	1.26
Oil B	4	CTO	Pine/spruce	714	38.0	16.4	80.4	9.0	822	195	10.4	1.34
Oil C	5	CTO	Birch	813	38.0	15.6	80.8	9.1	2635	215	9.8	1.34
Oil D	5	DTO	Birch	2084	37.7	18.6	80.2	9.2	2201	159	10.4	1.37
Oil E	5	DTO	Pine/spruce	3313	37.2	19.7	80.0	9.0	1447	104	10.9	1.34
Oil F	5	DTO	10% Bark in pine/spruce	1954	37.9	18.1	81.4	9.7	973	100	8.8	1.42

[a] Viscosity at 40°C.
[b] On dry basis.
[c] Micro carbon residue (ASTM D4530).
[d] Oxygen by difference.

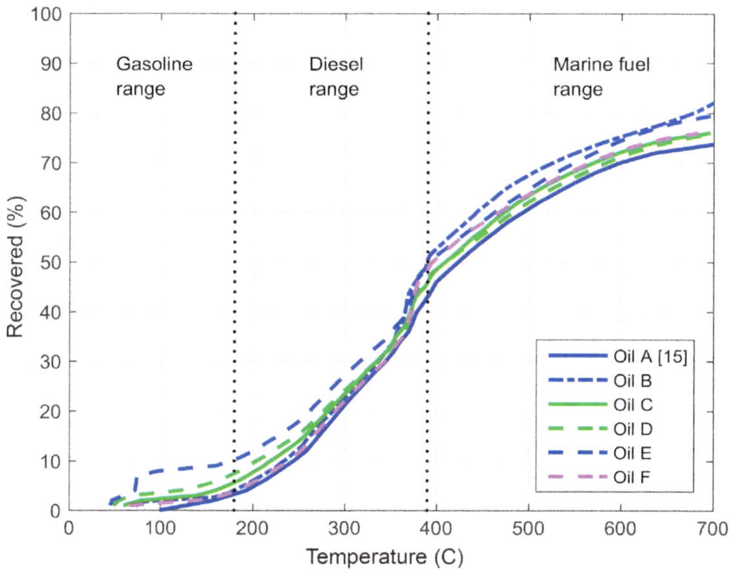

Fig. 10.3 Simulated distillation (ASTM D7169) of the oils given in Table 10.2 produced from pine/spruce *(blue)*, birch *(green)*, and 10% bark in pine/spruce *(purple)*.

10.2.2 Upgrading to drop-in biofuels (second stage)

A significant degree of biomass deoxygenation takes place during the first-stage Hydrofaction™ HTL, but the remaining oxygen content of the renewable crude oil affects the physiochemical properties that determine its compatibility with petroleum equivalents. In other words and similar to petroleum crude oils, the renewable crude oil can be referred to as an intermediate that requires further upgrading in order to gain transportation fuel value. Upgrading of the renewable Hydrofaction™ crude oil is done through catalytic hydrotreatment, which is a widely used process within conventional petroleum refining [22]. The process applies heterogeneous bimetallic catalysts and relatively high partial H_2 pressures to remove oxygen as mainly water through HDO and hydrogenate the renewable crude oil to remove aromatics and improve the diesel and marine fuel characteristics such as cetane number, density, and aromaticity. As indicated with a recirculation stream in Fig. 10.1, the hydrogen produced in situ during the HTL stage can be separated from the gaseous HTL product and utilised to cover part (~50%) of the hydrogen consumption in the second stage.

The current chapter presents hydrotreating experiments carried out in both continuous and batch reactor systems (BRS) and using both sulphided and nonsulphided catalysts. The objective is to show the product quality of wood-derived renewable oil after upgrading while also addressing catalyst activation and stability issues. Catalyst stability is crucial for the economic feasibility of hydrotreating; in petroleum refining, fixed-bed hydrotreating catalysts are generally in operation for several months or years before being regenerated or replaced. Most petroleum hydrotreaters apply sulphided bimetallic catalysts to remove sulphur and nitrogen in, for example, a middle-distillate fraction. Such sulphided catalysts stay active from the sulphur that is indigenous to the petroleum-derived feed. Hydrofaction™ renewable crude oil contains only 100–200 ppm sulphur, and as shown in the experimental results below, this affects the activity of a sulphided $NiMo/Al_2O_3$ catalyst. Based on these observations, two additional approaches were tested in the current study to improve catalyst activity:

(1) The renewable crude oil was spiked with butanethiol to a total sulphur content of 1 wt% S and upgraded in a continuous fixed-bed reactor system to test if a higher sulphur concentration improved the catalyst activity.
(2) Alternatively, the use of nonsulphided hydrotreating catalysts was tested as a potential upgrading path that reduces/eliminates the need of expensive gas cleaning operations downstream of the hydrotreater.

The two approaches have been tested experimentally, and the results are presented in the following.

10.2.2.1 Experimental procedure, Hydrotreating studies (second stage)

Feedstocks for the upgrading experiments presented below were different samples of renewable Hydrofaction™ crude oils produced at the pilot described in Section 10.2.1.6. The oils originate from oil production campaigns similar to that presented in Ref. [21].

10.2.2.2 Continuous reactor system

The CRS consists of a dual-piston syringe pump that ensures constant volumetric flowrate of oil. The pistons were kept at 90°C to reduce the viscosity and ensure a constant density, which was used to estimate the mass flowrate of oil. Oil and hydrogen were mixed in-line, and pressurised hydrogen was fed at a rate corresponding to 900 NL/L oil in all experiments. The fixed-bed reactor was made of ¾ in. stainless steel tubing, was operated in upflow mode, and was heated by electric heat cables. A 1/8 in. multipoint thermocouple facilitated 10 measurements inside the reactor. The feed mixture was heated at the entrance of the reactor, and the entrance region ensured a constant temperature of the mixture when reaching the catalyst bed. Reactor products were separated at 60°C and operating pressure in a cylindrical vessel. Gases were either collected for compositional analysis or passed through a wet test meter for volumetric flowrate. A back-pressure regulator at the gas-line outlet was used to set the operating pressure of the system.

The experimental results below present two different upgrading campaigns where the CRS was used. Both campaigns used a $NiMo/Al_2CO_3$ catalyst (activated in situ by sulphidation), but in the so-called spiked campaign, the renewable crude oil was spiked with butanethiol to increase the total sulphur content to 1 wt% S. The nonspiked campaign was carried out over 16 days (380 h), while the spiked campaign was running for 28 days (660 h) and mass balances were taken at least two times daily.

10.2.2.3 Batch reactor system

The BRS consisted of two 25 mL microbatch reactors build from 316L Swagelok fittings that enable pressure logging. The reactors were loaded with reactants, leak tested, and pressurised with H_2 prior to a rapid heat-up by introducing them into a preheated Techne SBL-2D fluidised sand bath with temperature control. Thorough mixing of the reactants was ensured by shaking the reactors at 450 Hz using a shaking device. All experiments were carried out in repeats to ensure accurate data. Gaseous products were quantified by weighing the mass loss during depressurisation into a gas sampling vessel for compositional analysis. Catalysts were filtered off the reactor liquid products, and the latter were centrifuged at 3800 g for 10 min to separate the aqueous and upgraded oil products. Additional details on the BRS upgrading methodology is given in Ref. [34].

Space velocity is a thought parameter in relation to microbatch reactors, but it is advantageous to adapt to the terminology of continuous systems to enable comparison with the CRS experiments. With that in mind, a WHSV equivalent (WHSV*) incorporating reaction time, oil, and catalyst loading for the BRS experiments was defined as $\text{WHSV}^* = m_{\text{oil}}/(m_{\text{catalyst}} \times t_{\text{reaction}})$. The hydrogen availability at initiation of the experiments was 525 NL/L oil.

10.2.2.4 Sulphided catalysts

A commercial $NiMo/Al_2O_3$ catalyst in the form of 3.2 mm extrudates was used for the sulphided catalyst tests. For the CRS, this catalyst was activated in situ using vacuum gas oil with 3.7 wt% dimethyl disulphide at a liquid hourly space velocity (LHSV) of $1.5\,h^{-1}$ and a pressure of 41 bar.

10.2.2.5 Nonsulphided catalysts

The nonsulphided catalysts have up until now only been screened in the BRS to evaluate the degree of deoxygenation and hydrogenation that these catalysts can provide. Two different catalyst types were tested, a $NiW/SiO_2/Al_2O_3$ catalyst and a Pd/Al_2O_3 hydrogenation type catalyst. The nonsulphided catalysts were activated by reduction ex situ. After reduction in H_2 atmosphere, the catalysts were stabilised using 1 vol% O_2 in N_2 for 5 h, in order to avoid reoxidation of the catalyst during loading into the BRS. The flowrate of different gases during activation was 225–250 mL/min.

10.2.2.6 Results: Hydrotreatment using a sulphided catalyst (second stage)

Continuous hydrotreatment of renewable Hydrofaction™ crude oil was successfully carried out on a sulphided $NiMo/Al_2O_3$ catalyst. Two different campaigns were carried out with 380 and 660 h on stream, respectively. A standard renewable Hydrofaction™ crude oil with a sulphur content of 309 ppm was used in the first campaign (non-spiked). Based on observations indicating potential desulphurisation of the catalyst (Fig. 10.4), a similar renewable crude oil (oil B) was spiked to 1 wt% sulphur in the

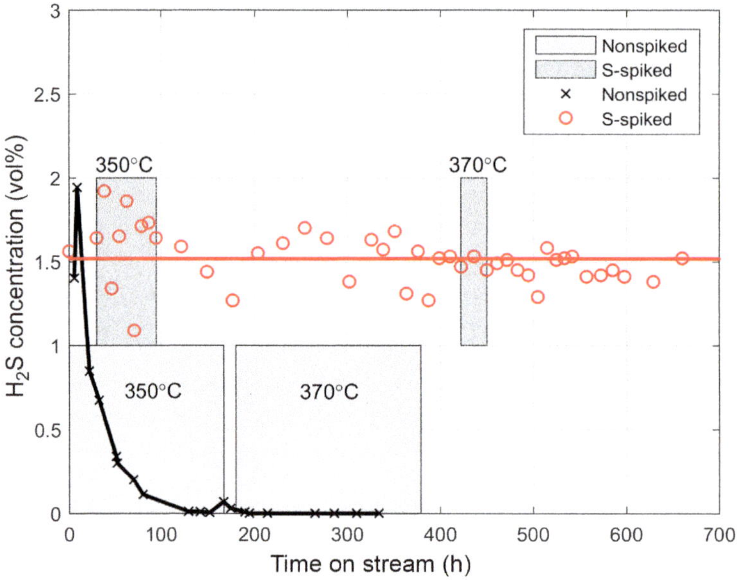

Fig. 10.4 H_2S concentration of the product gas during the two continuous hydrotreating campaigns.

second campaign (spiked). Additionally, data from two different hydrotreating temperatures will be presented for each campaign, adding up to four different datasets. In the spiked campaign, several other operating parameters were screened, but these are left out of the current study for simplicity.

Table 10.3 lists the hydrotreating conditions of the different experiments, together with mass balance data and the product gas composition. At first sight, the datasets appear rather similar with mass balances closing to 93%–96% and with volumetric oil yields between 96 and 100 vol%. However, it is worth focusing on the gas composition and in particular the concentration of H_2S, which is also pictured as a function of time on stream for the two campaigns in Fig. 10.4. It is clear that the H_2S concentration is stable throughout the *s*-spiked campaign, whereas it drops to below detection level during the first 150 h on stream in the nonspiked campaign. This depicts the reasoning behind testing the effect of crude oil sulphur content.

Table 10.4 lists physiochemical properties of the renewable crude oils used as feedstock for the two campaigns and the corresponding hydrotreated products. Complete deoxygenation and TAN elimination were achieved at both temperatures applied during both campaigns. Furthermore, the viscosity is reduced almost three orders of magnitude. Density, HHV, and H/C ratios indicate that the hydrogenation is improved during the spiked campaign. This observation is supported by the hydrogen consumptions (Table 10.3), which seem slightly higher for the spiked campaign as compared with the nonspiked campaign. This is an important finding, since a high degree of hydrogenation is desirable when producing diesel and marine drop-in fuels. Likewise, it is clear from both campaigns that the higher operating temperature (370°C) reduces

Table 10.3 **Hydrotreating conditions, mass balance results, and product gas composition at two operating temperatures for both the nonspiked and sulphur-spiked upgrading campaign**

	Unit	Nonspiked		Spiked	
Operational data					
Flowrate	(mL/min)	0.30	0.30	0.33	0.33
Temperature	(C)	350	370	350	370
Pressure	(bar)	62	62	66	66
LHSV	(h^{-1})	0.5	0.5	0.5	0.5
H_2 availability	(NL/L)	900	900	900	900
Mass balance	(wt%)	93.5	93.3	94.0	95.6
Gas yield	(wt%)	5.5	6.7	4.3	5.0
H_2 consumption	(wt%)	1.9	1.9	2.4	2.2
Water content	(wt%)	9.4	9.1	6.7	7.7
Hydrocarbon yield	(wt%)	81.0	80.0	85.5	84.6
Hydrocarbon yield	(vol%)	99.7	98.2	97.4	96.0
Gas composition					
H_2	(vol%)	94.6	93.0	96.1	93.9
H_2S	(vol%)	0.3	0.0	1.6	1.5
C_1	(vol%)	1.7	2.3	0.6	1.4
C_2	(vol%)	0.5	0.9	0.2	0.4
C_3	(vol%)	0.2	0.5	0.1	0.3
C_4	(vol%)	1.2	1.1	0.7	1.1
C_5	(vol%)	0.0	0.0	0.0	0.0
CO_2	(vol%)	1.3	1.3	0.7	1.4
CO	(vol%)	0.1	0.9	0.0	0.0

Table 10.4 **Physiochemical properties of the products obtained at different conditions as compared with the renewable crude oils used as feedstock**

	Unit	Oil G	Nonspiked		Oil B[c]	Spiked	
Op. temperature	(C)	—	350	370	—	350	370
Density at 15°C	(kg/m^3)	1103	989	989	1073	942	945
TAN	(mg/g)	55.7	<0.0	<0.0	48.0	<0.0	<0.0
Viscosity at 20°C	(cP)	80432	297	166	14429	34	33
HHV[a]	(MJ/kg)	37.2	42.1	42.2	38.0	43.3	43.2
Carbon[a]	(wt%)	80.6	88.1	88.4	80.4	88.6	87.4
Hydrogen[a]	(wt%)	9.1	11.9	11.6	9.0	13.4	12.6
Nitrogen[a]	(ppm)	1500	1175	986	822	424	402
Sulphur[a]	(ppm)	309	389	212	195	764	572
Oxygen[b]	(wt%)	10.1	0.0	0.0	10.4	0.0	0.0
H/C	(–)	1.34	1.61	1.57	1.34	1.84	1.71

[a] On dry basis.
[b] Oxygen by difference.
[c] Analysis before spiking with $C_4H_{10}S$.

hydrogenation as compared with 350°C. This is because hydrogenation is an exothermic reaction that is favoured by lower temperatures until reaction kinetics become rate limiting [22]. As a result, the continuous hydrotreating experiments suggest that the hydrogenation activity of the catalyst is reduced for a low-sulphur renewable crude oil compared with a sulphur-spiked equivalent.

HDN seems to be improved by both a higher operating temperature and a catalyst in its sulphided state (spiked campaign). Whereas the sulphur content is rather unchanged for products of the nonspiked campaign, it has increased in the products of the spiked campaign as compared with the biocrude. This is most likely due to H_2S trapped in the products, because the hydrotreating setup did not enable stripping of H_2S from the products.

Simulated distillation curves of the hydrotreated products are given in Fig. 10.5. True-boiling-point (TBP) distillation was also carried out and given in Fig. 10.5 of both the renewable crude oil and the 350°C samples from both the spiked and nonspiked campaigns.

The deoxygenation results in a significant degree of volatilisation and boiling-point reduction during hydrotreatment. In particular, products of the spiked campaign have been improved as compared with the simulated distillation of the renewable crude oil. The IBP-390°C distillate is improved with more than 30% in the spiked products compared with the renewable crude oil, whereas it is only around 10% for the nonspiked

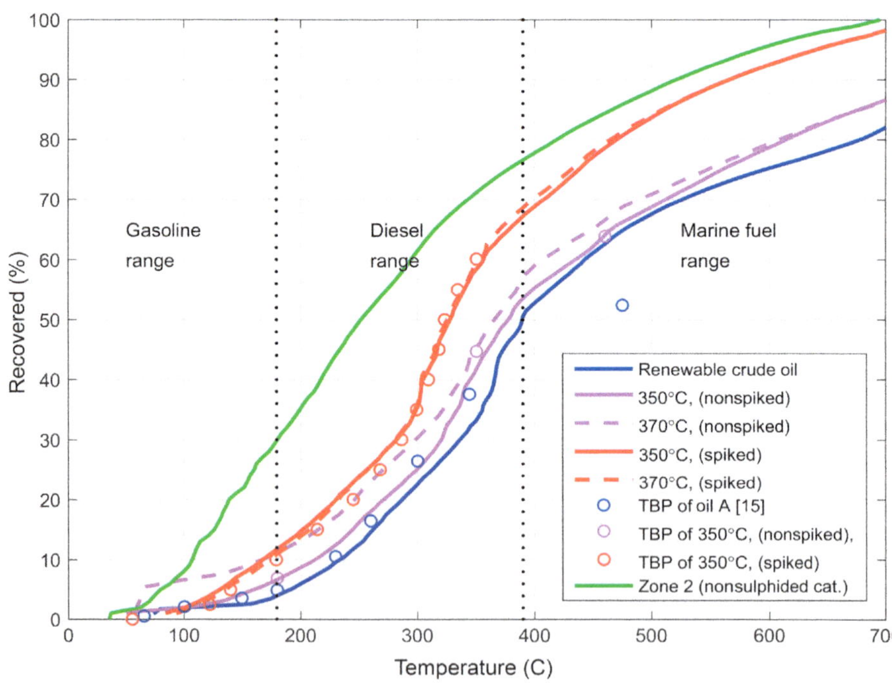

Fig. 10.5 Simulated distillation (ASTM D7169) profiles of the hydrotreated products compared with the renewable crude oil. TBP refers to a true-boiling-point distillation (ASTM D2892).

products. This indicates that cracking of heavy ends is improved when the catalyst is kept in sulphided state by sulphur spiking. The operating temperature of a specific campaign seems to have less effect on the boiling-point distribution. A higher degree of cracking is observed at 370°C for both campaigns but at the cost of a reduced degree of hydrogenation.

10.2.2.7 Results: Two-zone upgrading using nonsulphided catalyst (second stage)

The use of nonsulphided hydrotreating catalysts for upgrading of a low-sulphur renewable Hydrofaction™ crude oil may prove beneficial, because expensive gas cleaning equipment may be avoided when H_2S is reduced/avoided in the gaseous products. In Section 10.2.2.6, it was found that the hydrogenation and cracking activity of a sulphided catalyst is reduced when processing a low-sulphur renewable crude oil. Hydrogenation is however very important, when the objective is to produce diesel and marine fuel drop-in products. Based on a desire to improve the H/C ratio, a two-zone upgrading process was proposed, where deoxygenation and reduction of the boiling-point distribution is the objective of the first zone, while deep hydrogenation to improve the H/C ratio is the objective of the second zone. A schematic of such process is depicted in Fig. 10.6.

Initial screening of the nonsulphided catalysts was carried out to evaluate their ability to deoxygenate and hydrogenate the renewable crude oil. The screening included variation of temperature, pressure, and space velocity. The approach of the study was to initially screen the nonsulphided catalysts in microbatch reactors and

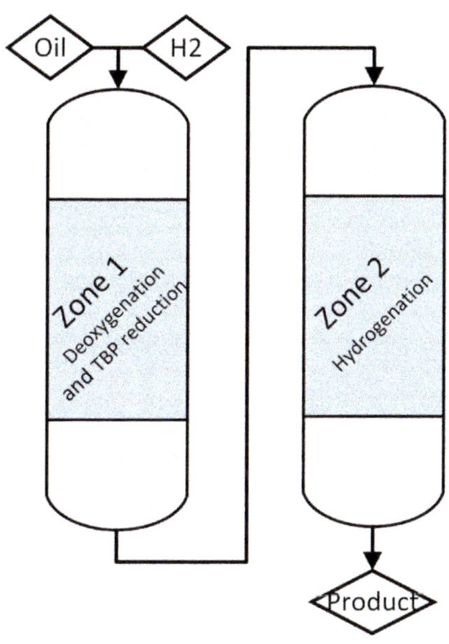

Fig. 10.6 Schematic of two-zone upgrading approach that is tested using nonsulphided catalysts.

then batchwise upgrade enough oil at the optimum conditions of zone 1 and zone 2 to analyse both the intermediate zone 1 product and the final zone 2 product from such two-zone upgrading. An initial parameter screening resulted in a preferred set of operating conditions for each catalyst shown in Table 10.5. Multiple zone 1 experiments at these conditions were carried out to obtain sufficient oil to enable analysis and have enough feedstock for the zone 2 experiments that were also carried out a number of times to get enough samples for analyses.

The amount of sample from the BRS limits the amount of possible analyses, but density, HHV, and elemental composition are given in Table 10.6 for the renewable crude oil, the intermediate zone 1 product, and the final zone 2 product. The oxygen content is reduced to 0.7 wt% in zone 1 and further reduced to below detection level in zone 2. The hydrogenation facilitates an overall increase in H/C molar ratio from 1.25 to 1.58, which indicates a noticeable degree of dearomatisation. This is less than the H/C ratios achieved during the spiked campaign in the CRS (Table 10.4). However, the noble metal loading of the Pd catalyst was only around 500 ppm, and the H/C ratio is likely improved further by increasing the metal loading of this catalyst. The purpose of each reaction zone or catalyst is effectively visualised by the FTIR spectra given in Fig. 10.7. The spectra show that absorption from oxygenated functional groups in the Hydrofaction™ oil is nearly eliminated during the zone 1 reaction. Likewise, the olefinic and aromatic absorption are reduced during the zone 2 reaction, which indicates hydrogenation reactions.

Additional quality improvements are reflected by a 16% increase in HHV and a 15% reduction of the bulk density. Both the HHV and especially the density are improved

Table 10.5 Preferred set of operating conditions based on the initial screening of the catalysts

	Catalyst	Temperature	Pressure	WHSV*
Zone 1	$NiW/SiO_2/Al_2O_3$	350°C	~100 bar	$0.5\,h^{-1}$
Zone 2	Pd/Al_2O_3	300°C	~100 bar	$1.0\,h^{-1}$

Table 10.6 Product quality before, between, and after the two reaction zones using nonsulphided catalysts

	Density[a] (kg/m^3)	HHV[b] (MJ/kg)	Elemental comp.[b] (wt%) C	H	N (ppm)	S (ppm)	O[c]	H/C (–)
Oil A [21]	1051	38.6	81.4	8.6	1124	100	9.8	1.26
Zone 1 product	914	42.4	89.0	10.3	66	61	0.7	1.38
Zone 2 product	892	43.5	88.3	11.7	[d]	[d]	0.0	1.58

[a] Density at 15°C.
[b] On dry basis.
[c] Oxygen by difference.
[d] Insufficient sample.

compared with the products obtained using sulphided catalysts (Table 10.4). Finally and perhaps most importantly, the simulated distillation profile of the zone 2 product (Fig. 10.5) shows a remarkable improvement of the amount of distillates compared with the products obtained using the sulphided catalyst. More specifically, the IBP-390°C distillate fraction increased by 50% compared with the renewable crude oil. As a result, the initial screening of a two-zone upgrading process with nonsulphided catalyst appears as a promising route for upgrading of Hydrofaction™ crude oil. Though, long term catalyst activity and stability need to be studied in a continous reactor system.

10.3 Life cycle GHG emissions of Hydrofaction™

The sustainability of a renewable drop-in fuel is highly dependent on the energy efficiency of the production and the GHG emission reduction associated with use of the particular fuel. This section presents a calculation of the life-cycle GHG emission reductions associated with Hydrofaction™ products. The calculation is based on a 2000 barrel/day (BPD) Hydrofaction™ plant using forestry residues as feedstock and producing diesel and marine drop-in renewable fuels. Fig. 10.8 visualises the overall mass and energy flows for the 2000 BPD Hydrofaction™ plant assessed in the

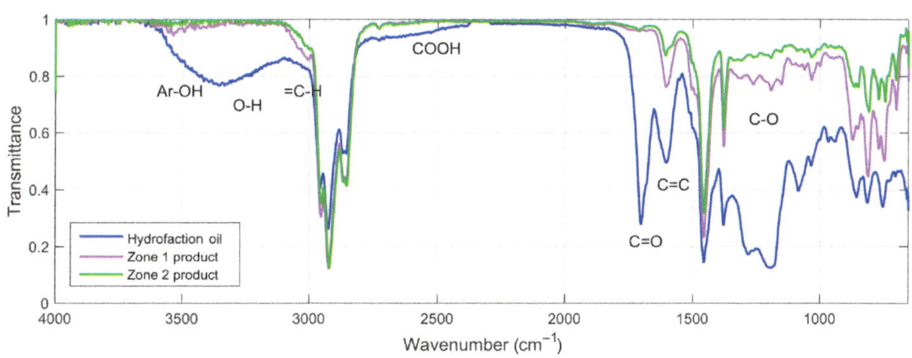

Fig. 10.7 Change in FTIR spectra during the two-zone upgrading process.

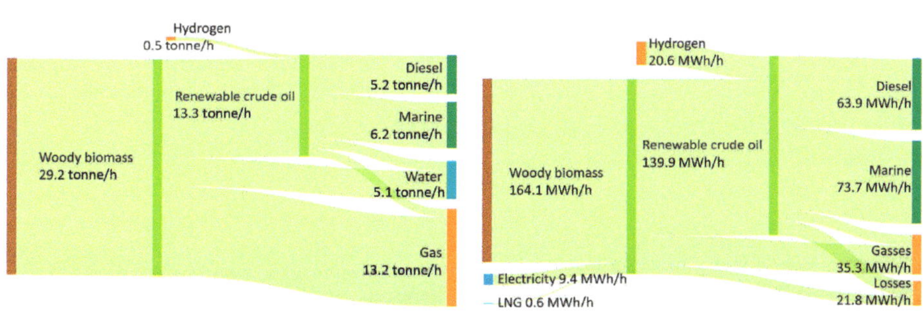

Fig. 10.8 Sankey diagrams of mass and energy balances in a potential 2000 BPD Hydrofaction™ plant.

GHG emission calculation. Mass and energy balances are based on the yield figures presented for the first and second Hydrofaction™ stage. More specifically, mass and energy balances for the first-stage Hydrofaction™ HTL are based on Ref. [21], while mass balances for the second stage are based on the *s*-spiked continuous upgrading experiment presented in Section 10.2.2.6. Fig. 10.8 illustrates the significant heating value that is available through the gaseous products from Hydrofaction™. In steady-state operation, combustion of these gaseous products is more than sufficient to make the overall process self-sustained with heat. However, an external LNG stream is included to cover start-up operations. Based on the energy balance in Fig. 10.8, the energy efficiency of the process can be calculated to 71%, which means that 71% of the energy added to the process is recovered in the drop-in fuels.

The location of the potential Hydrofaction™ plant assessed is the province of Alberta, Canada. Forestry residues are abundant in the region (Section 10.1.1), and fuel distribution infrastructure is widely available because extraction, refining, and distribution of oil sands and derived products is the largest industry in Alberta. Table 10.7 lists the emission intensities used in the GHG analysis of a Hydrofaction™ plant located in Alberta, Canada. Emission factors given in Ref. [15] by the government of Alberta have been used whenever possible. The plant applies forestry residues as feedstock that would have otherwise been disposed in permanent wood waste stockpiles. The methane emissions from decomposition of such wood waste stockpiles in the baseline scenario have been calculated based on the guidelines in Ref. [15]. It is assumed that the feedstock is transported on average 200 km from the harvesting site [16,35]. The gaseous Hydrofaction™ products are combusted to provide process heat, and the emission intensities of CH_4 and N_2O combustion products are assumed equal to that from combustion of liquefied natural gas (LNG). LNG is purchased to cover any additional heat requirements related to process start-up. The effect of using standard grid electricity versus renewable grid electricity has been tested. Heat, electricity, and make-up HTL catalyst requirements are based on design studies for a 2000 BPD Hydrofaction™ plant. Upgrading catalysts is assumed to be changed once every year, and transportation of all make-up catalysts to site is assumed to be on average 1000 km by heavy truck. Hydrogen consumption during upgrading is defined to a conservative 4 wt% (oil feed basis), higher than shown in Fig. 10.8, in order to accommodate for improved hydrogenation and losses from process gas recirculation. The upgraded Hydrofaction™ oil is fractionated into 50 vol% renewable drop-in diesel fuel used to displace petroleum diesel and 50 vol% drop-in marine fuel used to displace heavy fuel oil. Emissions related to distribution and dispensing of the finished fuels are left out of the calculation, because it is assumed to be the same and independent of the carbon origin of the fuel. Finally, GHG emissions related to construction of the plant are assumed negligible over the lifetime of the plant.

Table 10.8 lists the GHG emissions related to a 2000 BPD Hydrofaction™ plant producing diesel and marine fuel as compared with a baseline scenario where petroleum-derived diesel and heavy fuel oil products are refined and used in Alberta, Canada. The GHG emission reduction associated with using Hydrofaction™ products as compared with petroleum-derived products is 77%. This is comparable with the 70% GHG emission reduction determined for hydrotreated HTL fuel by Ref. [5], where emissions related to plant construction are also neglected.

Table 10.7 Emission intensities used in the GHG emission analysis of Hydrofaction™ products

Emission source	Unit	CO_2	CH_4	N_2O	CO_2e[a]	Reference
Feedstock harvesting and transportation (200 m), 15.9 L diesel/tonne	(kg/tonne)	4.2E+01	2.1E−03	6.4E−03		[15,16,35]
Combustion of liquefied natural gas (LNG)	(kg/m^3 LNG)	1.9E+00	3.7E−04	3.3E−05		[15]
Emission factor for grid electricity	(kg/kWh)	BNA			6.4E−01	[15]
Emission factor for renewable grid electricity	(kg/kWh)	BNA			0.0E+00	
Hydrogen emission intensity	(kg/kg H_2)	BNA			8.9E+00	[36]
K_2CO_3 emission intensity	(kg/kg)	BNA			2.7E−01	[37]
NaOH emission intensity	(kg/kg)	BNA			4.7E−01	[38]
Upgrading catalyst emission intensity	(kg/kg)	BNA			5.5E+00	[39]
Catalyst transportation by truck, 1000 km	(kg/kg)	BNA			4.9E−02	[40]
Emission intensity of fuel extraction and production	(kg/L)	1.4E−01	1.1E−02	4.0E−06		[15]
Combustion of diesel	(kg/L)	2.7E+00	1.3E−04	4.0E−04		[15]
Combustion of heavy fuel oil	(kg/L)	3.1E+00	3.4E−05	6.4E−05		[15]
Emission intensity from decomposition of wood waste stockpile	(kg/tonne)	Biogenic	2.0E+01			[15]

[a] 1 kg CH_4 = 25 kg CO_2e; 1 kg N_2O = 298 kg CO_2e; [15] BNA, breakdown not available.

Table 10.8 **GHG emission reduction associated with displacement of diesel and heavy fuel oil by renewable Hydrofaction™ products**

Emission source	Amount		CO_2 (kg/h)	CH_4 (kg/h)	N_2O (kg/h)	CO_2e* (kg/h)
2000 BPD Hydrofaction™ plant						
Feedstock collection and transport	29.2	tonne/h	1.2E+03	6.2E−02	1.9E−01	1295
Combustion of HTL process gas	3436	m^3 LNGe/h	Biogenic	1.3E+00	1.1E−01	66
Combustion of purchased LNG for heat	60.9	m^3/h	1.2E+02	2.3E−02	2.0E−03	118
Purchased electricity	9362	kWh/h	BNA	BNA	BNA	5992
Purchased hydrogen	530	kg/h	BNA	BNA	BNA	4717
Drop-in diesel fuel usage	6.5	m^3/h	Biogenic	8.6E−01	2.6E+00	791
Drop-in marine fuel usage	6.5	m^3/h	Biogenic	2.2E−01	4.1E−01	129
Make-up catalysts	1087	kg/h	BNA	BNA	BNA	443
CO2 sequestration	0	tonne/h	BNA	BNA	BNA	0
Hydrofaction™ GHG emission total						13550
Petroleum baseline						
Extraction and refining of petroleum diesel and fuel oil	12.9	m^3/h	1.8E+03	1.4E+02	5.2E−02	5313
Diesel combustion	6.5	m^3/h	1.7E+04	8.6E−01	2.6E+00	17974
Heavy fuel oil combustion	6.5	m^3/h	2.0E+04	2.2E−01	4.1E−01	20287
Wood waste stockpiling, decomposition	29.2	tonne/h	Biogenic	5.9E+02		14642
Baseline GHG emission total						58216
GHG emission reduction						44666
GHG emission reduction relative to baseline						77

Table 10.8 lists the details of a base-case scenario where the Hydrofaction™ process purchases all the hydrogen required for upgrading and uses standard grid electricity (coal-intensive in Alberta) and where the biogenic CO_2 from the HTL gaseous products is all emitted. It is clear that the major contributions to GHG emissions from Hydrofaction™ are related to the purchased electricity and hydrogen, covering 44% and 35% of the total emissions, respectively. Based on that, the following three additional scenarios were tested in the LCA.

Case 1: The base case is modified so hydrogen produced in situ during first-stage Hydrofaction™ is applied during upgrading of the renewable crude oil to lower the amount of hydrogen that need to be purchased. No additional LNG needs to be purchased, because there is sufficient heating value in the remaining gaseous products after hydrogen separation. Additional electricity demand is not accounted for.

Case 2: The base case is modified so renewable grid electricity is used to reduce the emission intensity of electricity but at the expense of a higher cost.

Case 3: The base case is modified so all the biogenic CO_2 from the HTL gaseous products is extracted (before combusting the gaseous products with air) and sequestered. Additional electricity demand is not accounted for.

Case 4: The base case is modified to include all of the above.

Fig. 10.9 compares the GHG emissions from the different scenarios and the resulting emission reduction. The effect of utilising in situ produced hydrogen in case

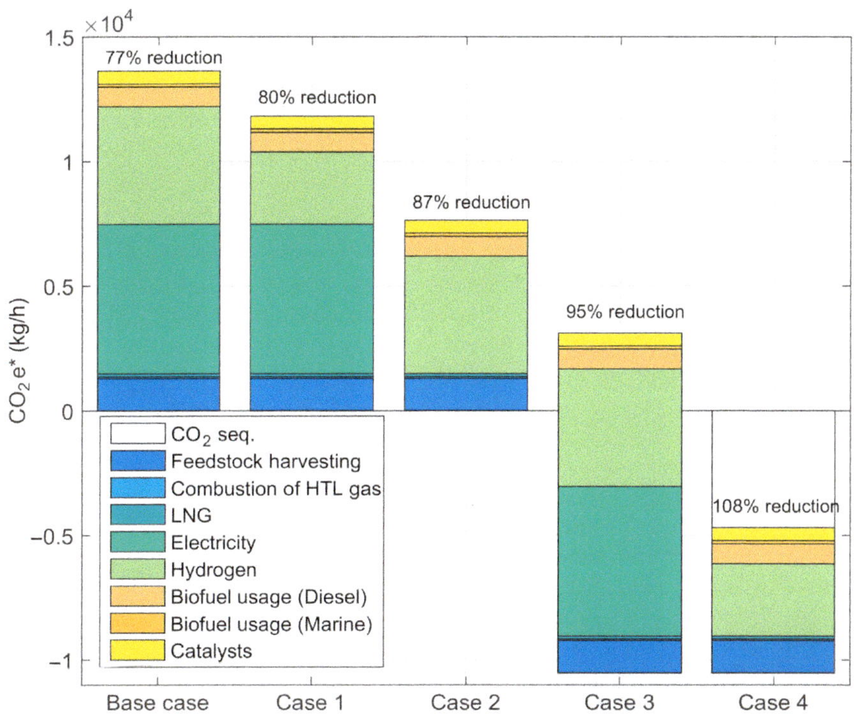

Fig. 10.9 GHG emission intensity of different Hydrofaction™ cases.

1 is minor, but this process modification is likely to be cost-efficient, since hydrogen is an expensive utility. The use of renewable electricity instead of standard grid electricity in case 2 has a major effect on the GHG emissions, which is because power generation in Alberta is relatively carbon-intensive due to coal-fired power plants. The cost-efficiency of case 2 depends on carbon legislation and the additional price of renewable electricity. The possibility of sequestering CO_2 from the HTL gaseous products is also worth considering as illustrated with case 3, since CO_2 is a major product from the first-stage Hydrofaction™ HTL and sequestration can significantly enhance the project's GHG emission savings. The rightmost column of Fig. 10.9 includes all the modifications to show the potential GHG emission reduction associated with Hydrofaction™ products. The result is a 108% GHG emission reduction, which implies a process that is not only CO_2 neutral but actually CO_2 negative.

10.4 Conclusion

Hydrofaction™ aims to fill the need for renewable heavy-transport fuels by converting forestry residues into drop-in renewable diesel and marine fuel using supercritical HTL and subsequent hydrotreating. Process descriptions and details are provided and supported by experimental data from continuous pilot operations. A comparison of six renewable Hydrofaction™ crude oils produced by supercritical HTL from different pine, spruce, birch, and bark mixtures shows similar quality independent of the woody feedstock. The characteristic fuel properties of these oils were significantly improved during two continuous hydrotreating campaigns with 300 and 700 h on stream. The low sulphur content (100–200 ppm S) of the crude oils was however found to inhibit the activity of the sulphided catalyst. A two-zone deoxygenation and hydrogenation reactor system using nonsulphided catalysts was screened in batch reactors, and improved devolatilisation relative to the sulphided catalysts was observed. Finally, mass and energy balances for a 2000 BPD Hydrofaction™ plant indicate that 1 tonne forestry residues can be converted to more than 400 L renewable diesel and marine fuels with an energy recovery of 71% and a GHG emission reduction of 77%–108%, reflecting an energy- and resource-efficient technology for the production of renewable fuels for heavy transportation.

Acknowledgements

The authors are thankful for the collaboration with Professor Lasse A. Rosendahl, Aalborg University, Denmark, and Professor Pedro Pereira Almao and his team at University of Calgary, Canada. The authors acknowledge the funding provided by EASME Horizon 2020 (Grant No. 666712), Danish Energy Technology Development and Demonstration Program (Grant No. 64013-0513), and Innovation Fund Denmark (Grant No. 4135-00126B).

References

[1] IPCC (2014), Climate Change 2014: Synthesis Report. Contribution of Working Groups I, II and III to the Fifth Assessment Report of the Intergovernmental Panel on Climate Change, (Core Writing Team: R.K. Pachauri and L.A. Meyer (eds.)). IPCC, Geneva, 151 pp.
[2] United Nations/Framework Convention on Climate Change (2015) Adoption of the Paris Agreement, 21st Conference of the Parties, Paris, France, United Nations.
[3] IEA (2016), World energy outlook 2016, Executive summary, Tech. rep., International Energy Agency. Paris.
[4] European Commission. Communication from the commission to the european parliament, the council, the european economic and social committee and the committee of the regions. A european strategy for low-emission mobility; 2016. SWD(2016), 244 final, Brussels, Belgium.
[5] I. Tews, Y. Zhu, C. Drennan, D. Elliott, L. Snowden-Swan, K. Onarheim, Y. Solantausta, D. Beckman (2014), Biomass direct liquefaction options: Technoeconomic and life cycle assessment, Tech. Rep., Prepared for U.S. Department of Energy, Springfield.
[6] De Jong S, Faaij A, Slade R, Mawhood R, Junginger M. The feasibility of short-term production strategies for renewable jet fuels is—a comprehensive techno-economic comparison. Biofuels Bioprod Biorefin 2015;9(6):778–800.
[7] Jensen CU, Hoffmann J, Rosendahl LA. Co-processing potential of HTL bio-crude at petroleum refineries. Part 2: A parametric hydrotreating study. Fuel 2016;165:536–43.
[8] FAO (2016), State of the World's Forests 2016. Forests and Agriculture: Land-Use Challenges and Opportunities. Food and Agriculture Organization of the United Nations, Rome.
[9] Ramachandran Nair PK, Mohan KB, Nair VD. Agroforestry as a strategy for carbon sequestration. J Plant Nutr Soil Sci 2009;172(1):10–23.
[10] Millar CI, Stephenson NL, Stephens SL. Climate change and forests of the future: managing in the face of uncertainty. Ecol Appl 2007;17(8):2145–51.
[11] O'Laughlin, Jay, and Mahoney, Ron (2008), Forests and Carbon. From University of Idaho Extension's 'Woodland notes', 19, 2.
[12] Bradley, D. (2007), Canada- Sustainable forest biomass supply chains. Prepared for IEA Task 40 by Climate change solution. Ottawa, Canada.
[13] Peter B, Niquidet K. Estimates of residual fibre supply and the impacts of new bioenergy capacity from a forest sector transportation model of the Canadian Prairie Provinces. Forest Policy Econ 2016;69:62–72.
[14] U.S. Department of Energy (2016), 2016 Billion-Ton Report: Advancing Domestic Resources for a Thriving Bioeconomy, Volume 1: Economic availability of feedstocks. M. H. Langholtz, B. J. Stokes, and L. M. Eaton (Leads), ORNL/TM-2016/160. Oak Ridge National Laboratory, Oak Ridge, TN. 448p.
[15] Government of Alberta (2015), Carbon offset emission factors handbook, ESRD Climate Change 2015, No. 1, Alberta Carbon Offset Program. Edmonton, Canada.
[16] Cambero C, Alexandre MH, Sowlati T. Life cycle greenhouse gas analysis of bioenergy generation alternatives using forest and wood residues in remote locations: a case study in British Columbia, Canada. Resour Conserv Recycl 2015;105:59–72.
[17] BTG BV (2002), Methane and nitrous oxide emissions from biomass waste stockpiles. Technical Report no. 12, Prepared for Worldbank PCFplus Research by Biomass Technology Group BV, Washington, DC.

[18] Macdonald C. Market pulp:overcapacity on the way. Pulp & Paper-Canada 2016; 117(3):9–11.

[19] RISI. China, Brazil paper and pulp production advances, even as growth slows globally, http://www.risiinfo.com/press-release/china-brazil-paper-pulp-production-advances-even-growth-slows-globally/; 2017. Accessed January 24, 2017.

[20] FAO (2009), State of the World's Forests 2009. Food and Agriculture Organization of the United Nations, Rome, Italy.

[21] Jensen CU, Guerrero JKR, Karatzos S, Olofsson G, Iversen SB. Fundamentals of Hydrofaction™: renewable crude oil from woody biomass. Biomass Convers Biorefin 2017; https://doi.org/10.1007/s13399-017-0248-8.

[22] Gary JH, Handwerk GE, Kaiser MJ. Petroleum refining, technology and economics. 5th ed. CRC Press; 2007, ISBN: 0-8493-7038-8.

[23] Lane, J. (2015), 4 minutes with… Taner Onoglu, Vice President, Altaca Energy. Biofuels Digest. http://www.biofuelsdigest.com/bdigest/2015/06/04/4-minutes-with-taner-onoglu-vice-president-altaca-energy/. [Accessed 21 February 2017].

[24] Genifuel 2015. Genifuel corporation closeout report for NAABB program, DE-FOA-0000123 http://www.energy.gov/sites/prod/files/2016/01/f28/naabb_genifuel_pilot_system_final_report.pdf. [Accessed 21 February 2017].

[25] Muradel Pty Ltd. (2017), Technology http://www.muradel.com/aboutus.asp#tech. [Accessed 21 February 2017].

[26] Licella Pty Ltd. Cat-HTR, http://www.licella.com.au/cat-htr/; 2017.

[27] Elliott DC, Biller P, Ross AB, Schmidt AJ, Jones SB. Hydrothermal liquefaction of biomass: developments from batch to continuous process. Bioresour Technol 2015;178:147–56.

[28] Elliott DC. In: Process development for biomass liquefaction. Volume 25:4, Conference 180, Report no. CONF-800814-P3, Pacific Northwest Lab., American Chemical Society, Division of Fuel Chemistry, Richland, USA; 1980.

[29] Harvey AH, Peskin AP, Klein SA. NIST/ASME Steam properties database, Software Version 3.0. In: Standard Reference Database. National Institute of Standards and Technology (NIST), US Department of Commerce; 2013.

[30] Akiya N, Savage PE. Roles of water for chemical reactions in high-temperature water. Chem Rev 2002;102:2725–50.

[31] Kruse A, Dinjus E. Hot compressed water as reaction medium and reactant properties and synthesis reactions. J Supercritical Fluid 2007;39:362–80.

[32] Sintamarean IM, Grigoras IF, Jensen CU, Toor SS, Pedersen TH, Rosendahl LA. Two-stage alkaline hydrothermal liquefaction of wood to biocrude in a continuous bench scale system. Biomass Convers Biorefin 2017; https://doi.org/10.1007/s13399-017-0247-9.

[33] Jensen CU, Olofsson G, Rosendahl LA. Impact of Nitrogenous Alkaline Agent on Continuous HTL of Lignocellulosic Biomass and Biocrude Upgrading. Fuel Process Technol 2017;159:376–85.

[34] Hoffmann J, Jensen CU, Rosendahl LA. Co-processing potential of HTL bio-crude at petroleum refineries. Part 1: Fractional distillation and characterization. Fuel 2016;165:526–35.

[35] McKechnie J, Chen J, Vakalis D, MacLean H. Energy use and greenhouse gas inventory model for harvested wood product manufacture in Ontario; 2014. Climate change research report, CCRR-39, Ontario, Canada.

[36] Ruether J, Ramezan M, Grol E. Life-cycle analysis of greenhouse gas emissions for hydrogen fuel production in the United States from LNG and coal. DOE/NETL-2006/1227; 2005.

[37] Pendolovska, V., Fernandez, R., Mandl, N., Gugele, B., and Ritter, M. (2014), Annual European Union Greenhouse Gas Inventory 1990–2012 and Inventory Report 2014. Technical report no. 09/2014 by European Environment Agency. Copenhagen.
[38] Biograce. Harmonised calculations of biofuel greenhouse gas emissions in Europe, list of standard values, http://www.biograce.net/content/ghgcalculationtools/standardvalues; 2017.
[39] Snowden-Swan LJ, Spies KA, Lee GJ, Zhu Y. Life cycle greenhouse gas emissions analysis of catalysts for hydrotreating of fast pyrolysis bio-oil. Biomass Bioenergy 2016;86:136–45.
[40] Railway association of Canada. Rail Freight Greenhouse Gas Calculator, http://www.railcan.ca/environment/calculator; 2017.

Index

Note: Page numbers followed by *f* indicate figures, and *t* indicate tables.

D

E

I

R

S

CPI Antony Rowe
Chippenham, UK
2017-11-15 21:16